To    STUART.

FROM    MUM o DAD.

CHRISTMAS    1993.

*Bird Life of Coasts and Estuaries* describes the bird life of the British coastline and adjacent off-shore waters from an ecological point of view, using information from the latest research to show how bird distribution and abundance are related to important environmental variables such as marine currents, weather, coastal landform and the influence of man.

Separate chapters look at the open sea, rocky coasts, estuarine shores and the coastal fringe. Typical birds from each of these habitats are introduced and their foraging and breeding behaviour, distribution and abundance described. In the final chapter, threats to coastal birds, such as habitat destruction and pollution are discussed, together with the positive action that can be taken to safeguard against these problems. The text is illustrated throughout by beautiful and informative illustrations of the birds themselves and the habitats in which they live.

The book will appeal to the layman who wants to know more about coastal birds, the birder who wants to find out how birds interact with their environment, and all those who are interested in the habitats that make up what is arguably Britain's most important natural asset.

Dr Peter Ferns is a senior lecturer and chairman of the Ecology and Environmental Management Group in the School of Biology at the University of Wales College of Cardiff. He has carried out research on coastal and estuarine birds in the Bristol Channel and Severn Estuary for over 20 years, and has been involved recently in assessing the impact of estuarine barrages proposed for various parts of the British coast as part of a renewable energy programme.

During his career, Dr Ferns has also studied breeding waders in the Arctic region, wildfowl on reservoirs, sandgrouse in deserts, as well as urban wildlife. The author of *Wading Birds in the Severn Estuary* (1977), he has also written numerous journal articles on the ecology and physiology of birds and mammals.

Bird life of coasts and estuaries

**Bird Life Series**

This series describes the bird life characteristic of each of the various major habitat types in the British Isles in a way that shows how the birds are adapted to the environments they inhabit, and how the environments themselves shape the birds' behaviour and breeding patterns. This beautifully illustrated series provides the reader with a clear understanding of the ecological and conservation aspects of the bird life of Britain and Ireland.

Published title

*Bird life of mountain and upland* by Derek Ratcliffe

Forthcoming titles

*Bird life of woodland*
*Bird life of towns, parks and gardens*
*Bird life of farmland, grassland and heathland*
*Bird life of freshwater wetland*

The editor of the series is Dr C.M. Perrins, who is the Director of the Edward Grey Institute of Field Ornithology at the University of Oxford.

# Bird life of coasts and estuaries

P.N. FERNS

*Senior Lecturer in Biology, School of Pure and Applied Biology, University of Wales College of Cardiff*

*With line illustrations by Chris Rose*

CAMBRIDGE
UNIVERSITY PRESS

Published by the Press Syndicate of the University of Cambridge
The Pitt Building, Trumpington Street, Cambridge CB2 1RP
40 West 20th Street, New York, NY 10011-4211, USA
10 Stamford Road, Oakleigh, Victoria 3166, Australia

First published 1992

Printed in Great Britain by
Butler & Tanner Ltd, Frome and London

*A catalogue record for this book is available from the British Library*

*Library of Congress cataloguing in publication data*

Ferns, P. N.
Bird life of coasts and estuaries / P. N. Ferns ; with line
illustrations by Chris Rose.
    p.   cm. – (Bird life series)
Includes bibliographical references and index.
ISBN 0–521–34569–3 (hardback)
1. Birds – British Isles.   2. Coastal fauna – British Isles.
3. Estuarine fauna – British Isles.   I. Title.   II. Series.
QL690.B65F47   1992
598.2941'0914'6 – dc20   92–14095 CIP

ISBN 0 521 34569 3   hardback

WV

*To Joanne*

Swan's trumpeting
was my diversion. Instead of mirth
I shared the Gannet's cry and Whimbrel's laughter,
in place of mead the music of the gull.

translated from *The Seafarer*
(late tenth century)

# CONTENTS

# PREFACE

Anyone writing a book about the coastal birds of the British Isles is in the fortunate position of having a wealth of material from which to select. Two of the earliest monographs on individual species were Grieve's on the Great Auk published in 1885 and Gurney's on the Gannet in 1913. Nowadays, monographs are published more frequently, with books on the Shelduck, Arctic Skua, Puffin and Manx Shearwater all appearing in the 1980s.

An important landmark in seabird studies was the regular national population survey of Fulmars started by James Fisher in 1939. Long-term studies of the productivity and mortality at particular colonies were later started by university researchers, notably George Dunnet on Fulmars in 1950, John Coulson on Kittiwakes in 1954 and Bryan Nelson on Gannets in 1960. Following the formation of the Seabird Group in 1965, a national survey of all breeding seabird populations was undertaken in 1969/70 (Operation Seafarer). Studies of a range of species at large mixed colonies such as Skokholm were also started. At the latter site, Chris Perrins' students began to put the pioneering natural history observations of Ronald Lockley into their proper scientific context. Meanwhile, Niko Tinbergen's work on seabirds at Ravenglass, Walney and the Farne Islands was revolutionising the study of animal behaviour.

A central role in widescale population surveys is now played by the British Trust for Ornithology (BTO). By harnessing the skills and enthusiasm of amateur ornithologists, the BTO is able to achieve the kind of comprehensive coverage that is the envy of most other countries. The Seabird Group, itself an offshoot of the BTO, was responsible, with the Nature Conservancy Council (NCC), for the most recent survey of our breeding seabird populations in 1985-87. The results of this survey (the Seabird Colony Register) have just been published (Lloyd, Tasker & Partridge, 1991). The NCC, which has now been split into English Nature, Scottish Natural Heritage and the Countryside Council for Wales, has also become involved in seabird research of its own. Its most notable recent contributions have been the study of the distribution of birds across the open ocean, through the work of the Seabirds at Sea Team, and the completion of an Estuaries Review containing a wealth of information about British estuaries (Davidson *et al.*, 1991). The Institute of Terrestrial Ecology (ITE), despite its name, has a number of staff working on coastal birds. Its best known seabird researcher is Mike Harris who worked on gulls in South Wales before moving to Scotland to study what some people would call real seabirds. A second generation of seabird researchers, trained by some of the people mentioned above, is now established in the universities, including Tim Birkhead, Bob Furness and Pat Monaghan.

Interest in waders has only recently become as strong as that in seabirds. Although breeding waders had been studied for many years by a few enthusiasts, such as Desmond Nethersole-Thompson and Bill Hale, wintering birds received little attention until the Wader Study Group was formed in 1970. This too was an offshoot of the BTO. Large numbers of waders began to be caught using cannon-nets on the Wash and elsewhere, thanks largely to the energy of Clive Minton. The tantalising ringing recoveries of these birds from abroad led to a number of British expeditions to places such as Iceland, Greenland, Morocco and Mauritania. University researchers also began to take an interest in waders, firstly at Aberdeen's Culterty Field Station on the Ythan Estuary under George Dunnet, then at Lindisfarne and Teesmouth under Peter Evans and in the Firth of Forth under David Bryant. The ITE also became involved through John Goss-Custard, who started work on Redshanks on the Ythan, and went on to study Oystercatchers on the Exe Estuary.

Wintering waders are now counted regularly by the Birds of Estuaries Enquiry, organised by the BTO. The population sizes of British breeding waders have also been estimated. The BTO, the Royal Society for the Protection of Birds (RSPB), the NCC and the Wader Study Group have between them organised surveys covering wet meadows in England and Wales, the machair, Scottish agricultural land, upland areas and saltmarshes. The Wader Study Group has now become a major international forum for work on waders and has organised several research projects such as that on the spring passage of Knots.

The RSPB has always provided financial support for research on birds, but has increasingly carried out work of its own. It has naturally enough concentrated on the biology of rare and declining species, and on management techniques for increasing bird numbers on reserves. The Wildfowl and Wetlands Trust has been actively involved in research for a great many years, and through the work of Myrfyn Owen, Malcolm Ogilvie and others has shown how ducks and geese exploit coastal habitats such as saltmarshes. Other notable work on coastal wildfowl has been undertaken by Harry Milne and Ian Patterson who studied the biology of Eiders and Shelducks at the Ythan. The International Waterfowl and Wetlands Research Bureau has a mainly co-ordinating function, but by organising meetings and workshops has produced a number of valuable reviews, particularly of the sizes of national and flyway populations of wildfowl and waders.

During recent years there has been something of a decline in the scope and quality of research on birds carried out in Britain. This is reflected most vividly by the fact that our journals are to an increasing extent filled with the excellent work being carried out by our friends on the continent, particularly the Scandinavians and Dutch. This decline is a consequence of the diversion of British resources away from publicly funded research and into private hands. Much less work is done as a result, and most of what remains is related to projects that threaten natural environments, such as housing and leisure developments, including marinas and barrages. Our few professional ornithologists, including those in the BTO, NCC, RSPB, ITE and the universities and polytechnics, have little choice but to spend more time collecting information on the effects of such proposals, and either presenting a case on behalf of developers or trying to prevent the developments taking place at all.

It is also galling, following a decade that has for the first time seen environmental issues placed firmly on the political agenda, to see more and more money spent by both government and industry on public relations, and less and less on environmental research and monitoring. For example, coastal birds are ideal species for monitoring the health of coastal and marine environments, yet no regular program exists to determine the levels of any of the toxic contaminants that they contain. It is hard to find anything encouraging to say about the changes that have taken place in recent years, but there is just a glimmer of hope. More public funds can be generated for research by the imposition of taxes and fines on those who pollute and damage the environment (which includes nearly all of us). This is already being done by the National Rivers Authority, which is introducing charges to cover the costs of its monitoring and research programmes. This principle needs to be strengthened and extended.

The references at the end of this book provide an introduction to the wealth of literature available to anyone wishing to pursue a particular topic further. Species of birds that do not occur in Britain are occasionally referred to, but only if their biology is of general interest. One or two exotic habitats, such as mangroves, are also mentioned because they are important to British birds at certain times of the year.

I am grateful to the research students who have shared studies of the gull and wader populations of the Bristol Channel with me, especially Greg Mudge, Dave Worrall, Judy Anderson, Hanna Siman and Ryland Talbot. Needless to say, they did most of the hard work. It has also been a pleasure to study waders in the company of Harry Green and my colleagues in the Celtic Wader Research Group. The following people have helped in the preparation of this book by providing material: S. Angus, J. Cawley, P. Doody, G.R. Hosey, J.M. Edington, A. Fitchett, P. Fox, J. Frikke, J.D. Goss-Custard, H.E. Grenfell, M.S.C. Havard, P. Hawkey, G.R. Hosey, P. Howlett, F. Hunter, J. Laursen, D.R. Lees, C.J. Mettam, S.J. Moon, M.E. Moser, G.P. Mudge, H. Nicholls, H.Y. Siman, T.J. Stowe, R.J. Talbot, M.L. Tasker, L.M. Warren, M.J. Wells, K. Williams, V. Williams and D.H. Worrall. At Cambridge University Press, M.G. Walters, R.C. Smith and S. Trevitt have been encouraging and helpful. I would also like to thank Professors Denis Bellamy, Ray Beverton and Mike Claridge in whose departments I have worked whilst preparing this book. The following have read through parts of the text and I am most grateful for their comments: J.A. Fowler, R. Prys-Jones, P. Rothwell, R.W. Summers, L. Campbell and N. Harrison. Chris Perrins, as series editor, has made a great many improvements and corrections to the text. Shelley Hinsley also read through the whole of the text, and this book has benefitted from numerous discussions with her. Chris Rose has patiently coped with my requests for difficult combinations of features in some of the illustrations, and the results of his efforts speak for themselves. Any figures that do not have the originator's name at the end of the caption were prepared by myself.

I have used the following terms in accordance with their correct geographical definitions: Britain refers to England, Scotland and Wales; the United Kingdom to Britain and Northern Ireland; Ireland to the Republic of Ireland (Eire) and Northern Ireland; and the British Isles to the United Kingdom, Eire, the Channel Islands and the Isle of Man. Finally, I have used the flowering plant names recommended by Dony, Jury & Perring (1986).

# INTRODUCTION

From childhood the coast fascinates us. As children we love going to the seaside, playing in the sand, searching amongst rockpools and swimming in the sea. More people go to the coast for holidays than to any other kind of habitat, although nowadays it is as likely to be Benidorm or Palm Beach as Blackpool and Brighton. Finally, when we grow old we like to retire by the sea if we can afford to. What is it about the coast that fascinates us so much? Perhaps it is because the British are derived from a long line of invaders – Celts, Norsemen, Saxons – all of whom had to cross the sea to get here and all of whom had a seafaring tradition.

Our love of the coast means that we associate the cry of seabirds (notably the Herring Gull) with pleasant memories, yet our very presence on the shore places coastal habitats under intense pressure. We are fortunate in the British Isles in having over 15 000 km (9300 miles) of coastline (Houston & Jones, 1990) surrounding a land area of 302 500 km$^2$ (116 780 square miles). If this same area were contained in a single, completely circular island, it would have a coastline of only 1950 km (1212 miles). Part of the reason for this enormous difference is that the British Isles are divided up into a complex of over 1000 islands (no one has ever counted them all properly), varying in size from the mainland of Britain at 229 850 km$^2$ (84 200 square miles), to tiny Rockall at around a thousandth of a square kilometre.

A further factor that increases the length of the coastline is that none of the islands is circular. Even a roundish island like Pabbay in the Sound of Harris, has a coastline that is 22% greater than it would have if it were truly circular. Benbecula, also in the Outer Hebrides but with a very complex shape, has a coastline nearly 300% greater than that of a circular island with an equivalent area (see Fig. 1).

These two factors alone do not explain why estimates of the length of the coastline made by different observers vary so much. For example, the Reader's Digest Association (1965) gives a figure of 11 400 km (7100 miles) for the British Isles, whereas Moser & Summers (1987) give 17 800 km (11 100 miles) for the United Kingdom alone. The former estimate was probably made using just a few small-scale maps, whereas the latter was derived from many larger scale ones. Rather like a fractal, the shape of the coast remains relatively complex even at larger magnifications (Fig. 2), and so its apparent length depends upon the scale at which it is measured.

The length of available coastline is of direct importance to birds since the greater the length, the greater the area of habitat for both feeding and nesting. For example, the average density of Oystercatchers along Britain's beaches and rocky shores is five per square kilometre in winter (Moser & Summers, 1987) and any reduction in the length of such shore would almost certainly mean that fewer birds could be supported. A similar situation exists for those species nesting on cliffs.

The fact that we have so many islands is important not just because it increases the length of the coast. Islands usually have fewer predators than the mainland, and thus offer greater security during breeding.

**Fig. 1.** Influence of island shape on coastline length (redrawn from Ordnance Survey maps). All islands except Portland are in the Outer Hebrides.

Pabbay
Area = 8 km²
Length of coast = 12 km
(10 km if circular)

1 km

Portland
Area = 11 km²
Length of coast = 16 km
(12 km if circular)

1 km

Hirta
Area = 7 km²
Length of coast = 17 km
(9 km if circular)

1 km

Benbecula
Area = 76 km²
Length of coast = 120 km
(31 km if circular)

1 km

Ireland lacks some of the predators present in the rest of Europe, including Britain. This is because, at the end of the last Ice Age, the rise in sea level closed the Irish Channel at a time when the climate was still too cold to have allowed certain species such as the Weasel and Polecat to have penetrated very far north. Birds are not limited to overland dispersal, and reaching such islands is not a problem for them, but even they have been secondarily affected by such factors. For example, Ireland tends to have few Short-eared Owls and no Tawny Owls because it lacks several species of the voles and shrews that form an important part of their diet.

The absence of predators is even more marked in the case of the smaller islands. All of the 16 islands in the British Isles with an area of more than 100 km² are inhabited by the Norway Rat, Peregrine, Jackdaw, Carrion Crow and several species of large gulls. However, only the mainland of Great Britain has all of the following potential predators of adult coastal birds and their eggs and young – Grass Snake, Adder, Hedgehog, Red Squirrel, Grey Squirrel, Fox, Pine Marten, Stoat, Weasel, Polecat, Mink, Badger, Otter, Wildcat, Common Seal, Grey Seal, Red Kite, Hen Harrier, Sparrowhawk, Golden Eagle, Barn Owl, Little Owl, Tawny Owl and Magpie. Large islands and those close to the mainland, such as Ireland, Skye, Anglesey, Arran and the Isle of Wight, have 13–16 of these 23 potential predators, whilst smaller and more distant ones have far fewer. For example, Rhum has four, whilst Lewis and Harris (a single island) has only six, despite having the third largest land area after Great Britain and Ireland. The Orkneys and Shetlands contain some of the largest and most important breeding seabird colonies in the whole of the British Isles and their mainlands have only five and six of these predators respectively, whilst the smaller islands in their archipelagos have even fewer (Sharrock, 1976; Frazer, 1983; Lack, 1986; Yalden, 1991). Domestic cats and dogs are also predators which are absent from many of our smallest islands.

More than half of the 16 largest islands in Britain lack Foxes, which can be devastating predators of seabird colonies, and they are also absent from most small islands. Such islands offer even greater security when they are uninhabited by man and his domesticated stock and pets. Proving that the absence of predators contributes to the large size of Britain's breeding seabird colonies is difficult, but there is circumstantial evidence. For example, in 1934 the liner *Herefordshire* was being towed away for scrap but broke free in a storm. She grounded on Cardigan Island off the coast of Wales and the crew, plus a large number of Norway Rats, scrambled ashore. Although the exact status of the island's small colonies of Puffins and Manx Shearwaters was not known at the time, they were certainly extinct not long afterwards. The island

**Fig. 2.** A section of the coast on Benbecula in the Outer Hebrides seen at successively larger magnifications. Each of the maps *b* to *h* shows a tenfold enlargement of the section shown in the box on the preceding map. Filled areas = rock, stippled = sand, hatched = mixture of sand, rock and stones, MHWS = tide level at Mean High Water Springs, MLWS = tide level at Mean Low Water Springs. The shape of the coast remains complex at increasing magnifications, though it is roughly three

has now been cleared of rats and both species of seabird are beginning to return, the Puffins encouraged by wooden and concrete decoys placed near suitable nesting burrows, and the shearwaters by transplanting chicks into burrows, and broadcasting adult calls from loudspeakers (Saunders, 1986).

In addition to the length of the shore, its width is also important to birds because it determines the area available for feeding. Thus the Orkney Islands support many more waders (*c.* 51 000) on their wide shores than do the Shetland Islands (*c.* 12 000), even though the length of the coastline is less (Summers *et al.*, 1988). The width of the shore depends on its gradient. This may be steep in the case of a cliff or shingle beach consisting of hard rocks (as in the Shetland Islands), but very shallow in the cases of wave-cut platforms on softer rocks (such as the

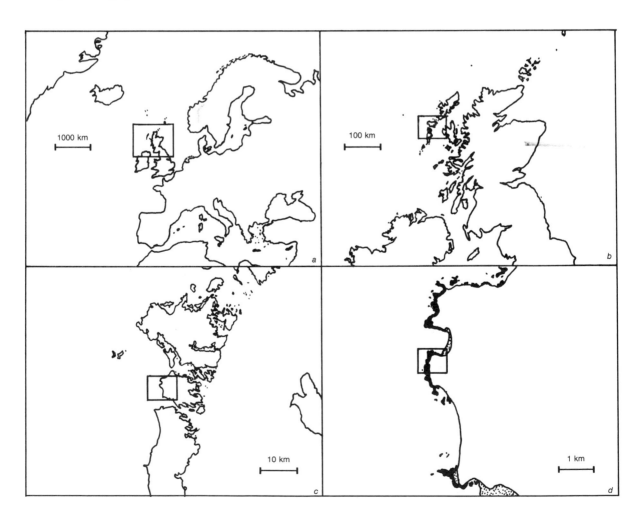

(Fig. 2 — *cont.*) times as convoluted in *a* to *c* than it is in *d* to *h*. This is one of the reasons why estimates of the length of the British coastline vary so much, depending upon the scale at which it is measured. For example, if the coastline measured on a map at the scale of *c* was 15 000 km, then maps at the scale of *b* would give a length of about 4000 km and maps at the scale of *d* about 20 000 km. An Oystercatcher can just be made out sitting on a nest in *h*.

Old Red Sandstone of the Orkney islands), or estuarine mudflats. Shore width also depends on the tidal range, which in Britain tends to be quite large. The average value for the whole country is 4.6 m at spring tides and 3.5 m at neaps, which compares with an overall average (springs and neaps) of 3 m or less for much of Europe, and under 1 m for the Baltic and Mediterranean. Available shore width may also be influenced by local factors. For example, exposed coasts with low tidal ranges, such as occur along the Atlantic coast of the Outer Hebrides and the west coast of Ireland, may have little of the intertidal zone exposed at all during storms because waves are driven ashore with sufficient force to sweep right up to the high water mark. Feeding opportunities for birds are reduced, but this may be offset to some extent by the fact that these same storms often bring ashore a large quantity of seaweed (especially

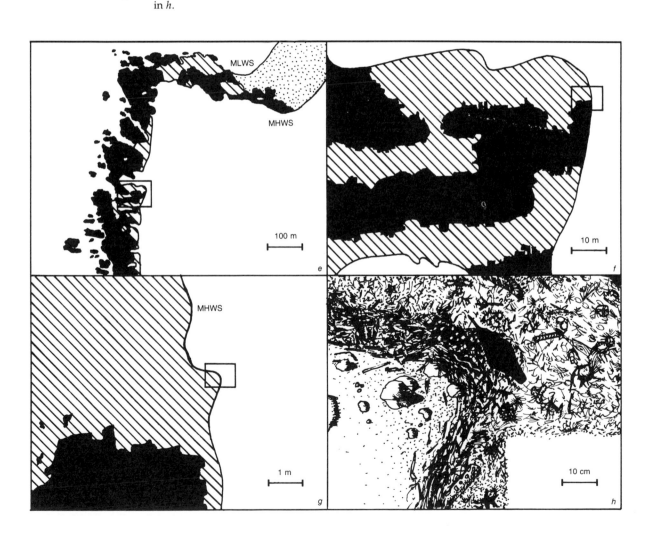

kelp) which piles up at the top of the shore and, as it decays, provides food in the form of sandhoppers, worms, maggots and flies.

The great length and diversity of our coast combine to provide a host of feeding and nesting opportunities for a wide range of bird species, including petrels, shearwaters, auks, gannets, cormorants, gulls, terns, skuas, waders and wildfowl. It is the task of this book to describe how these birds exploit coastal habitats. Britain itself is a small part of a wider biogeographical unit that includes the breeding and wintering habitats of many species, as well as the sites used during the migrations between them. For waders and wildfowl, the biogeographical unit is often referred to as a flyway and the major such unit in western Europe is the East Atlantic Flyway (Fig. 3). The major breeding area which this includes is northern Eurasia, but there are also several species that pass across into the New World to breed in Greenland and Canada, such as the Brent Goose and the Knot. These waders and wildfowl winter inland or along the coasts of Britain, Europe and northern and western Africa.

Most seabirds do not fit so conveniently into the flyway concept since they often disperse over a wide area of ocean during the non-breeding season (Fig. 3). We know little about their lives at sea, and population estimates are based on counts made at the breeding colonies. This is the complete reverse of the situation in waders and wildfowl where the breeding areas are remote and dispersed (e.g. across vast areas of tundra or sub-arctic wetlands), and counts can only be made at the relatively

**Fig. 3.** The East Atlantic Flyway and its associated breeding and wintering areas. The filled arrows show the direction of autumn movement from the former to the latter, whilst the open ones show autumn movements along the four other major flyways (P = Pacific, W = West Atlantic, E = Eastern European, A = Eastern Asian). Some typical species are shown as examples – breeders to the north of the equator and winterers to the south. Based on information in Harrison (1983), Mead (1983) and Davidson & Pienkowski (1987).

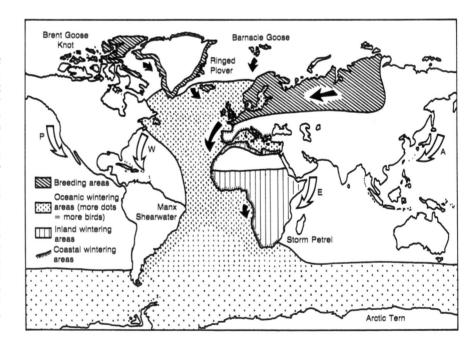

more concentrated and accessible wintering areas. Combining counts obtained by these different methods allows us to make an assessment of the overall importance of individual sites for coastal birds. This is based on the concept, enshrined in several international conventions, that an area which supports 1% or more of the population of a biogeographical unit or flyway for a particular species is of 'International Importance' for that species. If this criterion cannot be applied, for example if the size of the flyway population is unknown, a threshold value of 20% of the north-west European population is becoming increasingly accepted as a qualification for international significance (e.g. Batten *et al.*, 1990).

The fact that the list of internationally significant coastal species is so long in the case of the British Isles (70 species; see Appendix) is a measure both of the significance of our coast for birds and of our responsibility to conserve it.

# 1

## The coastal environment

The form of the coastline today is the result of geological processes which have taken place during the course of millions of years. A knowledge of the most important of these is helpful in understanding why, for example, our largest areas of rocky shore are situated in Scotland, and why most of our estuaries occur in the south. The structure of the coast is influenced by a complex array of factors including the nature of the bedrock, climate, waves, currents, tidal range, sea level change plus a number of biological factors, including the influence of man. This complexity has made it very difficult to devise a universally acceptable system for classifying coasts. However, nearly all such systems recognise the subdivision into erosional and depositional types. On erosional coastlines, material is constantly being removed by wave action and on depositional ones it is constantly being added. We will deal with these two types later, after considering some 'primary' forms of coastline which have been left over from previous geological events.

One of the most important factors influencing coastal morphology is changing sea level. This can result from changes in the volume of water in the oceans or changes in the level of the land itself. The latter may be due to tilting and/or folding of the earth's crust or may be a result of adjustments following the formation or removal of an ice sheet. In Britain, many of the sea level changes that have affected the coast are a result of movements associated with glaciation.

At the height of the ice ages, much of the world's oceanic water was locked up in ice sheets as the water evaporating from the surface of the oceans fell as snow over the expanded polar regions. Consequently, sea level was lower in absolute terms. However, the weight of the ice caused the land beneath it to sink, whilst that around the ice sheet bulged upwards in response to the squeezing pressure of the ice. When the ice melted, the sea level rose, but at the same time the land was freed from its burden of ice and also began to rise. Thus in those areas covered by ice sheets (and Britain was one), there was a complex interplay between the two factors – rising land and rising sea.

Approximately 13000 years ago, part of Scotland near Stirling was about 87 m lower than it is at present, but because world sea level was about 50 m lower than it is now, a beach formed at what is currently a height of 37 m above sea level. This is the highest raised beach in Britain, but there are many other lower ones further from the centre of glaciation where the land did not sink quite so far. In the vicinity of the English Channel, which was too far south to be ice-covered, the reduction in sea level due to glaciation was the dominant effect and thus Great Britain was joined by a land bridge to the rest of Europe.

Changes in sea level continue to take place at the present time. World wide, it is rising at a rate of a little over a millimetre per year, but this is more than offset in much of Scotland and Scandinavia by the fact that the land is rising (Shennan, 1987). Relative to southern England, Scotland is rising by at least 1.5 mm per year, and the south-west may be rising relative to the south-east, but by a smaller amount. These interactions mean that the relative sea level is rising in some places (see Fig. 4), and falling in others. More worryingly, there is increasing evidence that the greenhouse effect, caused by man's accidental interference with global heating and cooling processes in the atmosphere, is causing an acceleration in the rate of sea level rise. The implications of this for some coastal bird populations (and for man himself) could be serious and are considered in more detail in Chapter 6.

The bird fauna of coastal areas was obviously rather different during the height of the glaciations from what it is today. Although rather few fossil birds have been found from these periods, there are enough to know that while Woolly Rhinoceroses, Woolly Mammoths and Cave Lions roamed the tundras of Britain about 18000 years ago, Eagle Owls, Willow Grouse and cranes flew overhead. Moreover, during the interglacials (some of which were warmer than the present one), Dalmatian

**Fig. 4.** Relative changes in the level of spring high tides in north-west England. Curves from the south east show a similar rising trend, whereas those from the north show a declining trend because the land has risen faster than the sea. Redrawn from Tooley (1978).

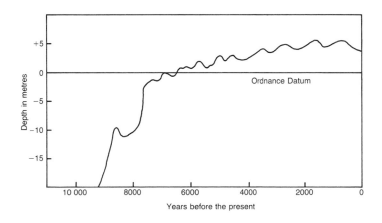

Pelicans, White Storks, Spoonbills and Jungle Fowl were present (Stuart, 1982).

Raised beaches are not the only evidence of past changes in sea level. Raised wave-cut platforms and cliff-lines are other features indicating relatively higher sea levels in the past. Such features are not usually of much relevance to coastal morphology today, but some submarine features exert a strong influence on the present coastline, giving rise to rias and fjords. Ria coastlines in Europe are often found in areas that were to the south of the ice sheet during the last glaciation and in which river systems developed as glacial meltwater streams above the lowered sea level. When the sea rose, it flooded these systems, producing long sea inlets (following the old river valleys) with narrow entrances, such as Milford Haven (Fig. 5). Such coastlines occur in southern Cornwall, South Wales and southern Ireland, as well as Brittany and northern Spain.

Many of our major estuaries are derived from rias in which sediment has accumulated because currents are slower than in both the rivers that feed them and the sea outside. This sediment gives rise to mudflats and saltmarshes, and the whole process may be accelerated by the formation of a sand spit across the entrance to the inlet. Mud and sand flats

**Fig. 5.** Geology of Milford Haven and south-west Wales. About 8000 waders and wildfowl winter in the Haven. In addition, 4000 or so pairs of auks, mainly Guillemots, nest on the Carboniferous Limestone cliffs to the south, together with a few pairs of Choughs. Based on Steers (1954), Evans (1986) and Lewis (1986).

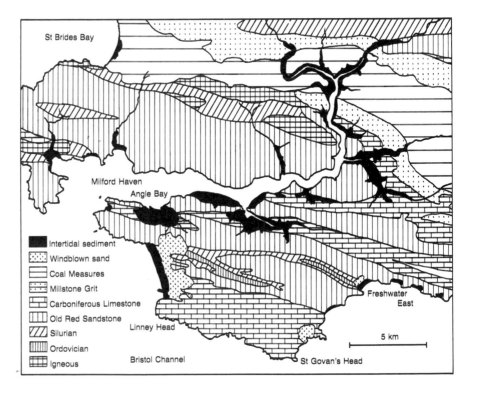

sometimes form in bays and inlets that have no major rivers associated with them, e.g. Portsmouth, Langstone and Chichester Harbours. These too are drowned land areas, which only later became silted up. The origins of all major British estuaries, such as the Wash, Morecambe Bay, Ribble, Dee and Severn, as well as many of the smaller ones, which altogether support around a million wading birds each winter, can be traced directly to the changing sea levels since the last Ice Age.

Fjords are glacial valleys, i.e. they have been gouged out by slow moving masses of ice rather than liquid water, and have been subsequently flooded by the sea. Typical fjords are valleys cut through solid rock, with a U-shaped profile and high valley sides. Typical fjord coasts occur in Norway. The firths and sea lochs of the west coast of Scotland are also fjords, but their valley sides slope less steeply than those in other parts of the world. Subsequent inundation of areas dissected by repeated glaciations has produced the complex network of islands and lakes that is so typical of much of Scotland (Fig. 6). The sheltered waters of this deeply indented coast support good numbers of Great Northern Divers, Black-throated Divers and Grey Herons (Moser *et al.*, 1986). The firths on the east coast of Scotland, however, are not fjords but are essentially rias, being more akin to the estuaries of England and Wales. Fjords fill up with sediment far more slowly than rias since they are usually fed by tiny rivers and because they are usually very deep. Areas of mudflat do eventually accumulate, but are much less extensive than those which form in rias and so are nowhere near as important to birds.

**Fig. 6.** Glaciated scenery of part of north-west Scotland showing numerous lochs and islands. All the named lochs have floors deepened by glaciers. The Sound of Raasay, the Inner Sound and several of the sea lochs are fjords. The arrows indicate the directions in which the glaciers flowed (deduced from Bickmore & Shaw, 1963).

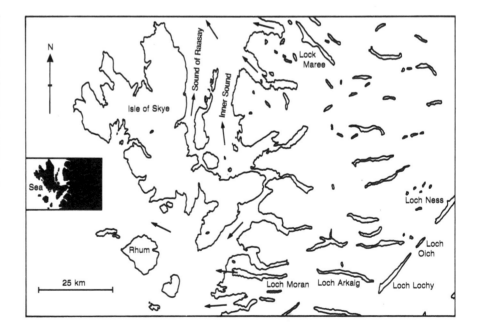

In ria and fjord areas, the nature of the bedrock (hardness, resistance to erosion) and its orientation (folding and faulting) often exert a strong influence upon the form of the coast. In Milford Haven (Fig. 5), softer areas of Carboniferous Limestone have been more quickly eroded than the Old Red Sandstone, and consequently the inlets of Angle Bay, Pembroke River, Cresswell River and Carew River have formed. These have filled with muddy sediments and they support most of the wading birds in the Haven. The lower reaches follow the east–west fold axis of the rocks. Many of the small bays along the Bristol Channel coast, such as Freshwater East, have formed along small faults providing points of weakness in the rock which has thus succumbed to wave action. Other areas of Carboniferous Limestone are more resistant to the force of the sea, for example those which form Linney Head and St Govan's Head.

While all these primary coastlines are the result of geological activities in the past, their present structure is a result of the present action of the sea, which tends to produce the four main types of coastline described below. In the first of these, erosive forces predominate, while in the remaining three it is deposition and accumulation of material that is most important.

### Sublittoral zone and the open sea

Around the British Isles, the seas over the continental shelf reach a depth of about 200 m. At the edge of the shelf, the continental slope dips down at an angle of about 5° until it reaches a depth of about 2000 m. The region between 2000 and 6000 m forms the ocean depths, which comprise about 83% of the world's sea floor. Although the continental shelf occupies only about 8% of the sea floor on a world-wide basis, and the continental slope about the same amount, they represent the majority of the submarine surfaces around the British Isles. Old river valleys, formed during the periods of lower sea levels during the ice ages, cross the shelf and in many places the sea bed is covered with deposits of glacial and periglacial materials. These deposits have been moved about by tidal currents where the latter are powerful enough. For example, just offshore in the North Sea beyond the Straits of Dover, and off the Norfolk coast, there are lines of sandbanks, such as the Goodwin Sands, which were formed in this way. Many of these once supported economically important fisheries for sole and skate, but the stocks are now depleted.

Solid rock outcrops occur over some areas of the continental shelf, usually where currents are high, as in the case of the English Channel. Mud occurs offshore too, as do materials of organic origin such as calcareous fragments from algae and shellfish off the coast of Brittany and in the Mediterranean. The zone immediately below the low water mark of spring tides is an area of transition from the extreme fluctuations in temperature, wave action and salinity that are characteristic of the intertidal zone, to the more consistent conditions of the sea itself. The subtidal zone is of course free from any risk of dehydration, which is one of the most significant influences upon intertidal animals. The only birds capable of exploiting benthic animals in the subtidal zone for food are those capable of diving to the bottom. These include Eiders and Black Guillemots on rocky shores and Long-tailed Ducks and scoters on sandy ones. Such species are able to forage in the intertidal zone at high tide, but as the tide recedes they can also exploit any subtidal shoals and sandbanks that may be within their diving range.

Further out to sea, the nature of the sea floor becomes less important to most seabirds since they increasingly rely upon plankton and pelagic fish as a source of food. However, those which depend on human fishing activities may still be influenced by the nature of the sea bed. For example, Norway Lobsters (scampi) occur where the sea floor is muddy and Lesser Black-backed Gulls feed extensively on the bycatch (i.e. discarded fish) from this fishery in the Irish Sea.

By the time the Gulf Stream, or North Atlantic Drift, reaches the British Isles, its waters are relatively unproductive because most of the nutrients have been used. Oceanic phytoplankton, the green plants at the base of the marine food chain, are limited to the surface layers of the sea through which light can penetrate – a maximum of 100 m under ideal conditions. As well as light, they need minerals and nutrients (just as land plants do) if they are to grow and reproduce. Particularly important nutrients in the sea, because they may become limiting, are phosphates, nitrates and silicates. On land these minerals are usually quite freely available because they are recycled from decaying organic matter in the soil, but in the sea they fall to the sea bed and become unavailable to the plankton. Only where water rises from the sea bed in the form of an upwelling, or in coastal waters where fresh water from rivers, rich in nutrients, floats on top of the denser saline sea, do these materials become abundant (see p. 80). The temperature of the water has little direct influence on the growth of planktonic organisms since they have evolved enzymes that function at whatever sea temperatures they are regularly subjected to. In fact, most of the world's productive marine waters are cold. In the tropics, the sea surface heats up, expands and, being less dense, remains at the surface. Its nutrients are soon exhausted and it thus becomes unproductive. The relatively low productivity of tropical waters is a direct consequence of this process.

There are regions of the sea where winds diverge – for example, between the trade winds blowing towards the equator and the westerlies towards the poles – and this causes water to rise to the surface to replace that which is pushed away by surface currents. The rising water is cold, because deeper water is normally colder and denser, but it brings nutrients to the surface from the sea bed and consequently enables plankton to thrive. This phenomenon explains the zones of high productivity present in the Arctic and Antarctic Oceans. Where ocean currents diverge, for example where water piled up at the equator by the trade winds begins to flow back again, there may also be upwellings (see Fig. 30) and these in turn explain the zones of high productivity that occur in these regions.

Nearly all coasts are regions of enhanced productivity because of nutrient-rich river run-off, tidal mixing and upwelling. However, there are some especially productive regions along the west coast of South America and southern Africa in places where few rivers enter the sea. This is explained by the prevailing currents which sweep cold, nutrient-rich Antarctic waters northwards past the projecting tips of these continents. These are known as the Humboldt and Benguela Currents respectively. Some of the world's largest seabird colonies are situated in these areas and consequently some of the world's thickest guano deposits occur there. This guano is derived from the nutrients that birds have removed from the sea in the form of krill, squid and fish, and not surprisingly it makes an excellent agricultural fertiliser. Any onshore wind that blows across these cold marine currents picks up very little moisture so that very little rain falls, resulting in the formation of extensive deserts inland. Occasionally, the normal pattern of water circulation is disrupted and warmer water stays at the surface with devastating consequences for seabirds and fisheries. Off the Pacific coast of South America these are called El niño or ENSO (El niño southern oscillation) years. During the last decade, ENSO events occurred in 1982/83, 1986/87 and 1991/92. The effects on seabirds can be extremely widespread, reaching as far away from the South American coast as Johnston Atoll (west of Hawaii) and Christmas Island. Seabird population sizes over a wide area may actually be directly limited by these periodic catastrophes (Schreiber & Schreiber, 1989).

A zone of high productivity around the British coast (see p. 78) is important to the immature stages of many fish that become commercially important as adults. However, the immature fish themselves are being increasingly exploited for fish meal manufacture. The young fish feed directly on the zooplankton, that in turn feed on the phytoplankton, though the feeding relationships between the many thousands of different species in the food chain of the open sea are complex. Unlike the green plants that form the basis of the food chain on land, most planktonic plants are minute in size, usually consisting of just one or two cells. Their total mass at any one time is small, to such an extent that there may actually be a greater bulk of zooplankton feeding upon them. This seeming paradox is explained by the fact that they grow and reproduce very rapidly. Occasionally, larger photosynthetic plants are found in the sea. For example, in the intertidal zone, seaweeds can anchor themselves to the sea bed and still receive adequate amounts of light, and in the Sargasso Sea in the Caribbean the free floating weed *Sargassum natans* is able to survive.

While the Gulf Stream's nutrients are mostly exhausted by the time it reaches the British Isles, its waters are still relatively warm. Thus the prevailing south-west winds are warm and damp, and this exerts a major influence on the general climate of the British Isles, making it much milder than, for example, Newfoundland which is situated at the same latitude on the other side of the Atlantic. Hebron on the coast of Newfoundland is at the same latitude as northern Scotland, but has average monthly temperatures below freezing for eight months of the year. The absence of long periods of freezing conditions makes Britain and Ireland an ideal wintering site for waders and wildfowl whose ability to feed would be impaired by such conditions.

The average productivity of the sea (Table 1) is only about 20% of that of the land, but intertidal ecosystems are nearly as productive as the most productive terrestrial ones. A tropical rain forest forms about 2 kg of new tissue per square metre per year, but needs 20 kg of existing tissue with which to achieve it. Most of this is in the form of the wood needed to support trees in their constant struggle to reach the light. A kelp bed can achieve the same productivity with less than 5% of the biomass. In fact, the total biomass of plants in the whole of the marine ecosystem is only 0.2% of that of the plants on land and yet they contribute one third of the world's annual total productivity. This is mainly due, of course, to the large size of the world's oceans compared with the land – they occupy about 70% of the earth's surface.

### Cliffs, stacks and rock platforms

During the formation of a cliff, wave action first cuts a notch at the base. The overhanging material then collapses onto the beach where it either remains or is carried away, sometimes accumulating below the low tide mark where wave action is less powerful. The net result is a cliff (the steepness of which depends upon how rapidly the rock crumbles and falls off) above a wave-cut platform. Waves also erode caves, frequently along lines of weakness in the rock such as faults or softer strata. The roofs of caves eventually collapse, leaving isolated offshore pillars or stacks. The latter can provide ideal nesting conditions for birds, for they

Table 1. *Productivity of the oceans compared with some terrestrial ecosystems. From Whittaker (1975)*

| Ecosystem | Average net primary production[a] (g dry matter/m²/year) |
|---|---|
| Open ocean | 125 |
| Continental shelf | 360 |
| Upwelling zones | 500 |
| Estuaries | 1500 |
| Algal beds and coral reefs | 2500 |
| Average for the oceans | 152 |
| Desert | 3–90 |
| Tundra | 140 |
| Grassland | 600 |
| Cultivated land | 650 |
| Coniferous forest | 800 |
| Temperate deciduous forest | 1200 |
| Marshland and swamp | 2000 |
| Tropical rain forest | 2200 |
| Average for the land | 773 |

[a]This is the total amount of new growth by plants each year, regardless of whether it dies or is eaten by herbivores.

offer sheltered conditions on their lee side, as well as security from terrestrial predators.

Most of the erosion that forms these features takes place during storms. Soft deposits, such as clay and volcanic ash, erode quickly and because of constant slumping do not produce a steep cliff with nesting ledges suitable for birds. The section of cliff between Spurn Head and Flamborough Head in Yorkshire is of this type. During the last century, before coastal defences were erected, this area was receding at a rate of about 2 m a year, with whole villages being lost to the sea. Some sections of this coast are still eroding at a high rate. The sediment arising from this erosion is carried south past Spurn Head and contributes to the infilling of the Wash and thus the formation of the mudflats and sand banks in one of Britain's largest estuaries.

Harder rocks retreat less rapidly and tend to produce vertical cliffs unless the strata dip towards the sea. Calcareous rocks, columnar basalts, sandstones and harder shales produce some of the best cliffs for nesting seabirds. The fallen material that accumulates at the cliff base during autumn and winter storms also provides suitable nest sites for species such as Shags and Storm Petrels if the gaps are of the right size and it is far enough above the splash zone (Fig. 7). Cliffs with no wave-cut platform, such as occur in many fjords, do not provide anywhere for scree to pile up above the water mark and so they lack nest sites of these types. Furthermore, since so much of our coast has been subject to sea defence works, there are many areas where scree no longer accumulates.

Fortunately man's activities are not entirely negative since some species have taken to nesting on buildings, notably several species of gulls. Black Guillemots sometimes nest in holes in breakwaters and harbour walls. Dry stone walls and ruined buildings resemble scree and are used for nesting by Storm Petrels. Swifts, Swallows and House Martins, which were limited in distribution to natural cliff sites during the Medieval period or earlier, now nest almost exclusively on man-made structures.

**Fig. 7.** Carboniferous Limestone cliffs and rocky beach near St Govan's Head, Dyfed.

Cliffs do not necessarily have to be close to the sea in order for them to be exploited by seabirds. Some Fulmar and Kittiwake colonies are found well inland. There is a Cormorant colony on cliffs at Towyn in North Wales situated about 8 km from the sea, and this species also nests in trees occasionally. In the rest of Europe, Cormorants often nest well inland and feed in freshwater areas. The majority of auks, however, are limited to sites within a short distance of the sea, because their young leave the nest before they are fully grown and capable of flight. Two exceptions to this are the Puffin, which is almost fully grown and capable of flight at fledging, and Brünnich's Guillemot, which can glide for considerable distances at fledging, even though it is only about a quarter of the weight of an adult. Both species occasionally nest as much as 1 km from the sea.

Cliff ledges may become colonised by salt-tolerant vegetation, such as Thrift. At the cliff top, salt-tolerant species grade into ordinary plants, although the transition may occur some distance away from the cliff edge if salt-laden air drifts inland during winter storms. The quantity and luxuriance of the vegetation on the cliff face itself depends on two main factors – the amount of soil, and the aspect. More soil accumulates where the strata dip landward, since pockets form naturally under such circumstances. Southerly and westerly facing cliffs receive more sunlight, which both warms the soil and favours photosynthesis, but they are also exposed to more intense driving rain and salt spray from the prevailing wind. This provides conditions similar to those in a salt-marsh. Since cliff soils usually occur in shallow pockets, drying out is also a problem in the summer. The net result of all these factors is that northerly facing cliffs tend to offer the most favourable conditions for plant growth. There have been no systematic studies of the uses made by birds of cliff vegetation, but the lack of grazing pressure usually means that a good seed crop is produced, thus providing good feeding opportunities for species such as Linnets, Chaffinches and Skylarks.

On the wave-cut platform situated at the foot of most cliffs, an inter-tidal rocky shore flora and fauna typically develops. This shows a well-marked zonation with different and characteristic communities of plants and animals occurring at different levels of the shore according to the amount of tidal exposure. These communities generally have to withstand more severe wave action than soft shore ones, but they do have the advantage of a secure surface upon which to attach themselves. Such shores support several species of waders and gulls which forage at low tide, as well as wildfowl and other diving species which forage in the intertidal zone at high tide and in the subtidal zone at low tide.

Table 2. *Definition of sedimentary materials on the Wentworth scale. From Bird (1984)*

| Material | Diameter range (mm) |
| --- | --- |
| Boulders | >256 |
| Cobbles | 64–256 |
| Pebbles | 4–64 |
| Gravel | 2–4 |
| Sand | 0.0625–2 |
| Silt | 0.0034–0.0625 |
| Clay | <0.0034 |

## Beaches, spits, barriers and dunes

Unlike cliffs, where continual destruction by the sea is the norm, beach and dune systems usually undergo a slow building process, albeit interrupted by phases of rapid erosion during storms. Terms such as 'gravel' and 'pebbles' tend to be used in imprecise ways, but they do have a precise meaning (Table 2). A real beach usually contains a complex and partially sorted mixture of several grades of material. The individual fragments are much rounder than similar materials found in fresh waters because of the more frequent and powerful abrasive forces acting upon them.

Where does the material that forms beaches and dunes come from? This is a difficult question to answer. Most boulders cannot be moved long distances without the aid of ice and so these have normally come from nearby cliffs. Lumps of soft, friable material derived from cliffs soon break down into fine sediments and are swept away. Sandstones may degrade directly into sand if the grains are not well cemented. Only the hardest materials remain intact for long enough to form the mass of pebbles and shingle on our beaches. As these rub together they slowly degenerate into silt, not sand, so rocks do not necessarily pass down through all the size grades indicated in Table 2 as they erode. Not all beach materials originate locally. Sand, mud and silt may come from rivers or be swept shoreward from the sea floor. With the rise in sea level since the last Ice Age, the shore has migrated landwards over a considerable distance and may have carried with it a large amount of periglacial material.

One of the most characteristic pebble-forming materials is flint. This falls from the joints in the chalk rocks as they are eroded by the sea. However, this is not the only route by which they reach the sea. Flints exposed on the wave-cut platform below the beach provide ideal anchorage points for the holdfasts of seaweeds. When these eventually break away (often in storms) they get carried shorewards by the action of the waves on the weed. Many of the shingle beaches of south-east Britain are derived from flint, and the fact that many chalk cliffs are protected from erosion without a noticeable diminution in the amount of shingle suggests that the latter mechanism, involving seaweed, provides the major source of flint pebbles. In other parts of the country, shingle beaches may be dominated by shale, slate or sandstone fragments. Beaches on granite coasts are often sandy since this rock weathers directly into its sand-forming mineral crystals.

Waves and currents are the dominant factors causing the movement of beach materials. Most waves are the direct result of wind action and they obviously travel in the direction in which the wind pushes them. Swell is the term used for waves that travel beyond the area where the wind is actually blowing. The direction and other properties of waves change once they reach shallower water. The coast modifies wave direction in three ways – refraction, reflection and diffraction. Reflection occurs when a wave hits a vertical or steeply shelving object and bounces back. Diffraction occurs when waves pass the end of an obstruction and they peter out in the lee of the obstuction in a variety of different directions. Refraction occurs in shallow water which slows down the waves and makes them take up a curved shape in bays and inlets (Fig. 8).

**Fig. 8.** The Rhins – an example of a tombolo. Wave refraction in Luce Bay modifies the direction of the approaching waves, transporting beach material northwards and increasing the size of the neck of land connecting The Rhins to the mainland. The bayhead beach conforms to the shape of the approaching waves.

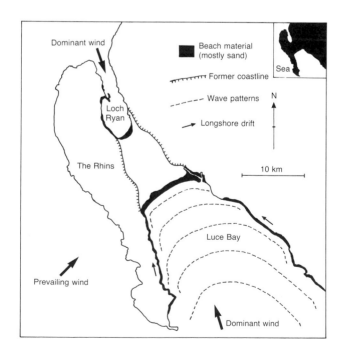

While the period between waves remains unaltered as they approach the shore, the wavelength and speed decrease whilst their steepness increases, especially at the front. Whether they break or not, a large part of the energy they contain is dissipated in moving material up and down the beach at the water's edge. If they approach the beach at an oblique angle, this material is carried up the beach at an angle (Fig. 9) and the net result is that it gets transported along the beach in the direction of wave movement. This is known as longshore drift. If the waves mostly travel in one direction, the beach gradually moves in that direction too. In the case of bays, longshore drift carries material to the head of the bay where it accumulates and adjusts its shape to the dominant wave pattern as illustrated in Figs. 8 and 10. The dominant wind is not necessarily the same as the prevailing wind (which is south-westerly in most of Britain), since if the latter is offshore it causes no significant wave action.

One of the reasons why shingle and pebbles accumulate at the top of the shore, and finer sediments lower down, is that the forward movement of waves up the beach (swash) is more powerful than the backward movement (backwash). The swash can thus carry large stones, but the backwash only small ones. A good deal of the higher level beach material is deposited in this way during storms.

Longshore drift leads to the formation of a number of important coastal features, many of which are significant for birds. It can form spits of shingle, sand or silt by simply extending a beach out into open water where there is a sudden change in the direction of the coast, such as near the entrance of a bay or estuary. A good example of such a structure is the Spurn peninsula (Fig. 11). The dominant wind and wave action in this part of the North Sea is from the north east, the prevailing south westerly winds causing no significant waves near the shore on the east coast. Thus longshore drift carries material from the rapidly eroding

**Fig. 9.** How longshore drift works. The dominant wind is the one that produces the waves which determine the net direction of movement of beach material (usually because it has the greatest fetch).

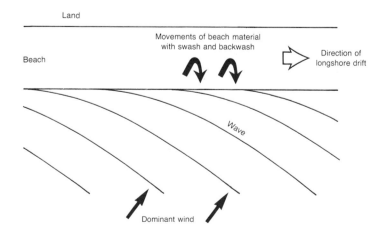

cliffs in the north, southwards past the mouth of the Humber. The spit that is formed in this region is recurved, i.e. deflected inwards, towards the Humber Estuary. It is the fifth spit for which historical records exist. Each of its predecessors was formed slightly further out in the North Sea at a time when the cliffs to the north had a more seaward location. Each was in turn destroyed when it became detached from the beach that was its source of sediment, usually in a storm (Boer, 1988).

Birds flying south along the North Sea coast during autumn (and spring) concentrate along the Spurn peninsula and it was for this reason that a bird observatory was established there in 1945 – the first on the British mainland. Why so many passerines migrate south past Spurn in the spring remains something of a mystery, but the waders that also do so are presumably returning to breeding grounds in Eurasia from wintering grounds in northern Britain, perhaps via gathering areas in the Wash and Wadden Sea (Cudworth, 1976). The remoteness of the peninsula, and of parts of the south side of the estuary, have enabled Little Terns to continue to breed there (Goodall, 1988).

**Fig. 10.** Sandy bayhead beach near St Govan's Head, Dyfed.

Sometimes a spit may form a connection between an island and the mainland, and such a structure is known as a tombolo. In the Orkneys, Shetlands and Isles of Scilly there are islands which are connected together by tombolo-like structures. The Rhins in Scotland (Fig. 8) is a tombolo. Longshore drift continues to accumulate material in Luce Bay and Loch Ryan, and thus the area of fusion with the mainland is still increasing. The dominant wind in Loch Ryan occurs less frequently than that in Luce Bay, and the northern side of the neck thus grows more slowly. Both bays support wintering waders and wildfowl (about 10 000 in total), although Loch Ryan has more than half of these, perhaps because of its greater shelter.

When longshore drift from two different directions meets on the open coast there can be large accumulations of material just as there is in a bayhead beach. Dungeness is such a structure, representing a massive accumulation in a succession of shingle ridges (Fig. 12). At one time, a continuous line of shingle probably enclosed a lagoon which developed into the Walland and Romney Marshes, and cut off the port of Appledore from the sea. This shingle has grown into its present eastwardly migrating, projecting form. As on most mobile shorelines, man has taken steps to control this process (by constructing sea walls and by artificially feeding the beach with shingle), in this case to protect the nuclear power stations that are situated close to the southerly tip (May, 1985).

Dungeness is located close to the narrowest part of the English Channel and this made it an ideal site at which to establish a bird observatory in 1952. The cooling water from the power stations, which wells up at

**Fig. 11.** The Spurn peninsula. Although the prevailing wind is from the south-west, as in most of Britain, the dominant wind is from the north-east.

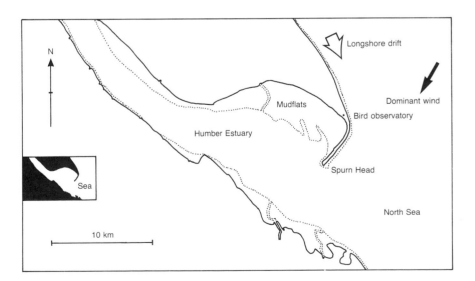

the 'patch', bringing food to the surface, is one of the best known sites in the country for seeing seabirds at close quarters. Kentish Plovers once bred on the shingle ridges in a colony that reached 44 pairs in 1906 after being protected from egg collection, but soon disappeared when the Dungeness to Hythe miniature railway was constructed through the middle of the colony and bungalows were built in the area. Common and Little Terns continue to breed, but in much smaller numbers than in the last century, and on islands in flooded gravel pits rather than on open shingle. Much of the shingle has now become colonised by scrub in the absence of sheep and goat grazing (Scott, 1985).

One of the most poorly understood geomorphological features of the coast is the barrier beach. Very extensive barrier beaches, broken by tidal entrances, occur along much of the Atlantic coast of the USA. The Wadden Sea in the Netherlands is enclosed by the barrier island chain of the Frisians, which is merely an enlarged and very segmented barrier beach. One way in which such structures may have formed is by the elongation of spits. This seems to be the explanation in the case of Orford Ness in Suffolk where the mouth of the river Alde is deflected some 18 km to the south by a long shingle spit. A number of islands have formed along the course of the river behind the spit, one of which is Havergate. This island was embanked to protect the grazing land from tidal inundation, but when a sluice was destroyed by a stray shell during the Second World War, the embankments retained some brackish water and Avocets bred. In 1949 the RSPB acquired Havergate as a reserve and it has been managed for breeding Avocets and other birds ever since (Cadbury *et al.*, 1989).

**Fig. 12.** Dungeness. Redrawn from Bird (1984) after Lewis (1932).

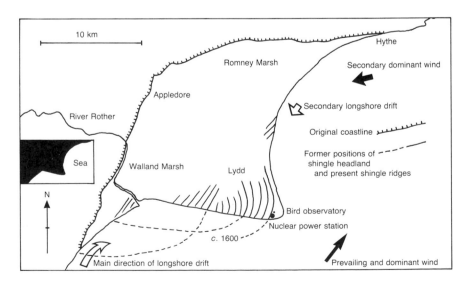

Chesil Beach (Fig. 13) is rather more of a mystery than Orford Ness. It extends as an unbroken ridge of shingle from Bridport to the Isle of Portland. Wider and higher at the Portland end, it rests on a bed of clay just above the low water mark. The lagoon behind the beach (the Fleet) is brackish and its landward margin consists of a coastline which has never been subjected to the erosive power of the sea and hence has no cliffs. The beach is still moving slowly towards the land as storms carry shingle over the top and deposit it in the Fleet. Eventually, as at the west end today, the whole beach will come to lie against the land behind the Fleet and cliffs will then begin to form during storms.

Chesil Beach may have been formed in the same way as the great North American barrier beaches, i.e. by a shoreward sweeping of material previously deposited by glacial meltwater rivers on land that is currently below sea level. This material accumulated at the high tide mark and, as the sea level rose following the end of the last ice age, gradually rose with it. Water levels in the Abbotsbury swannery at the western end of the Fleet have been controlled for some 600 years in

**Fig. 13.** Chesil Beach from the Isle of Portland, with Portland Harbour to the right and its connection with the Fleet in the middle distance.

order to maximise the breeding productivity of the Mute Swans. The young swans were marked by nicks in the webs of their feet to indicate ownership at the annual swan-upping, as well as being pinioned to make them easier to catch for the table (see also p. 192).

At the other end of Chesil Beach, the Isle of Portland is a massive block of Portland Limestone. The bird observatory on Portland Bill is ideally situated at one of the major points of departure and arrival of migrants from the western half of Britain. Projecting as it does into the English Channel, it is also an excellent place for the observation of seabird migration with large eastward movements during the spring (Clafton, 1976).

Sand and shingle spits and barrier beaches sometimes wholly or partly enclose bodies of water. These vary in character from freshwater lakes to estuaries, depending on their degree of communication with the sea outside. The Ley at Slapton in Devon is an example of a relatively isolated lagoon. Depending on rainfall and the state of the tide, salt water may seep in through the barrier beach, or brackish water from the lagoon may seep out. Such areas gradually silt up and develop extensive reedswamp and willow carr. The birds that occur there are a mixture of coastal and freshwater species. The Hampshire and Sussex Harbours (Portsmouth, Langstone, Chichester and Pagham) have something of the character of coastal lagoons about them because they have very narrow connections with the sea. However, only small freshwater streams flow into them, and consequently they are very saline. Coastal lagoon complexes such as the Camargue in France and the Coto Doñana in Spain provide some of the world's finest wetlands for birds.

Almost 7% of the coastline of England and Wales (500 km) is backed by sand dunes. Obviously, these only occur where there are sandy beaches and adequate supplies of incoming sand. Another feature that facilitates dune building is an onshore wind, and consequently the largest dune systems in Britain are found on west and south facing shores. When an onshore wind is blowing, sand at the top of the beach dries out and blows inland to form dunes (Fig. 14). Foredunes build up immediately behind the beach but can be repeatedly cut back again by storms. Dunes can build in other ways. For example, if the prevailing wind is offshore, and sand is supplied by longshore drift, they can build in a seaward direction.

Plants play a vital part in trapping and stabilizing wind-blown sand. Marram and Sea Wheat Grass are two of the most important species. The former is slightly tolerant of salt and actually requires a continuous supply of sand in order to thrive. As the foredune begins to migrate inland, so a new one forms in front of it, and eventually a whole series of landward moving dune ridges arise. These ridges become broken wherever the vegetation is destroyed by fire, overgrazing or erosion. Wet areas (dune slacks) may form in the depressions between dunes. The plant communities of the dunes furthest from the sea become more complex and may include trees, so these dunes are more stable (Ranwell & Boar, 1986). Rather few birds exploit the unstable foredunes, but the fixed dunes often support thriving communities. Many of our largest colonies of Eiders, Black-headed Gulls, Herring Gulls and Sandwich Terns (Fig. 15) occur in this kind of habitat.

Machair is a special type of sand dune system found on the Atlantic facing coasts of some of the Hebridean islands and in a few places in Ireland. Instead of the stable dunes being relatively undulating, they are replaced by a flat grassy plain (Fig. 16). The mechanisms of machair formation are not fully understood but they appear to have involved the erosion of a previously higher sand surface (formed at a time of rather different climatic conditions), or the infilling of water-filled depressions by wind-blown sand. The sand that makes up the machair is almost entirely shell-sand derived from the rich communities of bivalve mol-

**Fig. 14.** Formation of sand dunes. Three successive stages are shown in which the dunes are built in a landward direction by a prevailing onshore wind.

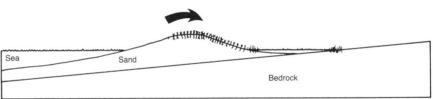

Initially, a single dune forms at the top of the beach, colonised by Marram, with brackish dune slacks behind it

The original dune then migrates inland and becomes colonised by more dense vegetation such as Sea-buckthorn while a new dune forms in front of it

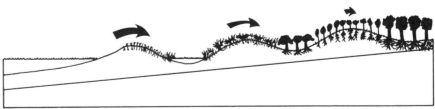

The process continues and the most landward dunes become colonised by White Poplar, with sallow in the damper hollows. Eventually, other trees characteristic of mature woodland take over and there may be many more than three dune ridges

luscs that occur offshore. These communities live in the sandy under-water sediments formed from the broken shells of their own ancestors. Rocky outcrops also occur along with much wider tracts of rocky sea bed further out to sea. These support submarine forests of kelp. Red-throated, Black-throated and Great Northern Divers feed on fish amongst these kelp forests, whilst Long-tailed Ducks feed on the bivalves in the sandy bays.

Being made up of shell-sand, the machair is rich in calcium. When fertilised with kelp collected from the beach after winter storms, it makes a productive soil which has been farmed for centuries. Both

**Fig. 15.** Sandwich Terns nesting on a sheltered sand dune at the top of the shore. Chris Rose.

cultivated and uncultivated machair support large populations of breeding waders – Ringed Plovers, Lapwings and Oystercatchers on the drier areas, and Redshanks, Dunlins and Snipes where it is wetter. Inland from the machair are less productive grasslands and moorlands supporting fewer birds.

### Estuaries, marshes, deltas and reefs

In the relatively sheltered conditions of deep bays and inlets, and behind spits, finer sediments are able to settle out giving rise to mudflats. The largest areas of mudflats occur in estuaries, which can be defined as partially enclosed coastal waters with a free connection to the open sea, but within which the sea water is significantly diluted with fresh water from a river system. The salinity of sea water is about 35 parts per thousand of salt, whilst that of river water is less than 0.5 parts per thousand. Thus the range of salinities found in estuaries lies between these two extremes.

Many British estuaries are rias, and sedimentation is gradually filling them in. If the sea level were to remain constant for long enough, many such systems would end up as dry plains with rivers winding their way through them to the sea. Sediments in estuaries may originate from three different sources – material washed down by rivers, material washed in from the sea and material eroded from the adjacent shorelines. The amount in suspension in estuarine waters is usually much greater than in the sea outside or in the rivers feeding it, which makes the problem of determining the origin of the accumulated sediment quite difficult. A good deal may have been swept in from the sea during the postglacial rise in sea level by the same sort of processes which formed barrier beaches. Where rivers are rich in sediment, they may constitute the major source. This is the case in many Mediterranean estuaries where removal of tree cover and agricultural over-exploitation of river basins since Classical times has resulted in rapid soil erosion. Mining operations may likewise contribute to estuarine sediments, as for example in the River Fal in Cornwall which has much china clay waste in its mudflats.

The general form that mudflats take depends on an estuary's overall

**Fig. 16.** Section through machair. The prevailing onshore winds in this case are very strong and do not permit high dunes to persist. Instead they are replaced by a flat grassy plain, very close to the water table. Redrawn from Ritchie (1979).

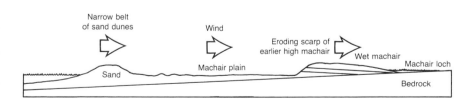

shape, but there is a continuous process of change. The positions of creeks and river channels move and meander across the mudflats much more rapidly than in the case of a river crossing its floodplain on dry land. The Severn Estuary has a large tidal range and therefore considerable quantities of sediment are held in suspension. The estuary narrows near the Severn Bridge (top right hand corner of Fig. 17) where there is a hard bedrock of Carboniferous Limestone. Because of this narrowing, the velocity of the water is high and little sediment settles out, leaving an almost bare rock platform. Just downstream of the narrows where the velocity drops slightly, the heaviest materials that the water can carry (gravel and stones) settle out. Much of the central part of the estuary is occupied by a large area of mobile sand. Around the periphery of the estuary, where the tidal currents are reduced, muds have settled out. The sand banks at the top of the shore at Weston-super-Mare, Sand Bay and Berrow probably formed as typical bayhead beaches at a time when there was much less mud in the estuary. At a later date, the sand on the lower shore became covered by mud as the estuary filled up with finer sediments. Even the top of the shore is now becoming more muddy, much to the consternation of those who depend on the holiday trade at Weston-super-Mare. Here, sand is being added to the top of the shore artificially in order to preserve the sandy beach.

**Fig. 17.** Intertidal zones of the lower Severn Estuary. More shorebirds feed on the peripheral mudflats than on the relatively mobile central sandbanks. Based on Evans (1980).

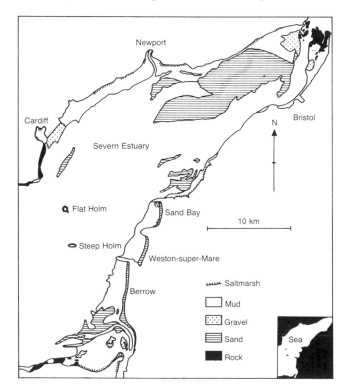

Estuaries are characterised by high turbidity in the water column, by low and fluctuating salinities, and by the dominance of tidal currents over wave action. Fewer invertebrate animals survive under these difficult conditions compared with sandy areas outside estuaries. Filter feeders do not thrive in areas of high turbidity because their filtering systems become clogged with silt. Consequently they are replaced by species feeding on the deposited organic materials which are abundant on estuarine mudflats (McLusky, 1981). In turn, a whole community of different species of birds, notably waders with long legs and long bills, has evolved to exploit these animals as food (Fig. 18).

Seabirds do not usually penetrate very far into estuaries because most of them require clear water in which to locate and capture their prey. The Cormorant is an exception in being able to catch fish in murky water. However, estuaries are important nursery areas for young fish such as flatfish, Bass and Sprats, and these become important seabird prey at a later stage in their lives.

Towards the top of the shore, vegetation plays an important part in stabilizing mudflats. In temperate areas, there is a distinct zonation of different species up the shore, from diatoms on the open mudflats (especially in the summer months), through carpets of eelgrass (*Zostera*) and *Enteromorpha* (which are important sources of food for estuarine grazers such as Brent Geese and Wigeon), to the larger plants of the saltmarsh proper. In Britain, many of our saltmarshes are dominated by *Spartina anglica* (a modified hybrid between the native *Spartina maritima* and the accidentally imported *Spartina alterniflora*) which developed spontaneously in Southampton Water during the nineteenth century. It has since been deliberately spread all over the country by man. This is because its vigorously growing shoots trap sediment and hence raise and stabilise the upper shore, minimising the impact of wave attack on vulnerable sea defences. It is also used as an aid to the reclamation of intertidal areas for agriculture and industry. It has largely replaced glasswort (*Salicornia*) in the lower shore marshes of most British estuaries, though it gradually becomes replaced on the upper marsh by native plants more tolerant of long periods of exposure (Long & Mason, 1983). In some parts of southern England where *Spartina* was first introduced, it is dying back for reasons that are completely unknown (Tubbs, 1984). The breeding bird communities of *Spartina* areas are poor compared with botanically more diverse saltmarshes, but they do sometimes include populations of Lapwings, Redshanks and Dunlins.

In tropical areas, mangroves often take the place of smaller saltmarsh plants. There are many species of mangrove, all adapted to life in tidal conditions. They have an elaborate system of prop roots to anchor the plant in the soft sediments and support it against the force of the waves.

**Fig. 18.** Mudflats provide a safe anchorage for boats and a rich feeding area for shorebirds. At least four species of waders and two of wildfowl are visible in this case (Grey Plover, Dunlin, Redshank, Curlew, Brent Goose and Shelduck). Chris Rose.

Mangroves in some parts of the world, for example in West Africa, provide important wintering areas for coastal birds from Europe such as Whimbrels and Common Sandpipers. Along the coast between the Gambia and Equatorial Guinea, for example, there are 19 000 km$^2$ of mangroves which are used by coastal birds. Unfortunately they are very difficult to census effectively (Altenburg, 1987).

When the amount of sediment carried by a river exceeds the capacity of the currents at its mouth to carry it away, it accumulates there in the form of a delta. Rivers in some parts of the world have enormous deltas, such as those of the Ganges and Brahmaputra which together cover 75 000 km$^2$. Active delta regions undergo considerable change as material is deposited and new channels are formed. As a result of these processes the delta gradually progresses seawards. Spits and bars of material may migrate away from the delta under the action of longshore drift. Since tides and currents are strong in Britain, and because most rivers discharge into rias, there are few deltaic structures around our coasts. However, delta-like systems do occur on both sides of the entrance to Pagham Harbour in Sussex. In this case, the sediment is coastal rather than riparian in origin and the deltas are formed by tidal rather than river action. Similar double tidal deltas occur in the gaps between the Frisian Islands in the Netherlands.

Coral reefs are a distinctive coastal feature, composed solely of the accumulated calcium carbonate skeletons of coral polyps. The reefs increase in thickness as successive generations of polyps build on top of each other. Coral reef shorelines are found only in the tropics (within 30° of the equator) in clear, shallow water, since the algae which live within the coral tissues and are important for their survival, require light for photosynthesis. Fringing reefs are formed adjacent to the shore, barrier reefs offshore, and atolls around islands which have sunk. A fall in the level of the land relative to the sea is necessary for the formation of both barrier reefs and atolls. Some reefs have bases which are so deep that even the largest fall in sea level during the ice ages would not have uncovered them indicating that they must be situated on parts of the earth's crust which have sunk. Many of the world's seabirds breed on coral structures that have emerged from the sea and become dry land. This is sometimes the only land available in large tracts of open ocean that have suitable food supplies but would otherwise be unexploited by breeding birds.

The 'reclamation' of land from the sea is in some cases a natural process, for example when saltmarshes or sand dunes are formed and then undergo succession to eventually support woodland. Man has greatly accelerated this process by planting *Spartina anglica* and by the construction of sea walls, dykes and ditches. These completely exclude the tide and yet allow rain water to drain away at low tide through sluices. Many thousands of hectares of land that are below the high water mark of spring tides have been reclaimed in this way. On the south side of the Severn Estuary, the Somerset Levels are an example of this, representing the second largest area of fenland in the country. In its original state this area must have been a spectacular wetland with populations of Grey Herons, Bitterns and various marsh-dwelling warblers. The bones of pelicans, cranes and White-tailed Eagles are also known from deposits dated at around 1000 BC (Coles & Coles, 1986). The value of such areas to birds in the more recent past has also been considerable. Since they are prone to flash flooding, they can provide feeding opportunities for a number of coastal species at high tide in winter. The high water table has slowed down the rate of agricultural change and the rough, damp pastures have supported a high proportion of the inland breeding wader populations of southern Britain. Unfortunately, most of the value of these areas is now disappearing as drainage 'improves' them and they come to resemble farmland elsewhere in the country.

# 2

# The open sea

In order to exploit the food resources of the open sea, animals that have evolved from land-based ancestors, such as the birds, have had to become highly specialised. Only a few groups have managed this, and it is worth reviewing the differences between them.

The penguins (Order Sphenisciformes) comprise 18 of the most highly specialised species of seabirds on earth. They have lost the power of flight and the wings have become modified as flippers for underwater propulsion. They are limited to the colder parts of the southern hemisphere except where the productive Humboldt and Benguela Currents penetrate into the tropics. Like marine mammals, they have a thick layer of blubber to insulate the body from the cold and the feathers are modified to produce a thick, dense pile, almost like that of a carpet. The divers or loons (Order Gaviiformes) comprise only four or five species, all limited to the northern hemisphere. The legs are set well back on the body, so far in fact that they can only shuffle or walk awkwardly on land. This limits them to nesting close to the water, usually on islands in remote lakes, lochs and tundra pools. They moult all the flight feathers simultaneously and hence become flightless for a period in the autumn. The front toes are webbed and the tail is short and stiff. The divers are probably the most efficient of the foot-propelled underwater swimming birds. They spend most of their time on the water, coming ashore only to breed. All species of divers winter at sea in inshore areas where they feed on fish and a range of marine invertebrates.

The grebes (Order Podicipediformes) comprise about 20 species which show several parallels with the divers. The legs are used for underwater propulsion and are thus set well back on the body where they are unsuited for efficient walking. They nest close to the water, often on floating nests. They need some space to get airborne and are flightless during the moult. Once aloft they are fast, efficient fliers. The tail is vestigial and the front toes are lobed and only slightly webbed. In the breeding season, all species inhabit fresh water, but several overwinter in estuarine or inshore marine waters (Fig. 19).

One of the largest and most successful groups of marine birds is the tubenoses (Order Procellariiformes) which includes the petrels, shearwaters and albatrosses. Some species are very abundant, for example Wilson's Petrel is reckoned to be the most abundant seabird on earth with a population numbering some hundreds of millions (Wood, 1976). In contrast, other species have a very restricted breeding distribution. Moreover, new species are still being discovered, rediscovered or created by the separation of what were previously thought to be only races. For example, it has recently been accepted that the Mediterranean races of the Manx Shearwater (the Balearic and Levantine Shearwaters) should once again be considered as a separate species, the Mediterranean Shearwater, as suggested by Bourne *et al.* (1988). One of the most surprising ornithological events of recent years was the capture in 1989 of three all black storm-petrels at Tynemouth. These, and similar birds from Madeira and France, are thought to be Swinhoe's Petrels (a race of Leach's Petrels) from an as yet undiscovered breeding colony in the North Atlantic (Bretagnolle *et al.*, 1991).

The order includes a large size range, from the diminutive storm-petrels to the Wandering Albatross with a weight of 12 kg and a wingspan of 2.5 m (the largest wings of any living bird). All these species are long-lived, taking several years to reach sexual maturity, and they all spend a large proportion of their lives on the wing. As the name tubenose implies, they often have long tubular nostrils. The bill is usu-

**Fig. 19.** Slavonian Grebe in winter plumage. This is our most maritime species of grebe, but still prefers sheltered coastal waters. Harold E. Grenfell.

ally large and hooked to facilitate grasping slippery prey, and the individual horny plates of which it is constructed are often clearly visible. The latter are present in all birds but the separate plates are not usually very distinct. The stomach of the tubenoses is enlarged and modified to produce and carry oils which are used to feed the young. The order is divided into four families – albatrosses (Diomedeidae), shearwaters (Procellariidae), storm-petrels (Oceanitidae) and diving-petrels (Pelecanoididae).

There are 13 species of albatrosses and apart from a couple in the Pacific Ocean, they are restricted to the southern hemisphere for breeding. They spend much of their time in the zone of westerly winds in the southern ocean, one of the most consistently windy areas in the world. The existence of fossil albatrosses in Lower Pleistocene deposits in East Anglia shows that there were albatrosses in the northern hemisphere about two million years ago, but there is no firm evidence that they bred in the north (Stuart, 1982). Black-browed Albatrosses, which normally inhabit the southern oceans as far north as the Tropic of Capricorn, are occasionally recorded in the North Atlantic and one individual has summered (and even built a nest) at the Hermaness gannetry on Unst in Shetland since 1974. A solitary female also summered for 34 years on the Faeroes during the last century. With the exception of the Sooty Albatross, all members of this family breed colonially, usually on exposed slopes and cliffs. Exposure is important because they need plenty of space, and a strong breeze, in order to take off and land. They lay a single egg as do all of the tubenoses.

The second family of tubenoses includes two species of giant petrels, two fulmars, two petrels, six prions, more than a dozen shearwaters and well over a dozen gadfly petrels (the taxonomy and number of species in the latter group is uncertain). The giant petrels are similar in appearance to the albatrosses, but they have even more massive bills, as they are carrion eaters and predators of other seabirds. The fulmars and prions are plankton feeders, the latter having comb-like plates on either side of the bill which act as a filtering device. Shearwaters are named after their habit of gliding at an angle close to the surface of the water, seeming to cut into it with their wings, a feature common to many members of the tubenose family. They mostly breed in underground burrows and only enter and leave these at night. They also use their wings to 'fly' underwater when searching for food, which consists mostly of surface shoaling fish and cephalopods. The gadfly petrels are named after their habit of flying in what has variously been described as a bounding, swinging, impetuous, bat-like, swift and careening fashion. They typically fly like this when the wind is strong. In calmer conditions a mixture of more leisurely flapping and gliding flight is used. Most species breed on remote islands in the southern hemisphere and consequently rather little is known about most aspects of their biology. They feed largely on squid and fish (Harrison, 1983).

The storm-petrels, with about 20 species, form the third family. They have shorter wings than the other groups of tubenoses, and this is associated with their buoyant, fluttering flight. The name petrel is either a corruption of pitterel (one who pitter-patters), or less probably of St Peter, both referring to the habit of these birds of 'walking on the water'. In fact they hover over the sea, dipping their feet into the water and pattering across the surface. As they do so, they scan the sea and reach down with the bill to take food, which consists of very small planktonic items. The precise function of the 'walking' has been the subject of much speculation but remains something of a mystery. Storm-petrels can swim and dive as well. With the exception of the Galapagos species (which visits its breeding colonies by day), they are all nocturnal visitors to their nests in burrows or crevices (Lockley, 1983).

The final family of tubenoses includes four species of diving-petrels. These remarkable southern hemisphere birds are very similar in size, appearance and movements to the Little Auk of the northern hemisphere. Their flight is low, swift and 'whirring'. They sometimes 'fly' straight through the crest of a wave, seeming to pass with ease from air to water and back again, using the wing for underwater propulsion. They collect planktonic food in throat pouches for transport back to the young.

The Order Pelecaniformes includes seven species of pelicans, nine gannets and boobies, 32 cormorants, two or more darters, five frigate-birds and three tropicbirds. Unlike the tubenoses, in which three toes are webbed, all four toes are joined by large webs of skin. The pelicans, gannets and boobies lack incubation patches on the belly and incubate their eggs beneath the webs of the feet which presumably transfer heat to them. The gannets and boobies have torpedo-shaped bodies and wedge-shaped tails. They feed by plunge diving from a considerable height and have modified wings and nostrils to permit this.

Cormorants are predominantly dark-plumaged birds with modified barbs on the contour feathers which permit air to escape when they dive, thus reducing buoyancy. This is in complete contrast to other diving seabirds in which a layer of air is trapped in the plumage at all times. Without the extra buoyancy, cormorants can remain submerged using less effort and can thus forage for long periods with less exertion. Propulsion underwater is by the feet alone, with the tail being used only as a rudder. Darters (anhingas) are similar in appearance to cormorants but have longer tails. They often float so low in the water that only the head and neck are visible (hence they are also known as snake-birds). Associated with this is the habit of stalking prey slowly underwater, rather than flushing and chasing it as do cormorants. They are tropical and subtropical in distribution and inhabit fresh waters and sheltered marine waters with wooded shores where they can roost and nest.

The frigatebirds have a very characteristic appearance in flight because of their adaptations for manoeuvrability, with large wings and a deeply forked tail. The feet have reduced webbing and sharp claws, and the plumage generally becomes saturated if they settle on the sea. This is because they obtain their food entirely on the wing, in the form of flying fish chased to the surface by shoals of tuna or dolphin, or by forcing other birds to disgorge food in flight by chasing and harrassing them. Thus they have no great need for waterproof plumage. They can also take items from the surface of the sea, or from the land with their beaks, but they are limited in distribution to oceans where flying fish and jumping squid occur (Nelson, 1975).

Tropicbirds are white, with greatly elongated central tail feathers. They are clumsy on land and cannot walk upright on their legs, but instead rest on their bellies and shuffle. They nest in hidden, shaded sites, often on cliffs from which they can take off without having to walk far. They feed by plunge diving for fish and squid.

Finally in this group there are the pelicans themselves. They are large birds with enormous distensible throat pouches. Their diet consists mostly of fish which are caught using the pouch as a net. Birds may fish alone, or in synchronised rings and arcs in order to round up shoaling fish. The Brown Pelican, which occurs in marine habitats, plunge dives like a gannet and then opens its pouch underwater at the last minute to engulf and entrap the targetted prey. Pelicans are mainly colonial tree-nesters.

In the mid-Tertiary epoch (about 30 million years ago), the Pelecaniformes gave rise to a family of wing-propelled flightless diving birds known as plotopterids. These include the largest species of diving birds that have ever existed – approximately 2 meters long from head to tail. They occurred in the northern hemisphere at a time when a number of giant penguins inhabited the south. Both these groups disappeared during the Miocene, probably when the seals and porpoises, which may have been better adapted to this particular underwater niche, first appeared (Feduccia, 1980).

The ducks, geese and swans (Order Anseriformes) include about 20 species which are regularly marine in habits. Most of these are diving sea-ducks – notably eiders, scoters and sawbills. They are fast-flying and usually undergo a well-marked moult migration to a secure area with reliable food supplies where they spend the flightless moult period. Some feed on bottom-dwelling bivalve molluscs whilst others pursue midwater fish.

The remaining species that inhabit the open sea belong to a single group which includes the auks, gulls, terns, skuas and waders (Order Charadriiformes). The most exclusively maritime of these are the auks (Family Alcidae) with 21 species. They have webs between the three forward pointing toes and have relatively small, pointed wings. This results in fast, aerial flight with rapid wing beats but also allows the wings to be used for underwater propulsion. All except the two smallest species become flightless during the annual moult. Auks remain at sea, often at a considerable distance from the coast, throughout the winter, coming ashore only to breed. Most breed in large colonies on sea cliffs or islands, though two of the murrelets (Pacific Ocean species) use coniferous forests (either in trees or on the ground) and visit their colonies only at night.

There are more species of auks in the North Pacific (18) than there are in the North Atlantic (6), but none of them occurs in the southern hemisphere. The smallest species, the Little Auk in the Atlantic and five species of auklets in the Pacific, are plankton feeders and transport food to the young in an enlarged throat pouch. Larger species feed on fish and plankton and usually carry food to the young in the bill. In the puffins and auklets, the chicks fledge only a little short of adult weight. In guillemots and Razorbills they fledge at 20–30% of adult weight, fluttering or gliding down to the sea on well-grown wing coverts (Gaston & Nettleship, 1981). The male parents then tend the chicks at sea for several weeks. This unusual system for the care of the young is presumably necessary because ferrying food back to the colony is too inefficient to allow an adequate growth rate, especially when one parent has to permanently guard the young at the exposed nest site. Four species of small murrelets, nesting in holes and rock crevices, go even further and lead the young directly to the sea a few days after hatching.

Only a few of the 45 species of gulls (Family Laridae) are true inhabitants of the open sea, the rest being mostly coastal. All species have the front three toes webbed and have relatively long, narrow wings. The group as a whole is commoner in the northern hemisphere than in the south and has a basically grey and white plumage. The feet, bill and orbital ring are often brightly coloured and the head in several species is black during the breeding season. A wide range of different feeding methods are used by different species, including fish and plankton feeding, scavenging and kleptoparasitism (food piracy). They also breed in a wide range of locations, but because their size and plumage makes them conspicuous, they usually choose sites that offer some protection from predators, such as cliffs, offshore islands, marshes and trees. The larger species of gulls take a number of years to reach breeding age and have a juvenile plumage that is quite different from that of adults. The young are fed near the nest for several weeks until they fledge at nearly adult size. Amongst the pelagic species, the Kittiwake is widely distributed across northern oceans in winter. Sabine's Gull moves south from breeding areas in the high Arctic to winter off the western South American and African coasts, and the Ivory Gull and Ross's Gull inhabit the Arctic sea ice edge during the winter period (Blomqvist & Elander, 1981).

The skuas (Family Stercorariidae), of which there are six species, are closely related to the gulls. Also known as jaegers (hunters), they are much darker than gulls, have longer claws and are mainly predatory and piratical. During the breeding season, the three tundra-nesting species (Pomarine, Arctic and Long-tailed) feed mainly on lemmings and their breeding success is consequently very dependent upon lemming abundance. Of these three, the Arctic Skua has the most variable diet during the breeding season and in British breeding areas such as Shetland, it is largely a food robber (as it also is in winter). The Great Skua feeds in a variety of ways, taking some fish for itself and some by piracy and scavenging from trawlers. It is also a predator of other seabirds. The name bonxie, which is often applied to this species, originated on Shetland and simply means 'dumpy bird'. The three southern hemisphere species are all closely related to the Great Skua and have a similarly wide range of feeding habits. The tundra-breeding skuas undergo long distance migrations to wintering areas off the coasts of Africa, South America and other parts of the southern hemisphere oceans. There they feed mainly by piracy, with the Arctic Skua tackling mainly auks, kittiwakes and terns, and the Long-tailed Skua mainly terns. The larger Great Skua kleptoparasitises mainly auks, gulls and gannets, whilst the Pomarine Skua appears to feed directly on small seabirds during winter, especially phalaropes (Furness, 1987*a*). Several species have different colour morphs which may differ in the details of their ecology. For example, pale phase Arctic Skuas tend to start breeding at an earlier age than dark phase ones (O'Donald, 1983).

The terns (Family Sternidae) include 42 species with similar plumage to the gulls but rather different habits. They are smaller and more delicately built than gulls and are more specialised for marine life. Many species have long pointed wings and forked tails, and remain on the wing for long periods. While the majority have black caps, the freshwater marsh terns and the noddies, which have a basically dark plumage with paler heads, can exhibit a complete reversal of the more typical tern coloration. Their most characteristic feeding method is shallow plunge-diving and the diet consists of small fish, squid and planktonic crustaceans, molluscs and worms. Terns usually have a smaller clutch size than gulls, often laying but a single egg, probably because their pelagic feeding habits make it difficult for them to provision more than a single chick. The juvenile plumage is much more similar to that of the adults than is the case in gulls. Most species are to some extent migratory and the Arctic Tern is usually claimed as being the greatest long distance migrant of any bird species, since it breeds in the Arctic and winters in the Antarctic.

The phalaropes also belong to the Order Charadriiformes, but are in the same family as most of the other waders (Scolopacidae, see Chapter 4). There are only three species and they all have slightly webbed feet and four toes which are lobed or fringed with broadened scales for swimming. Two species breed in freshwater marshes or coastal islands in the Arctic (Grey and Red-necked Phalaropes) or in temperate North America (Wilson's Phalarope). Only the former two species are marine in winter, both occurring in the Atlantic and Pacific Oceans, but concentrating in regions with abundant plankton such as the Humboldt and Benguela Currents. They float high in the water because the dense breast plumage traps a lot of air and acts as a raft. With their centrally placed legs they are able to turn easily as they swim. Their most characteristic feeding movement is to spin around pecking at the water surface with their fine, long, straight bills. This movement through the water is supposed to help stir up prey.

**Body form and flight**

Having introduced the main groups of seabirds, we will look next at how they are adapted to life at sea, starting with some size considerations which apply equally to all birds. Representative members of each of the main groups of seabirds are shown to scale in Fig. 20. In the case of tropical and southern hemisphere groups, the species illustrated are the ones most likely to be encountered in European waters, albeit very rarely. The relative dimensions of the wings and body are directly dependent on the method of flight employed, and there are two key factors to be considered in relation to this. The first is 'wing loading' – the average weight of bird supported by each square centimetre of wing. The wing area for this purpose is usually measured on the fully outstretched wings (birds are capable of varying their wing area considerably by extending or partially folding the wings). The mantle, i.e. the area of the back between the wings, is usually included in this measurement even though it generates no lift. The second factor is the 'aspect ratio' – which is the wing tip to wing tip distance (wingspan) squared and divided by the wing area. This is mathematically equivalent to dividing the wingspan by the average distance from the front edge to the back edge of the wings (Pennycuick, 1982). Birds with long thin wings tend to have high aspect ratios.

Wing loading tends to increase in larger species (Table 3), as a direct consequence of the fact that as the size of any object increases, its weight increases faster than its surface area. This is obvious if one considers a familiar object such as an apple. A typical eating apple has a diameter (a simple linear dimension) of about 6 cm. Its cross-sectional area when cut in half is $\pi r^2 = (22/7)3^2 = 28$ cm$^2$, and its volume is $(4/3)\pi r^3 = (4/3)(22/7)3^3$

**Fig. 20.** Flight outlines of representatives from 16 families of marine birds, drawn to approximately the same scale. The Procellariidae are divided into four groups – fulmars, prions, gadfly-petrels and shearwaters; whilst the Laridae are divided into two – gulls and terns. The only seabird families omitted are penguins (Spheniscidae), skimmers (Rynchopidae) and sheathbills (Chionidae).

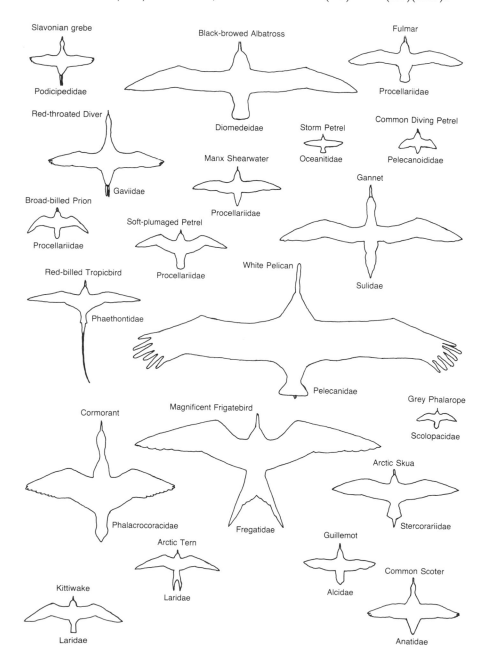

Table 3. *Estimates of wing loading and aspect ratio for the marine birds in Fig. 20. Compiled from information in Cramp & Simmons (1977, 1983), Cramp (1985), Harrison (1983) and Viscor & Fuster (1987)*

| Species | Weight (g) | Wing loading (g/cm$^2$) | Aspect ratio |
|---|---|---|---|
| White Pelican | 10 000 | 0.8 | 8 |
| Black-browed Albatross | 3500 | 1.1 | 13 |
| Gannet | 3000 | 1.1 | 14 |
| Cormorant | 3000 | 1.1 | 15 |
| Magnificent Frigatebird | 1400 | 0.5 | 13 |
| Red-throated Diver | 1240 | 1.1 | 12 |
| Common Scoter | 1100 | 1.3 | 11 |
| Guillemot | 950 | 1.8 | 12 |
| Fulmar | 800 | 0.7 | 12 |
| Red-billed Tropicbird | 760 | 1.0 | 17 |
| Arctic Skua | 420 | 0.5 | 12 |
| Manx Shearwater | 420 | 0.6 | 11 |
| Kittiwake | 410 | 0.5 | 11 |
| Slavonian Grebe | 390 | 1.1 | 10 |
| Soft-plumaged Petrel | 330 | 0.6 | 13 |
| Broad-billed Prion | 170 | 0.5 | 10 |
| Common Diving Petrel | 130 | 0.6 | 9 |
| Arctic Tern | 110 | 0.3 | 12 |
| Grey Phalarope | 55 | 0.3 | 10 |
| Storm Petrel | 25 | 0.2 | 11 |

= 113 cm$^3$. On the other hand, a very large cooking apple, with a diameter that is twice as great, would have a cross-sectional area of $(22/7)6^2 = 113$ cm$^2$, and a volume of $(4/3)(22/7)6^3 = 905$ cm$^3$. So when the diameter of an apple is doubled, its area increases fourfold (from 28 to 113 in the above example) and its volume eightfold (from 113 to 905). This is true for any measurement of length, area and volume, provided the same units of measurement are used as the scale increases. Wing area thus increases as the square of any linear increase in size in birds. In nearly all animals, weight is a property that is largely dependent upon volume. Thus weight increases as the cube of any linear increase in size in birds. So, if a bird were to double its body length, its wing area would increase fourfold and its weight would increase eightfold. The wing loading would therefore be doubled and in all probability the bird would fall out of the sky!

In reality, what happens is that larger birds have relatively larger wings in order to reduce the wing loading as much as possible, and relatively larger wing muscles to power them. This effectively sets an upper limit to the size that can be achieved by flying birds. The heaviest flying bird currently in existence is the Great Bustard in which the males may weigh as much as 18 kg, but this species flies relatively little. Mute Swans can be very nearly as heavy, with adult cobs reaching 15 kg, and they can fly well, though they require some space to get airborne.

The general increase in wing loading with increasing size (Table 3) has some exceptions. For example, the heaviest bird in the table, the White Pelican, has a rather lower than expected wing loading. This is because it has relatively big broad wings suitable for gliding slowly in thermals. It is difficult to flap such wings with much power since the available muscle mass for propulsion is relatively small. Two other species have a lower than expected wing loading, and these are the Magnificent Frigatebird and the Arctic Skua. In both cases this is partly an adaptation to achieve the manoeuvrability required to steal food from other seabirds during aerial pursuits.

Nearly all seabirds have a relatively high aspect ratio and this makes them more similar in 'design' to man-made gliders than many of the terrestrial birds we think of as being good soarers, such as vultures and storks. This is because the wings of many seabirds are adapted for gliding using some of the special air currents that occur over the sea. Most of the medium and large sized species in Fig. 20 have a high aspect ratio and a relatively uniform distance from the front to the back of the wing. A few have a more triangular wing shape which is broader at the base. The latter type of wing occurs in species that utilise flapping flight more regularly, such as the Cormorant, Guillemot and Common Scoter. Aspect ratio is not a measure that distinguishes between these two rather different wing shapes (i.e. triangular and uniform). The groups of seabirds that use mainly flapping flight are the divers, grebes, gadfly-petrels, storm-petrels, auks, phalaropes, terns and seaducks. The storm-petrels and prions have a fluttering, hovering flight and their small, lightly-loaded wings can be flapped rapidly. Albatrosses and shear-waters fly mainly by gliding rather than flapping, whilst the remaining species utilise a combination of both (Pennycuick, 1987). However, it should be emphasised that how a bird flies depends upon a whole range of circumstances, most importantly the weather conditions and whether or not it is attempting to make a specific journey (for example to feed young).

Not surprisingly, the flappers tend to have a larger muscle mass than the gliders. For example, the main flight muscles (the pectorals) represent 13% of the dry body weight in Puffins, 8% in Manx Shear-waters and only 5% in Fulmars. Puffins also have relatively large livers in order to process and store the energy necessary to power their large muscles (Osborn & Harris, 1984). Albatrosses and giant petrels have a wing lock mechanism which helps to reduce the muscular effort needed to hold the wings open and flat for gliding. This 'lock' consists of a sheet of tendon between the sternum and the humerus. It also prevents the wings from being raised above the horizontal and this accounts for the rather stiff-winged gliding position of these species (Pennycuick, 1982). In the Herring Gull, 6% of the flight muscle fibres are of the 'slow' type, which are relatively uncommon in birds. These slow fibres are efficient at remaining contracted for long periods and are probably responsible for holding the wings open during gliding.

One of the most important determinants of the overall shape of sea-birds' wings is their adaptation for gliding. Compared with resting levels of metabolism, gliding requires about twice as much energy, whereas flapping flight requires seven times as much (Phillips, Butler & Sharp, 1985). A high aspect ratio combined with a low wing loading, as in frigatebirds, allows a slow glide in weak air currents and thermals (Pennycuick, 1987). With a higher wing loading, as in albatrosses, a faster glide suited to stronger wind conditions is obtained. In contrast, the pelican's wing, with its low wing loading, relatively low aspect ratio (for a seabird), and slots between the tips of the primaries, is designed to cope with the problems of take-off and landing, rather than being an ideal gliding tool. This design of wing is essentially that of a species that exploits slow-moving terrestrial thermals and may have to take-off in relatively still air. Albatrosses, being fast gliders, are relatively clumsy when taking-off and landing, and require plenty of space or some wind assistance to get airborne. Similarly, frigatebirds can only perch awkwardly, and because they nest in trees, may occasionally crash land and become fatally entangled in the branches or trapped on the ground (Palmer, 1976*a*).

There are several types of air currents that occur over the sea and which seabirds can exploit (Fig. 21). First of all there are thermals. These only occur in a rather restricted range of conditions, notably when the air is colder than the sea and when the wind is less than 45 km/hour (28 mph). The North Atlantic in winter is one place where such conditions are moderately common. At wind speeds up to 25 km/hour (16 mph), rising air forms a series of polygonal columns or cells, rising in the centre and falling again around the edges as it cools. Gulls can sometimes be seen circling in the central rising air of these columns to quite a considerable height over the sea. The height of the columns is very variable, depending on both the thermal gradient and the horizontal wind speed. Birds glide between columns, but lose height as they do so. At wind speeds between about 25 and 40 km/hour (16–25 mph) these polygonal columns become extended into rectangular cells, and birds are thus able to soar in straight lines along the centres of the cells gaining height.

Unlike thermals over land, those over the sea do not cease at night (Welty, 1962).

Another type of thermal occurs in the zone of the trade winds where

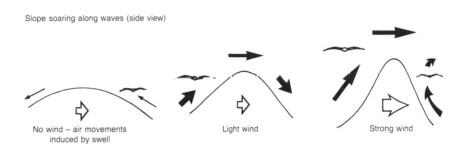

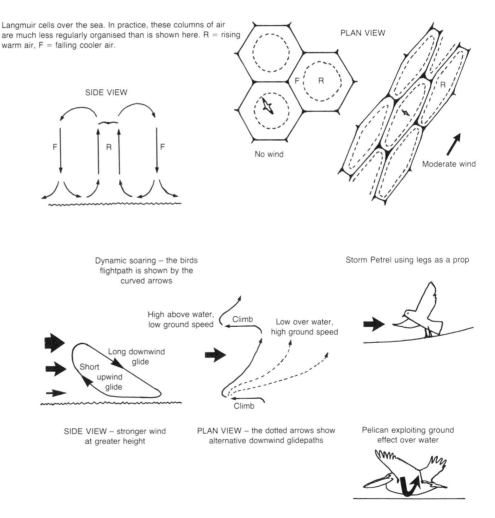

**Fig. 21.** Air currents and thermals over the sea, and the methods used by seabirds to exploit them. Filled arrows indicate the direction of air currents and open ones the direction of movement of the waves. Based on information in Welty (1962), Pennycuick (1975) and Withers (1979).

air is blown towards the equator over progressively warmer seas. Where the northerly and southerly trade winds converge at the equator, slowly rising air masses are found, forming characteristic cumulus clouds. Frigatebirds are particularly efficient at exploiting these thermals.

Wind velocities over the sea are quite high and seasonally consistent over much of the earth's surface. They produce waves, and the interaction between the waves and the air produces two effects that are of great importance to marine birds. Firstly, the air travels faster than the waves it produces and, consequently, there is an updraft on the upwind side of the wave. In the case of larger waves, this may also extend to part of the downwind side (see Fig. 21). Even if the wind drops completely, any waves that are still present will push against the air and create similar updrafts (this is one of the ways in which waves lose their energy and the sea gradually becomes calmer). These local air currents are perhaps the most important sources of lift exploited by albatrosses, petrels, shearwaters, gulls, terns and several other seabirds. It explains why they fly so close to the sea and spend so much time travelling in line with wave crests and troughs (Pennycuick, 1982). The smaller petrels tend to fly into the wind rather than across it, at least when foot pattering. The sea is moving more slowly than the wind and so their feet are 'pushed' back into the water acting like anchors whilst their extended wings face into the wind generating lift (Fig. 21) (Withers, 1979).

Air close to the sea surface travels much more slowly than air higher up because it is slowed down by friction. This effect is only noticeable in the zone 20 m or so above the sea – higher than this the wind velocity does not increase any further. Where wind speeds are moderately strong (50–90 km/hour; 30–56 mph), the velocity gradient is such that there is an increase of between 1 and 5 km/hour (up to 3 mph) for every extra metre of height. Albatrosses sometimes exploit this by descending downwind or crosswind gaining speed as they do so. They then turn into the wind and glide in level flight across the waves, gradually losing speed. While they are still moving quite quickly they adjust their glide path to a rising one, generally at an angle of about 20°. As they rise they begin to encounter faster and faster winds. This generates more lift and although it slows them down it enables them to prolong the glide path. Eventually they end up flying slowly relative to the 'ground' but at a height of several metres above the sea. They then turn downwind and begin the descending part of the cycle again. However, the importance of dynamic soaring, as this whole process is called, may have been exaggerated and the upwind turns and rises may simply be achieved as a result of speed gained by slope soaring and could thus be a means of gliding from one wave to another (Wilson, 1975).

One other mechanism is exploited by gliding seabirds and that is ground effect (Fig. 21). Close to the surface of the water, the air below the wings (which is always slightly compressed when the wings are steady) is unable to move away as easily as it does in mid air, and so more lift is generated than normal. The effect is exploited by pelicans which glide for surprisingly long distances low down over calm water without having to flap their wings, and this reduces their drag by about 50% (Hainsworth, 1988). Shearwaters and others exploit the same effect during calmer sea conditions (Rüppell, 1977).

In species with small tails, such as the divers, grebes and auks, the feet are used in flight for steering, braking and extra lift generation. The legs also aid in the take-off and landing in all species by means of jumping, running, paddling and aquaplaning. In those species with medium and large tails, the actual area of the tail can be varied by a factor of three or four simply by opening or closing it. Since it can also be raised, lowered and twisted, it provides an ideal control surface during flight, as well as generating extra lift and influencing the general direction of air flow across the wings. Species with rapid flight generally have fairly modest tails, e.g. the albatrosses, but species that fly slowly or manoeuvre rapidly have larger tails. The most extreme example of the latter is provided by the frigatebirds which have large and deeply forked tails providing the best design for rapid changes of direction during flight.

Most marine birds do not have long necks, but in those that do, such as the divers, grebes and cormorants, the neck is extended during flight. The slight forward shift in the centre of gravity of the body that this causes is offset by having the legs extended at the rear during flight. In the pelicans, the weight of the head, with its long bill and pouch, is so great that the neck has to be folded into an S-shape during flight. Even on the ground, or on water, the neck is folded to help support the weight of the bill and pouch.

The body of all flying birds is slung beneath the wings in flight, producing an inherently stable configuration. This stability can be increased still further if the wings are held in a V-shape, since if the bird is displaced it will automatically return to the horizontal position. Storm-petrels take advantage of this in order to maintain a stable position just above the water surface. A very flat wing configuration, such as that of tubenoses in the wing-locked position, is not as stable but does produce the maximum amount of lift (Fig. 22). A slightly more open V-shaped profile, as in gulls and terns (it is no coincidence that they are usually drawn this way in cartoons), is a compromise between the needs of gliding and flapping, since the inner parts of the wing (which provide most lift) are in a stable configuration, and the outer parts which bend down a little, help to trap air on the downstroke. The wings of frigatebirds have a very long humerus giving them a characteristic flight profile with the inner part of the M-shape seeming small compared with the outer sloping part. Such a configuration is unstable in flight (being almost the reverse of the V-shape), but this is just what is required in order to be able to change direction or initiate a dive quickly (Rüppell, 1977).

### Feeding and swimming

Most species of seabirds can feed by more than just a single method, though they are usually most skilful at only one or two. Aerial feeding is utilised by frigatebirds to catch flying fish, and by skuas, frigatebirds and some gulls to steal food from other seabirds. Large size, speed, agility and a powerful bill are essential for the latter. Frigatebirds also

**Fig. 22.** Fulmar slope soaring along the waves. Chris Rose.

take a good deal of food from the surface whilst flying (flight dipping), notably squid, and they are well adapted to exploit the thinly distributed food resources of the relatively unproductive tropical oceans. However, the difficulty of obtaining food means that their young are irregularly provisioned and grow slowly, limiting this group to breeding at intervals of greater than a year.

Other species that feed by flight dipping are the storm-petrels and some terns and gulls. Storm-petrels pick up crustaceans, tiny fish and squid as they patter across the surface. It has been suggested that the yellow or white areas on the webs of the feet of some species, such as Wilson's Petrel, might be used to lure or frighten prey, especially krill, thus making them easier to capture (Zink & Eldridge, 1980). However, this remains conjectural. Surface feeding whilst floating on the sea is employed by nearly all species of seabirds if conditions are suitable, e.g. if an abundance of food (natural or man-made) is present, but some species also feed routinely by this method. Albatrosses, for example, seize fish and squid whilst floating, both by day and at night. Broad-billed Prions swim along the surface with their heads underwater, scooping up and filtering plankton from the water with specially modified bill lamellae. This is often referred to as hydroplaning, to distinguish it from the surface filtering that is employed by many other species. During surface filtering, several small prey are grasped at the same time and water is expelled from the bill before they are swallowed. Much of the feeding of fulmars, shearwaters, petrels and gulls takes place while they swim on the surface, the prey being filtered or seized from the water (Ashmole, 1971; Harper, 1987).

Plunge-diving is also employed to varying degrees by many species. Terns specialise in shallow plunge-diving. This is usually done from a hovering position just a few metres above the water surface and the body does not usually submerge completely. The forked tail is valuable in changing direction quickly prior to diving, and in adjusting the dive itself. Common and Arctic Terns dive from a height of $1\frac{1}{2}$-8 m. The latter species often uses a series of 'stepped hovers' before committing itself to the plunge. Roseate Terns dive from somewhat higher (8–15 m) and execute rapid angled dives into the water, rather than hovering. Consequently, they probably plunge-dive to slightly greater depths and remain submerged for longer (*c.* 2 seconds) (Kirkham & Nisbet, 1987). Great Shearwaters plunge-dive from a few metres up in the air and then pursue prey underwater. The Brown Pelican dives from various heights and folds its wings right back just before entering the water in order to prevent impact stresses from damaging them. The bill is opened just above the target fish and the pouch rapidly expands as water, together with the fish, or other prey, rushes in. At the surface, the water must be drained out of the pouch before the bill can be lifted out of the water and the prey swallowed (a pouch full of water weighs more than the bird itself).

Tropicbirds plunge-dive from heights of 10 m or more, but it is not known how deeply they penetrate the water. Gannets and boobies are the most specialised of the plunge-divers. They may dive from heights as great as 100 m to depths of 10 m or so, though several species are adapted to more shallow dives than this. In order to withstand the considerable forces generated by impacts with the water at high speed (100 km/hour, 60 mph), they have evolved streamlined, torpedo-shaped bodies, and the wings fold backwards completely just before entry. The head and neck are cushioned to some extent by partially inflated cervical air sacs and the external nostrils are absent. The backward folding of the wings is achieved by the humerus folding to the rear to an extent not seen in other birds (except the Brown Pelican). The external nostrils are replaced by secondary 'nostrils' in the corners of the bill (Fig. 23) as is also the case in cormorants. These downward facing slits are roofed over by a flexible covering formed by the jugal bone and its overlying horny plate. This is hinged at the forward end in such a way that the force of the water upon impact closes it, thus preventing a sudden inrush of water (MacDonald, 1960). The pelicans have their nostrils partly occluded by a valve-like structure which also closes automatically when subjected to external pressure. When diving, gannets and boobies generally aim for a particular individual fish or squid, adjusting their dive through the air right up to the last moment. They can pursue prey underwater for some distance if it is not actually seized during the dive itself (Nelson, 1978b). Ainley (1977) has argued that plunge-divers ought to be commoner in clear water conditions because of the need to see prey clearly from the air. However, the reverse seems to be the case. Both plunge-divers and pursuit-divers are commoner in murky inshore waters than in clearer offshore ones (Haney & Stone, 1988).

Diving from the water surface (pursuit-diving) is used by divers,

**Fig. 23.** Side and front view of a Gannet's head to show the position of the flexible jugal operculum (hatched). The secondary nostrils open along the slit between this operculum and the lower jaw. The slit closes when the operculum is forced inwards by water pressure. Based on information in MacDonald (1960).

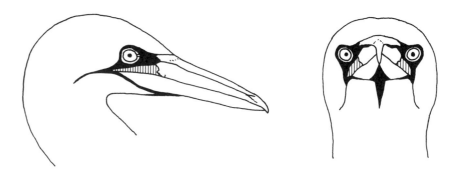

grebes, cormorants, auks, shearwaters, diving-petrels, seaducks and penguins. A variety of methods are used for underwater propulsion. Divers and grebes use their feet alone for both swimming and steering. The cormorants are similar, except that the tail, and to a limited extent the wings, are used for steering. Use of the wings for underwater propulsion can be more efficient and allows higher speeds to be achieved, depending on the compromises that have to be made with the species' flying ability. Shearwater wings are adapted primarily for flight and are rather inefficient underwater, allowing only shallow swimming dives to be made. The wings of auks, on the other hand, are much smaller, reducing their resistance to water on the recovery stroke. Both groups reduce the effective area of the wings still further by partly folding them underwater, turning them into a more sturdy paddle for forward propulsion. Those auks which dive and swim best, generally fly most poorly. Thus Little Auks are efficient feeders with a relatively low wing loading (0.5 g cm$^{-2}$) and so can forage at some distance from their breeding colonies. Typical foraging distances are 9–15 km, but they may go as far as 100 km, carrying food back to the young in the throat pouch. When feeding underwater, they are less efficient and for the most part take relatively slow-moving crustaceans which are individually pursued. Their diving times are about 30 seconds and they probably do not usually go much deeper than about 20 m. However, the plankton rises towards the surface at night, making it more accessible to those Little Auks that feed after dark.

In contrast, the diet of Brünnich's Guillemots consists mainly of fish, with crustaceans being taken as an alternative. The average dive time is about 60 seconds and the fact that they eat mainly benthic (bottom-dwelling) fish is in accordance with the idea that they are specialised for deep diving (see Table 4). They have a relatively high wing-loading as a consequence (*c.* 1.8 g cm$^{-2}$), but this does not seem to restrict their foraging range during the breeding season, for it is very similar to that of the Little Auk. Guillemots also take mainly fish and have diving capabilities slightly less good than Brünnich's Guillemots. However, they seem to be more adapted to pursuit than to deep diving (Bradstreet & Brown, 1985). Beak marks on Herrings, Sprats and whitings trawled in the Beauly Firth (at a maximum depth of about 30 m) suggest that many of these fish are attacked from below, but may escape when the bird tries to reposition them in the bill. The fish that escape are mostly more than 12 cm long, a size known to cause handling difficulties in captive Guillemots (Hislop & MacDonald, 1989). Guillemots tend to feed closer to their colonies during the breeding season (average distance about 10 km, maximum *c.* 50 km) than the previous two species. Puffins are also fish feeders, but appear to exploit schooling species. This can be

Table 4. *Diving depths of some species of seabirds for which information is available. Typical depths were derived from the median values of the normal depth ranges quoted in the literature (Cramp & Simmons, 1977; Cramp, 1985; Bradstreet & Brown, 1985; Burger & Simpson, 1986; Jury, 1986; Owen, Atkinson-Willes & Salmon, 1986; Barrett & Furness, 1990; Harris et al., 1990; Wanless, Burger & Harris, 1991)*

| | Depth of dives (m) | | |
|---|---|---|---|
| Species | Typical | Average deepest[a] | Maximum |
| King Penguin | 70 | | 240 |
| Brünnich's Guillemot | 25 | 46 | 100 |
| Guillemot | 18 | 43 | 180 |
| Common Scoter | 11 | | 30 |
| Black Guillemot | 10 | | 50 |
| Puffin | 9 | 30 | 68 |
| Eider | 8 | | 55 |
| Great Northern Diver | 7 | | 70 |
| Long-tailed Duck | 7 | | 62 |
| Red-throated Diver | 6 | | 21 |
| Razorbill | 5 | 24 | 140 |
| Shag | 5 | 28 | 42 |
| Cormorant | 5 | 33 | 37 |
| Little Auk | 5 | | 35 |
| Great Crested Grebe | 3 | | 30 |
| Slavonian Grebe | 5 | | 25 |
| Gannet | 2 | | 15 |
| Sandwich Tern | 0.8 | | 2 |
| Kittiwake | 0.5 | | 1 |
| Arctic Tern | 0.3 | | 0.5 |
| Common Tern | 0.2 | | 0.5 |

[a]Based on small depth gauges attached directly to the bird and later recovered and read. They record only the maximum depth reached.

inferred from the fact that, in the breeding season, they carry a number of fish of the same species, closely similar in size. There is, incidentally, no truth in the idea that these fish are sequentially orientated in opposite directions in the mouth (Harris, 1984).

Puffins probably do not dive very deeply and have relatively short dive times. They feed quite close to the colonies in the breeding season (average *c.* 5 km, maximum *c.* 50 km) but spread out well offshore during the winter (see Fig. 29). Razorbills feed on fish, but forage in relatively shallow coastal waters (up to 10 or 15 m), though there is a record of one having been seen from a manned submersible swimming around at a depth of 140 m (Jury, 1986). Black Guillemots take a wider range of prey than most other auks, but fish still make up the majority of the diet. They feed closer to the breeding colonies than all the other species but seem to dive deeper than Razorbills on average (Table 4). They swim underwater at a speed of 7 km/hour (4½ mph) and yet they are agile flyers, needing only about 6 m of open water in which to take off. Consequently they are able to forage amongst quite dense pack ice (Bradstreet & Brown, 1985).

It thus appears that flying ability in auks is only broadly related to the distance they will travel from the breeding colonies in order to collect food for the young. If the distance is very great, however, the young may receive infrequent feeds and hence grow slowly. The now extinct Great Auk was a species that underwent the final stage of adaptation to efficient underwater, wing-propelled swimming, by becoming flightless. Great Auks probably dived to depths of 75 m to feed, since this is the average depth of the Grand Banks off the coast of Newfoundland where they were abundant. They were probably able to forage at a considerable distance from their breeding colonies, just as penguins do today (in excess of 50 km), but compensated for this by bringing back extremely large amounts of semi-digested food to the young (Bradstreet & Brown, 1985). The limitations that flightlessness placed upon their choice of breeding sites made them exceptionally vulnerable to human predation which was the major factor that drove them to extinction.

The divers show the most extreme form of specialisation for foot-propelled underwater locomotion amongst living birds. This places almost as many constraints upon their choice of breeding sites as does the flying ability of the auks. In this case, it is the rearward placement of the legs that provides the problem, since it means that divers walk awkwardly. The nest must also be within a few metres of the water and the adults require a considerable length of open water to take-off and land. Similar constraints apply to grebes, though in this case the smaller species, such as the Little Grebe, are able to exploit small water bodies because of their greater aerial manoeuvrability.

Both divers and grebes can adjust the buoyancy of the body by varying the amount of air trapped in the feathers. They often 'sleek' the plumage before diving to reduce its air content. They also have an air sac system which is reduced in size compared with other birds, and this facilitates underwater activity. This is also why they float relatively low in the water (Campbell & Lack, 1985).

Cormorants show a number of adaptations for underwater activity, but do not dive particularly deep – 10 m or so being typical. They differ from most other diving species in retaining relatively broad wings. This enables them to soar quite slowly, and gives them a degree of flexibility in take-off and landing not possessed by divers and grebes. For example, they can nest in trees, which would be difficult for these other groups to land in. Cormorants have even been recorded catching Swallows in flight and eating them. Darters have flying abilities which are marginally better than those of cormorants. In both these groups the plumage becomes wetted to some extent during swimming. This is because modified barbs on the contour feathers allow air to escape when placed under pressure from the surrounding water. This reduces buoyancy and makes for greater manoeuvrability underwater, but does require that the plumage be dried out after a period in the water. This is often achieved by spreading the wings and allowing air to circulate around the body (Fig. 24). The plumage remains waterproof in rain because there is no water pressure to force out trapped air. Darters in fact have plumage that overall is about three times more wettable than that of cormorants. However, their wings and tail repel water and thus they are able to fly without having to dry these feathers first (Rijke, Jesser & Mahoney, 1989). The tail is large and stiff, and has an important

**Fig. 24.** Cormorant drying its plumage. Harold E. Grenfell.

role in helping to execute sharp underwater turns. One of the mysteries of cormorant foraging is how they are able to catch prey in the murky waters of rivers and estuaries, where visibility is very low. The answer may be that they explore the bottom at very close range in order to flush prey with the bill, which they can then grab quickly rather than actively chasing it through the water.

Divers, grebes, cormorants and darters are all mainly fish-feeders and are just as much at home in fresh waters as they are in the sea, if not more so in the case of many species. The webbing on the feet varies from all four toes being webbed in the cormorants and darters, to just the front three in divers. In the grebes, the three front toes are only slightly webbed, but also have broad lateral lobes of skin to aid underwater propulsion. Those on the outer sides of the toes are quite stiff. On the recovery stroke, the whole foot is rotated through 90° to reduce water resistance and allow for efficient swimming at the surface. In all these species, the feet are thrust backwards simultaneously for propulsion when diving at speed, but they may alternate when manoeuvring or moving more slowly. Alternate thrusts may also be used to assist the natural buoyancy of the body when returning to the surface. The feet are also used for braking and turning (Campbell & Lack, 1985).

Most of the seaducks use both the feet and wings in underwater swimming. This contrasts with the freshwater species and the mergansers in which the feet alone are used. Eiders, scoters and Long-tailed Ducks employ half-closed wings (with the primaries tucked under the secondaries) as additional underwater paddles (Snell, 1985). Along with the auks and divers, this enables the seaducks to reach considerable depths (Table 4). Most species feed on bottom-dwelling invertebrates, taken from rocky or sandy bottoms. Since Long-tailed Ducks apparently feed in parts of the American Great Lakes where the depth is 160 m, it is possible that they can dive as deep as this. Many diving birds dive to depths at which the bends, caused by the formation of nitrogen bubbles in the blood, would occur in a man making repeated dives from the surface holding his breath, so birds must also have some mechanism that prevents this (Schmidt-Neilson, 1990).

The rearward placement of the legs in proficient divers means that a very upright posture has to be adopted when walking on land in order to get the weight of the body balanced over the legs. In the extreme case of divers, grebes and tropicbirds, forward progression on land is achieved by shuffling forward with the belly resting on the ground, although divers and grebes can also lunge or run forward for short distances with the whole of the body lifted clear and very upright. In the auks, gannets and many of the tubenoses, the difficulties of balancing on land are alleviated by walking on their shank bones as well as their feet. The puffins are the only auks not to do this, and consequently have a characteristic upright, penguin-like posture.

Rearward placement of the legs also restricts take-off ability, hence the need of many species to first paddle and then scamper across the surface of the water to increase speed. Albatrosses and gannets have difficulty in taking off at low wind speeds and require an even longer run at such times. For example, a large flock of Gannets once drifted ashore at Sennen Cove in Cornwall when they were unable to take off in calm weather. These birds were probably gorged with fish and their attempts to take off were also impeded by a heavy swell which slowed them down (Nelson, 1978b). Many shearwaters are unable to take off from flat surfaces at their breeding colonies and have to shuffle on their bellies to the top of hillocks in order to launch themselves into the air. In such circumstances they are very vulnerable to predators and this is the main reason for their nocturnal visits to the colonies (Brooke, 1990).

The different methods of obtaining food require different types of bills in order to be effective. One of the most characteristic features of the food taken by marine birds is that it is usually wet and slippery. This requires that the bill be capable of exerting a good grip, and serrations or ridges along the opposing surfaces of the horny outer covering of the upper and lower mandibles is the means by which this is achieved. In plankton feeders there are often horny papillae on the inside of the bill and tongue in order to manipulate food in the mouth. Several species have a pointed, spear-like bill, including divers, grebes, gannets, darters, terns and some auks and gulls. This offers minimal water resistance during lunges or dives after prey and can also mean that a near miss, i.e. a failure to grasp the prey successfully between the mandibles, may none the less result in successful capture by stunning or even piercing it with the bill. Once caught, the prey is held by the serrated bill edges or by sharply pointed oral papillae. Puffins are able to carry several fish at a time because the upper part of the oral cavity exerts a grip via oral papillae against the muscular, flexible wedge formed by the tongue, which is pressed upwards by the lower jaw. Gripping the fish just behind the gills prevents them from slipping out and having them orientated both ways evens up the weight distribution of the load (Harris, 1984).

Most of the remaining seabirds have an elongated bill with a terminal hook on the upper mandible. This is ideal for grasping active prey such as fish and squid when they are picked out of the water, rather than being lunged at. Great Shearwaters have a stouter and more strongly hooked bill than Sooty Shearwaters, and where the two species occur together, e.g. along the Atlantic coasts of Canada, Greenland and Iceland during the austral winter (May–September), the former take tough-bodied prey such as squid and Mackerel, whereas the latter take more crustaceans and soft-bodied prey such as Herring (Brown *et al.*, 1981). The hooked tip is not only useful for exerting a gaff-like grip, but also for tearing apart larger prey which cannot be swallowed whole.

There are many other adaptations that greatly enhance the efficiency of seabird feeding. The coordinated movements of some species, such as the pelicans, may trap shoals of fish. Common Terns may defend sections of shoals of fish against conspecifics (Safina & Burger, 1988). Many species, especially in the unproductive tropical oceans where food is thinly distributed, depend on feeding groups of tuna, sharks, dolphins, whales or seals to drive prey from deeper water to the surface and to concentrate them into dense shoals. For example, Sooty Terns and frigatebirds are almost totally dependent on Yellowfin Tuna to provide them with feeding opportunities in the eastern equatorial Pacific Ocean (Au & Pitman, 1988). Underwater predators generally concentrate prey by swimming round and under the shoals, but Humpback Whales blow curtains of air bubbles around them. Herring Gulls and Kittiwakes tend to associate with Humpback Whales in the western Atlantic and feed on the small fish which they round up (Pierotti, 1988). Even fish like Mackerel may drive whitebait into a shoaling frenzy near the surface as they feed, thus making them readily available to birds. Seabirds must be constantly on the lookout for feeding opportunities of this sort and presumably converge on promising signs of activity such as spouting whales. Even whales that do not concentrate prey or drive it to the surface may be useful indicators of high prey abundance. No wonder then that seabirds have adapted so readily to following ships to feed on man's waste, especially that of the fishing and whaling industries.

Another factor that has an important influence on seabird foraging is the diurnal migration of the plankton. Zooplankton as small as a millimetre in diameter may undergo a regular vertical movement of the order of 10 m a day, coming closer to the surface at night to graze on phytoplankton, and sinking to greater depths during the day to avoid the chlorophyll they have consumed (inside the phytoplankton) becoming energised by the light shining through their thin-walled bodies, and releasing potentially lethal oxygen radicals (Hendry, 1988). Their aquatic predators have to move with them if they are to feed, and consequently the larger zooplankton eaten by birds are available closer to the surface at night. This would be of little use to visual foragers were it not for the fact that a great many planktonic organisms are luminescent. At least one of the functions of luminescence is camouflage (rendering the body invisible to predators from below by appearing as light as the sky), so presumably the risk of predation by birds is less than that by fishes, squids and other aquatic predators. However, there are many other suggested functions, including defence, prey attraction, mate attraction and general illumination, and we simply do not know enough about the phenomenon yet to know if birds are able to exploit it when foraging. A high proportion of the prey brought by some petrels and albatrosses to their young at night are luminescent (e.g. Imber, 1973). However, the remains of squids in the guts of several species of shearwaters, petrels and albatrosses off the coast of South Africa (the White-chinned Petrel being the commonest species), came mostly from species that are believed to float to the surface when dead, whilst the remains of those which sink when dead were rare (Lipinski & Jackson, 1989). The latter authors thus argued that most squid were therefore scavenged. On the other hand, Harper (1987) saw many luminescent species being taken by seabirds at the surface whilst they were still alive.

Many species of seabirds are generalist and opportunist feeders, rather than specialists. Notable in this respect are the shearwaters, fulmars and gulls. There may also be regular seasonal switches in diet, and a good example of this is the Puffin which takes mainly large fish during the breeding season and mainly smaller planktonic organisms in winter (Harris, 1984).

## Plumage

On land, most birds spend a significant proportion of the time (usually the night) resting on the ground or in the trees. Only a few, such as Swifts, have adapted to a completely aerial existence – regularly remaining on the wing both by day and by night. In the case of seabirds, the only resting place available away from land is usually the sea itself – always wet, usually cold, and constantly in motion. Nearly all seabirds have to be able to rest on the sea at times, whilst most make frequent and prolonged contact with it during feeding. The exceptions are the frigatebirds which avoid contact with the water and become sodden if they do land on it for long. While at sea they must remain airborne by day and night just like Swifts. Other groups, such as gannets, some shearwaters and terns, seem to avoid prolonged periods of contact with the water. Divers, grebes, diving-petrels, phalaropes, gulls, auks and seaducks on the other hand habitually rest on the sea for long periods of time and in many cases remain virtually immersed for hours at a time.

In order for the latter groups to remain in contact with the water for so long, it is essential for them to possess dense, insulating and waterproof plumage. Water conducts heat away from the body 20–26 times faster than air of the same temperature, and so some means of preventing heat loss from the uninsulated legs and feet is also needed. The latter is achieved by means of a countercurrent heat exchange mechanism which is used by many birds and mammals, but is especially important in the legs of birds that swim for long periods in cold water. This mechanism consists of a blood capillary network, at the base of the leg or other extremity, in which cool blood in the veins coming from the limb is brought into close contact with warm arterial blood from the body (normally at about 42 °C in birds). This cools the blood going into the limb as well as warming the blood passing back into the body, thereby reducing the overall heat loss (Phillips, Butler & Sharp, 1985). For example, those species of gulls which habitually walk or rest on sea ice or on snow, such as Ross's Gull and the Ivory Gull, are able to do so without damage to the feet and without losing large amounts of heat. In these cases, the temperature of the foot may be only just above freezing, and there are insulating pads on the bottom of the feet to prevent them from actually dropping below zero. In such birds the body surface temperature below the feathers is about 38 °C, even when the environmental temperature drops as low as −16 °C. The top of the leg has a surface temperature of about 20 °C under these conditions, and the lower section about 7 °C. This gives some idea of the relative efficiency of the countercurrent heat exchanger, but it also indicates that considerable heat is still lost from the legs. This explains why roosting birds so often withdraw one leg beneath the belly feathers where heat loss is greatly reduced. It has been demonstrated experimentally in the Mallard that the capillary network reduces heat loss in the immersed legs by up to 30% (Midtgård, 1980).

For the legs to function at all at temperatures so much lower than the rest of the body requires that they have a number of special adaptations. For example, the lipids (fats) in the membranes of cells would normally solidify at such temperatures and the cells eventually die. In fact, bird and mammal leg lipids are much more unsaturated and oily than normal ones. This is one reason why foot oils (such as calves' foot oil) are used by man as lubricants in very cold working conditions (Irving, 1966).

Just as the countercurrent heat exchange system can be used to minimise heat loss, so it can help lose heat when the need arises. All that happens is that the capillary network shuts down and the blood instead passes directly from the body to the extremity where heat can then be lost to the cooler environment. Flying Herring Gulls can lose as much as 80% of their heat production through the feet by removing them from the plumage and exposing them to the air stream (Baudinette *et al.*, 1976). This can also be important in birds with dense plumage following bouts of intense activity. For example, the wings of penguins have a countercurrent heat exchanger, but this can be bypassed and the wings used as cooling radiators during activity on land.

The plumage of seabirds in prolonged contact with the water is particularly dense. The total number of feathers on the body is large, often exceeding 10 000, whereas such a number is rare even in large terrestrial birds (Markus, 1985). Moreover, the down feathers are well developed and very efficient at trapping plenty of insulating pockets of air. It is no coincidence that the species exploited by man for the most effective down products is a seabird, namely the Eider duck. The female lines the nest with down plucked from her own breast and this is harvested on Eider 'farms' in Iceland and elsewhere. The Eiders involved are wild, but continue to use the traditional farm nesting sites because they are afforded protection from predators and achieve good breeding success once they have relined the nest with more down.

In order to prevent water from saturating the down, the outer or contour feathers in most seabirds have hair-like outgrowths on the barbules. These produce a very hydrophobic, or water repellant, surface. Preen gland oil does not directly produce a waterproofing effect, but probably helps maintain the flexibility of these outgrowths so they do not break off and thus render the waterproofing ineffective.

## Coloration

Of the world's 300-plus species of seabirds, the commonest type of overall plumage pattern is dark above and light below (Table 5). The main functional significance of these pale undersides is usually considered to be that they are less conspicuous against a light sky when viewed from beneath, as suggested for aircraft attempting to attack submarines in the Second World War (Craik, 1944). This has been experimentally demonstrated using painted models and fish in an aquarium (cited by Tinbergen (1974)). More recently, Black-headed Gulls dyed black underneath have been shown to suffer a 31% reduction in success rate when hunting schools of Bleak in an indoor pool (Götmark, 1987). Thus pale undersides would be expected in all those species which are surface feeders or plunge-divers hunting prey with good vision, and this does broadly seem to be the case. Exceptions are species that forage mainly by night and are therefore rendered less conspicuous by being dark underneath, e.g. Sooty Shearwaters, or that forage in murky waters, e.g. Cormorants.

In general, dark plumage is more resistant to wear than light plumage and might therefore be selected for in the absence of other factors. The better wearing quality of heavily pigmented feather keratin explains the tendency of many light plumaged birds to have dark wing tips (e.g. gulls), since the tips of the primaries are subjected to particularly heavy wear. The amount of black, however, decreases with age in the Black-headed Gull (Allaine & Lebreton, 1990). The fact that dark plumage tends to absorb more heat may also be of advantage to those species,

Table 5. *Dominant plumage patterns in 294 species of adult seabirds, ignoring wing bars, wing tips, throat patches, rump patches, etc. Species that have different plumage phases have been omitted. Based on information from Harrison (1983)*

| | Number of species in each plumage category | | |
|---|---|---|---|
| Group | All white or grey, or mostly so | All dark | Mostly dark above and paler below |
| Penguins | 0 | 0 | 16 |
| Divers | 0 | 0 | 4 |
| Grebes | 0 | 0 | 9 |
| Albatrosses | 1 | 3 | 8 |
| Petrels | 4 | 14 | 25 |
| Prions | 0 | 0 | 3 |
| Shearwaters | 0 | 4 | 15 |
| Storm-petrels | 1 | 12 | 7 |
| Diving-petrels | 0 | 0 | 4 |
| Tropicbirds | 3 | 0 | 0 |
| Gannets and boobies | 4 | 0 | 4 |
| Cormorants and shags | 0 | 17 | 9 |
| Pelicans | 4 | 1 | 2 |
| Frigatebirds | 0 | 5 | 0 |
| Phalaropes | 0 | 0 | 3 |
| Skuas | 0 | 4 | 2 |
| Gulls | 31 | 2 | 12 |
| Terns | 32 | 4 | 3 |
| Auks | 0 | 4 | 18 |
| Total | 80 | 70 | 144 |
| % of species | 27 | 24 | 49 |

such as cormorants and darters, that need to dry their plumage after swimming. It has been suggested that the Scandinavian race of the Lesser Black-backed Gull is much darker as an adaptation to wintering in more tropical regions than other races. Individual feathers treated to six months of UV radiation and then scraped with a knife, show more damage in white and grey areas than they do in black ones (Bergman, 1982).

Birds with white upperparts are far more conspicuous than those with dark upperparts. Species that derive some advantage from feeding communally might therefore be expected to have evolved white plumage for this reason. This may be true of White Pelicans, for example. However, such plumage would not be advantageous if it attracted other competing species which interfered with feeding. The Blue-footed Booby, in which groups plunge-dive together in a coordinated fashion, seems to contradict this, in that it is a dark species, whereas the gannets which do not dive synchronously (but do feed communally) are basically white. The explanation may be that it is easier for gannets to catch fish if the latter are confused by large numbers of conspicuous white birds attacking them in a mêlée, whereas Blue-footed Boobies rely on taking schools of fish by surprise in a sudden co-ordinated attack (Parkin, Ewing & Ford, 1970). It may be advantageous for the generally pale gulls to attract other gulls, and other species, to a concentration of food because they fare better in a chaotic feeding mêlée, or because they can then rob other birds. An even more ingenious idea is that their white plumage may attract aquatic predators, such as whales, which may then concentrate the prey and drive it to the surface (Pierotti, 1988).

The presence of food robbers may also have been a significant factor influencing the evolution of plumage colours. Boobies may gain an advantage by not being conspicuous to frigatebirds from above, and thus they nearly all have dark upperparts, whereas there are no frigatebirds in most areas where gannets are common. Small, dark species that sit on the water are at least twice as difficult for a human observer to spot, even at distances of less than 300 m, than large pale flying ones (Powers, 1982). Auks, which are vulnerable to food robbery when they bring prey to the surface prior to swallowing it, may be dark for this reason. The food pirates themselves may tend to have dark underparts if this enables them to get closer to potential victims without being spotted. This may be one reason for the plumage polymorphism of Arctic Skuas in which dark bellied forms are commoner in the south and indulge in more food robbery than the northern pale form (Furness, 1987a). One of the few species that is small, all white, solitary feeding and that carries food to its young in its bill (all of which should make it particularly vulnerable to piracy), is the Fairy Tern. This species does a lot of foraging at night when the chances of robbery are presumably lower, but while this may be how it gets away with being conspicuous, it does not explain why it is white in the first place! The only other all-white (as opposed to all-grey) seabirds are polar in distribution and clearly gain an advantage by being camouflaged amongst snow and ice, e.g. Glaucous Gull, Snow Petrel.

The plumage changes that occur during the course of growth and maturation are of considerable interest because of the light they shed on the selective factors that may influence plumage coloration. Juvenile birds are presumably much more vulnerable to food robbery, and in many species where the adults are conspicuous, such as the Gannet, the young have dark upperparts. This may tend to make them less conspicuous when they are attempting to swallow fish at the surface after successful plunge-dives, thus minimising the chances of piratical Herring and Great Black-backed Gulls spotting them. It usually takes five years for the complete transition to adult plumage to occur in gannets. Juvenile gulls generally have browner plumage than adults, possibly because they feed along the shoreline to a greater extent than adults and are better camouflaged there by being brown. Presumably they have yet to master the skills needed in order to forage in the same way as their parents. The larger gulls take 3–5 years to achieve adult plumage, though they may be capable of breeding whilst they still have some juvenile feathers (Tinbergen, 1974). Juvenile Razorbills and Brünnich's Guillemots have an unexplained plumage dimorphism in which the majority of fledging chicks resemble summer plumage adults in having a black chin and throat, but a significant minority resemble winter plumage adults in having a white chin and throat (Gaston & Nettleship, 1981).

Some species show seasonal changes in plumage coloration, notably a number of gulls which develop a black head during the breeding season. In these cases, the head usually plays a part in breeding displays, especially those associated with territorial defence. Both sexes contribute to this defence and both have black heads. In fact, the sex of neighbouring individuals may be indistinguishable to territory owners (van Rhijn, 1985). Most species with a black head tend to forage inland on invertebrates during the breeding season (presumably a conspicuous face would be disadvantageous foraging at sea). Many of the terns have some black on the head, but this is restricted to the crown which is invisible to prey looking from below and in front of the foraging bird. Perhaps significantly, the black cap tends to recede further back across the forehead during the winter. Some of the marsh terns, such as the Black Tern, lose their black underparts during the breeding season and become much whiter.

Other species show marked seasonal changes in plumage. For example, Black Guillemots become all black except for white wing patches (and a scarlet gape), during the breeding season. The Cormorant also develops more contrasting plumage in summer, with the white throat and thigh patches standing out very clearly. In general the winter plumage is much duller. For example, the head of many gulls develops brownish speckles. This makes them slightly less 'bright', but the significance of such a subtle change is difficult to determine. In Puffins, the bright eye ornaments (consisting of coloured, cornified skin) and several of the bill plates are shed in the autumn (Fig. 25), and the legs and feet become much duller.

Several species have prominent wing bars and/or rump patches which may facilitate species recognition. This is well developed in the wading birds (see Chapter 4), but a few seabirds show it too. For example, the prions or whalebirds have a conspicuous black wing bar which extends across the back of the body, and storm-petrels have a variety of differently shaped white rump patches. Wing and tail bars may also serve to distinguish juvenile birds. Young Kittiwakes have a black wing bar and nape patch, which is completely lacking in the adults. Most juvenile gulls have black terminal tail bands. Wing bars may also facilitate rapid aerial manoeuvring of flocks, as in waders.

The coloration of areas of bare skin in seabirds is usually unspectacular. Such skin is often darker in birds that live in areas of intense solar radiation in order to prevent tissue damage from ultra-violet radiation (the same thing occurs in mammals). This applies to the orbital ring (the narrow area of bare skin surrounding the eye) in many tropical and polar terns and gulls. The bridled form of the Guillemot has a white orbital ring and a white line running backwards from it across the side of the face. The adaptive significance of the bridle is not known, though it is controlled by a single recessive gene (like blue eye colour in humans). Bridled birds are uncommon in the south, but may form as much as 70% of populations breeding in the north, so this feature is probably linked in some way with physiological adaptation to colder conditions (Jefferies & Parslow, 1976).

**Fig. 25.** Seasonal and age-related changes in the head plumage, bill plates and skin (gape and eye ornaments) of the Puffin. Adult winter in the centre, recently fledged juvenile at the top left, first summer (no bill grooves) top right, second or third summer (1+ bill grooves) bottom left, and adult summer (four or more years old, 3+ bill grooves) bottom right. Ageing characters based on Harris (1981). Chris Rose.

The feet of seabirds are often pink, red, yellow, or black, though in gannets and boobies they may be vivid red or blue. These, and other brightly coloured parts of the body, are usually visible only at short range and probably facilitate species recognition and thus prevent inter-breeding. Differently coloured eyes, bills, feet and areas of bare skin are often found in closely related species of penguins, tubenoses, pelecani-forms, gulls, terns and auks which nest close together in the same areas (Pierotti, 1987). Young birds imprint on the parental colour pattern and hence associate and mate with birds of their own species. In cases where eggs are deliberately or accidentally transferred to nests of the wrong species, the young may imprint on their foster parents, and eventually try to form mixed species pairs and possibly give rise to hybrid offspring (Harris, 1970*b*). These hybrids are often at a disadvantage compared with either parent, e.g. hybrids between Herring and Lesser Black-backed Gulls (Harris, Morley & Green, 1978), and this is why such species discriminating marks evolved in the first place. Leg colour is particularly important in seabirds because the young spend so much time close to their parents feet. The only superficial difference between the Yellow-legged Herring Gull, which is now considered to be a full species (Yésou, 1991), and the Herring Gull is the colour of the legs. The Herring Gull has pink legs.

Elaborate structures used for display are not common in seabirds, but in frigatebirds the males have a bright scarlet throat patch consisting of bare skin overlying the cervical air sacs. These sacs can be inflated to produce an enormous balloon-like structure used for display during the breeding season.

### Sense organs

Sight is one of the most important senses to all birds, since it provides vital information during both flying and foraging. Consequently, the eyes (Fig. 26) are relatively large – much larger than in mammals of comparable size. The function of the pecten, a structure unique to birds, is not known but it may be concerned with supplying nutrients to the retina and in regulating pressure within the eye. The third eyelid, or nictitating membrane, which is present in all birds and lies beneath the ordinary upper and lower lids, can be extended across the eye while it remains open. All birds can do this, but it is particularly useful in protecting the eyes of underwater swimmers, especially those which hit the water surface with some force. It protects the delicate corneal surface without the total loss of vision that closing the eyes completely would entail.

There appear to be no special adaptations in plunge-divers for seeing underwater, and presumably the plunge is directed to the right spot, with appropriate allowance being made for the different refractive properties of air and water by aiming to one side of the selected prey. The same is true of herons and other large waders which have to aim the bill at underwater prey (Lotem, Schechtman & Katzir, 1991).

For such species, there are two ways in which the problem of coping with both aerial and aquatic vision can be achieved. One is to adjust the shape of the lens of the eye by means of the ciliary muscles (to a much greater extent than is normally required in air). This method is used by cormorants and diving ducks. These species have an accommodative ability of about 50 diopters, which compares with less than 10 in many other birds (and humans). This is partly achieved by squeezing the soft lens with the ciliary muscles to such an extent that it bulges forward through the iris aperture. The other method is simply to have a very flat cornea, so that the differences in the refractive effects of air and water are minimised. This is the case in penguins, though their eyes are better adapted to underwater vision and they are somewhat short-sighted in air (Erichsen, 1985). In divers, auks and seaducks, the nictitating membrane has a clear central window which allows good underwater vision, though it does not provide any refractive correction as is sometimes claimed (Sivak, Bobier & Levy, 1978).

The cone cells in the retina of many terrestrial birds contain red, orange or yellow oil droplets. These filter out blue light, cutting down the effects of haze and forming part of a mechanism for efficient discrimination of reds, yellows, greens and browns. They also filter out some of the blue light from the sky, making it appear darker and there-

**Fig. 26.** Horizontal section through a Herring Gull's eye, showing the main features referred to in the text. The foveal regions are areas of high visual acuity. Redrawn from Welty (1962).

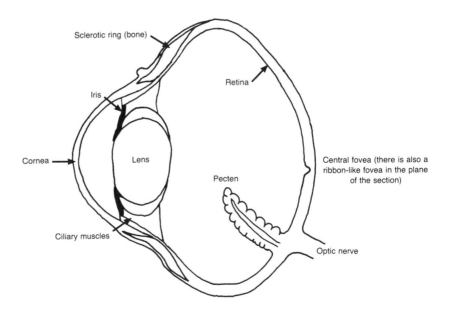

fore showing up white objects better. The white, communally feeding seabirds like gannets, terns and gulls also possess these oil droplets. The underwater feeders including penguins, shearwaters, cormorants and auks have very few such droplets, presumably because it is blue light that penetrates water furthest, and it is therefore essential that their sensitivity to it is not reduced (Burton, 1985).

Those species of seabirds which forage at night, such as the Laysan Albatross and the Swallow-tailed Gull, have enhanced levels of the pigment rhodopsin in their retinal rod cells (responsible for black and white vision) to enhance their nocturnal sight, just as is the case in owls. However, the Common Diving Petrel, which visits its breeding colonies at night and probably does some feeding at night too, has a visual threshold no more sensitive than that of the pigeon. Such sensitivity would probably be sufficient to guide it over a colony by star light, but not when stars are obscured by cloud (Brooke, 1989). The eyes of many birds have fovea. These are areas with densely packed cones and hence with heightened visual acuity. Ordinary fovea are absent in many seabirds, such as albatrosses, and instead there is a circular band of dense cones around the retina corresponding to the position of the horizon when the head is in the normal horizontal position. This band must serve to enhance the bird's ability to see objects on the horizon, or near it, from a considerable distance. Manx Shearwaters enhance their nocturnal vision by having eyes with a shorter focal length than those of diurnal birds. This means that less light is lost before it reaches the retina, though the image is smaller and thus some acuity is lost (Martin, 1990).

All birds have a sense of smell, but there has often been a tendency to underestimate the importance of this sense. The lining of the nasal passages, where the sense organs responsible for smell are situated, are particularly well developed in the tubenoses. This is probably the reason why so many members of the group have the elongated tubular nostrils that increase the surface area available for sensory tissue. The olfactory lobes of the brain that process the information from these organs are also relatively larger than those in any other group of birds except kiwis. Grebes too have quite large olfactory lobes. The sense of smell is important in two different aspects of the lives of tubenoses. Firstly, in those which breed in burrows and visit the colonies at night, it assists in locating the correct nesting hole (Brooke, 1990). Such species always approach their burrows in an upwind direction, and Leach's Petrels placed in a maze show a marked preference for air currents from their own burrow over that from another individual (Grubb, 1974). The characteristic musky smell imparted to the bird and its burrow is believed to derive from the preen gland secretion. Secondly, smell can be of great importance in locating food. Experiments using different food homogenates and oils placed on the surface of the sea attract more fulmars and shearwaters from a downwind direction than from elsewhere, provided that smells derived from potential foods are involved. There are differences in the attractiveness of different smells to different species too, with Fulmars coming to cod liver oil, but not to homogenised squid or krill (Hutchinson *et al.*, 1984). Very little work has been done on the sense of taste in seabirds, but no doubt it plays an important role in distinguishing edible from inedible materials.

Sense organs associated with touch are as widely distributed throughout the body in birds as they are in mammals. Tactile sensations are transmitted through the feathers and this may provide important feedback information for fine adjustment of attitude during flight. The bill, tongue and palate are all well supplied with touch receptors.

The hearing of seabirds appears to be similar to that of most other birds and has a range of sensitivity and response not very different from that of man. There is a complex inner ear system in all birds which supplies feedback on the orientation of the body and its acceleration in different directions, all of which represents vital information to any flying animal. It has been suggested that sense organs in the elongated nostrils of the tubenoses might aid in sensing differential wind velocities on either side of the head (which could help in gliding) but this remains to be proven.

## Other adaptations to marine life

One of the most characteristic features of tubenosed birds, present in all species and absent from all other groups, is the collection and storage in the stomach of oil. This oil is derived directly from the foods that they eat and is used both to feed the young and as a means of defence against predators by regurgitation and spitting. The oils accumulate in the anterior muscular stomach, known as the proventriculus, and while they are rich in lipids, they still contain about 80% water. Shearwaters may carry about 200 ml and fulmars about 300 ml. The oils accumulate because the acid conditions of the proventriculus favours protein diges-tion, but not fat digestion. On its own, this clearly would not provide a balanced diet for chicks and so it is given in a mixture along with regurgitated fresh fish and squid. It occurs in non-breeding tubenoses too, and in these it represents part of a cost-effective mechanism for dealing with food. Weight for weight, it is an efficient source of energy and so is the most sensible part of the diet to digest last. It provides a good reserve to tide the birds through periods (such as storms) during which they may have difficulty feeding. The chicks of Leach's Petrel fledge with a stomach full of oil which probably serves as an energy store during their first week at sea (Place, Sievert & Butler, 1991).

The oil also provides an excellent defence since it smells foul, and matts the plumage of avian predators. When it dries it becomes resinous. Peregrines and Gyrfalcons have been found with their plumage so matted that they were unable to fly after visiting Fulmar breeding colonies. Fulmars can aim quite accurately to distances of nearly 2 m and can spit five or six times. The chicks too have a well-marked spitting instinct which is even shown by newly hatched chicks before they have any oil in their stomachs (Jacob, 1982). Some seabirds, including Leach's Petrels, have the ability to digest about a third of the chitin they eat. This is of value because the exoskeletons of crustaceans, such as krill, consist mostly of chitin, representing a significant proportion of the amount of material ingested (Place & Jackson, 1990). Grebes deliberately eat their own feathers to aid in the formation of pellets, which are then voided orally to get rid of fish bones and gastric parasites (Piersma & van Eerden, 1989).

Pelagic seabirds seldom have any opportunity to drink fresh water and the avian kidney is incapable of excreting a salt solution more concentrated than seawater. This presents them with a major problem. The difficulties can be minimised by obtaining as much water as possible from the food, since this has a salt concentration lower than that of sea water, but it is still necessary to have some mechanism, more powerful than that afforded by the kidneys, for secreting salt. This is achieved by means of special salt secreting glands, situated between the eyes, and known as nasal glands (Fig. 27). These are present in most birds, but in those with regular access to fresh water they are quite small and inactive. They are morphologically and embryologically distinct from both the lachrymal (tear) glands which lubricate the eyes and eyelids, and the Haderian glands which produce a protective oily secretion which coats the surface of the eyes.

**Fig. 27.** Skull of Fulmar from above showing the depressions in the skull in which the salt glands are situated (stippled). E = eye socket, N = nostril.

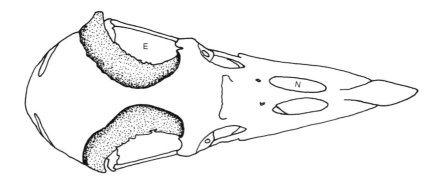

In seabirds and other species exposed to high salt concentrations in the diet, the paired nasal glands are large and usually lie in depressions in the skull immediately adjacent to the orbit of the eye, though there is some variation between species. Each lobe of the gland conveys a concentrated salt solution secreted by the glandular tissues into the nasal cavity from where it is 'sneezed' away, or trickles out onto the bill and runs down in a groove to the tip where it can be shaken off. The nasal glands only become active when the need to excrete salt arises, and in some species the size of the glands shows marked seasonal changes, especially if there is a change of habitat from sea to fresh water or vice versa (Hill, 1985). Work in the laboratory has shown that these size changes are simply a direct response to the amount of salt ingested (e.g. Staaland, 1968). Those species in which the external nostrils are partially or completely occluded have to get rid of their salt gland secretions through the mouth. Also, their salt glands are relatively smaller than in other seabirds and this may be because they are all fish eaters and their diet therefore contains somewhat less salt. They also appear to lack the enhanced sense of smell of the tubenoses.

Seabirds in general do have quite a large preen gland. Moreover, albatrosses and probably other species also have other sebaceous glands distributed over the whole body surface. As emphasised earlier, the secretions from these glands are not necessary for the plumage to be waterproof since this is just a function of the physical structure of the feathers, but they do help to maintain them in good condition, by lubricating them and thus maintaining their flexibility. The preen gland oil may also contain substances that help to control fungi and bacteria in the plumage. It also contains vitamin D precursors which may change to vitamin D under the action of sunlight, and can then be ingested during preening, though the latter point has yet to be proven. The reason seabirds have large preen glands is presumably just because they have a large number of dense feathers, and that these receive quite harsh physical treatment during swimming, diving and exposure to UV light.

The reason that oiled seabirds sometimes lose their waterproofing following washing with detergents to remove the oil, is that traces of the detergent remain in the plumage and act as a wetting agent. Thorough and repeated rinsing in clean water removes these traces and renders the birds waterproof again immediately. Birds with blocked or inactive preen glands remain waterproof but their plumage becomes dry and brittle (Jacob & Ziswiler, 1982).

## Breeding seasons

Seabirds living in temperate latitudes usually have relatively abundant food supplies, but they may be prevented from gaining access to them by adverse weather conditions, such as storms. In polar areas, where the sea freezes during the winter, they have to undergo seasonal migrations. In the tropics, seabirds are faced with the added problem that food supplies may be extremely thinly distributed. All birds breed at the time of year that enables them to produce the greatest numbers of surviving young at minimum risk to their own prospects of survival. This results in a well-marked seasonal peak of breeding in temperate and polar regions during the summer months. All seabirds are slow breeders, in the sense that incubation and the growth of young takes a long time, and this is simply a function of the long distances they usually have to travel between breeding sites and feeding areas. With very few exceptions, they are thus unable to raise more than a single clutch in a year. One such exception is probably Cassin's Auklet, which is able to lay early because it is a plankton feeder and does not accompany its young to the sea. Second broods may occur at some sites in Canada (Nelson, 1980).

Tropical breeders are often subjected to very little seasonal variation in either food supplies or weather conditions. Consequently, breeding success may vary little between one time of year and another. This has lead to the habit of breeding at intervals greater or less than a year. For example, Sooty Terns breed every six months on some islands in their huge breeding range in the tropical oceans, every nine months on others, and every 12 months at the majority of sites. Despite this variation, breeding is well synchronised at each of these colonies. In one colony in the Seychelles, for example, where half a million Sooty Terns breed on the ground with only a few inches between nest sites, 75% of the eggs are laid within nine days. This results in fewer losses to predators (skinks, rats, Turnstones and Cattle Egrets) than is the case in those eggs laid earlier or later in the season. This predator-swamping effect clearly outweighs the disadvantages of dense synchronised breeding, which include an increased risk of the transmission of parasitic infection and increased competition for food (Feare, 1976).

A breeding interval of 8–11 months is quite common in tropical seabirds and this has been attributed to the fact that in the absence of significant changes in day length to synchronise the bird's internal biological clock with the solar year, this clock seems to have an intrinsic period of nine months. When birds (of any kind) are kept in the laboratory under controlled and uniform conditions, this internal clock appears to cycle with a period of roughly nine months. For example, such captive birds undergo the annual moult at about this interval. The internal annual clock thus differs from the diurnal one in being shorter rather than longer than the external cycle with which it is normally synchronised (Gwinner, 1975).

Some seabirds breed at intervals greater than a year, and in these cases it appears that there is not enough time to rear the young and for the adults to get back into breeding condition again within 12 months. In markedly seasonal areas, such species breed in a two-year cycle, as is the case in a few of the larger albatrosses. In the tropics, it can result in repeated breeding cycles at some interval between 12 and 24 months. The seabirds of the Galapagos Islands are a good example of a group of species with a whole range of breeding strategies.

This complex system of breeding seasons on the Galapagos can be thrown into complete disarray by unusual climatic conditions. For example, during El Niño years, when warm waters replace the rich and productive cold waters from the Antarctic which normally surround the islands, there can be a complete breeding failure of all seabirds. Similarly, if food supplies suddenly become plentiful, there is an opportunity for any individuals not already breeding to start doing so (Nelson, 1980).

Amongst British breeding seabirds, all of which are annual, the timing of breeding does vary somewhat from colony to colony. In the Puffin, for example, egg laying is always earliest in eastern Britain and latest in the north and west. In Gannets, the mean laying date is two or three weeks earlier on the Bass Rock in the east than it is on Ailsa Craig in the west. The reasons for these differences are not known for sure, but the earlier availability of food, in the form of sandeels, in the east is often suggested. It could be related, on the other hand, to food supplies later in the season (Nelson, 1978a). For auks as a whole, in the wider Atlantic region, there is a highly significant relationship between sea-surface temperature and the timing of breeding (Fig. 28). Every extra degree centigrade in sea-surface temperature in August is correlated with a three-day earlier start to breeding, as measured by the median hatching date. Presumably this too acts through food availability, with birds breeding as soon as food supplies allow them to, as suggested by Perrins (1970).

### Distribution at sea

Counting birds at sea from a ship poses a number of problems, largely because of the tendency of some species to follow vessels. They do this because of the aerodynamic advantage they can obtain from the ship's aerial slipstream, and because of the food which may be thrown overboard or which may be brought to the surface in the ship's wake. Even if such birds are ignored when counting, flying birds are about twice as likely to be counted accurately as are those which are sitting on the water (Powers, 1982). Despite these difficulties, observations from ships remain the most important source of information about the distribution of birds at sea. At the very least they offer a means of assessing the relative densities of birds in different areas. Useful information can also be obtained from aircraft and from fixed structures such as oil production platforms.

In the North Sea, the distribution of birds has been intensively studied by the Seabirds at Sea Team of the Nature Conservancy Council. Between 1979 and 1986, over 1½ million bird records were obtained, 80% of these involving Fulmars, Guillemots, Kittiwakes and Gannets. The highest densities of birds were associated with midshelf fronts (see below), especially those off Shetland, Orkney, north-east Scotland and in the Skagerrak.

In the Fulmar, seasonal changes in distribution within the North Sea are less well marked than in other species, though there is a general tendency for numbers to increase in the south, when adults and young leave the breeding colonies in northern Britain. Moulting birds are seen earlier in the south, presumably because some high arctic breeders move into the area at this time. The density of Fulmars is sometimes correlated

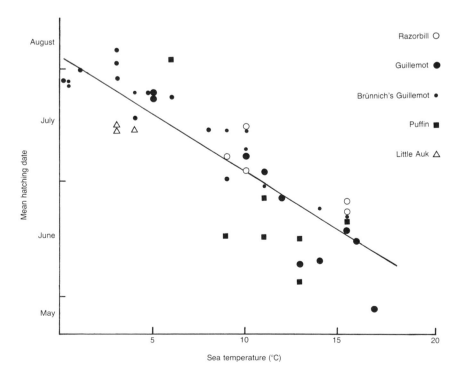

**Fig. 28.** Relationship between timing of breeding in auks (measured as the date by which half the clutch has hatched) and average sea-surface temperature in August. The latter is correlated with temperatures thoughout the spring and summer. Redrawn from Birkhead & Harris (1985).

with the numbers of trawlers in the same counting sectors, suggesting that human fishing activity is one of the factors influencing their distribution. Observations from the shore for the BTO's Winter Atlas confirm this general pattern, with the greatest concentration of Fulmars off the north-east coast, and off Ireland. There are probably around two million Fulmars in British waters during winter (Lack, 1986).

Gannets breed in the greatest numbers in the north and east of Britain. During winter, less than 2% of Gannets in British waters are immature. This is because most birds less than five years old winter off the coasts of France, Iberia and Africa. Wintering birds in the North Sea are thinly distributed, and there is no association with the breeding colonies between November and January as there is in the Fulmar. In February, March and April, however, numbers do begin to build up near the Shetlands and at Bass Rock. The proportion of immature birds rises to about 20% in May and remains high until November, but most of these birds occur well out to sea away from the breeding colonies.

The Kittiwake is by far the most regularly observed gull in offshore waters. In the North Sea, the main winter concentrations are well out to sea, with trawlers and other vessels becoming important sources of food in late winter, perhaps as the abundance of natural foods decline. During the breeding season, and into the autumn, Kittiwakes occur much closer inshore.

Puffins are largely pelagic in winter, being spread out over a considerable part of the North Atlantic, with relatively small numbers in the North Sea (Fig. 29a). During the spring, birds move closer inshore and begin to gather near the breeding colonies (Fig. 29b). They remain there during the main part of the breeding season (Fig. 29c), but gather in a dense concentration associated with the front off the Firth of Forth during the annual moult (Fig. 29d). Little Auks breed in the high Arctic and spend the winter throughout the North Atlantic. They penetrate into the North Sea in some numbers at this time of year, concentrating in the Skagerrak in January and February. Their limited diving depth means that they are vulnerable to prolonged winter storms which may force their prey (zooplankton and small fish) to remain out of reach in deeper water.

Both Guillemots and Razorbills are widely distributed around the British coast in winter. Visits to the breeding colonies occur sporadically at this time, though juveniles tend to move south to France, Iberia and the Mediterranean (Mead, 1974). At the end of the breeding season, birds move further offshore, though still in the general vicinity of the breeding colonies, in order to moult. Thus there are large numbers in the outer Moray Firth and Firth of Forth at this time of year. There are about two million Guillemots and 200 000 Razorbills in the North Sea in winter. Sprats appear to be a particularly important food for Guillemots and there is a general correlation between winter numbers in the southern North Sea and Sprat biomass (Blake *et al.*, 1984). Moreover, Guillemots are common in onshore waters where winter stocks of Sprats are present, such as the Scottish Firths. Local food shortages, caused by human overfishing of Sprat stocks (p. 264), have been responsible for a number of wrecks. For example, 10 000 unoiled Guillemot corpses in poor condition were washed ashore along the east coast of Britain in February 1983 (Tasker *et al.*, 1987).

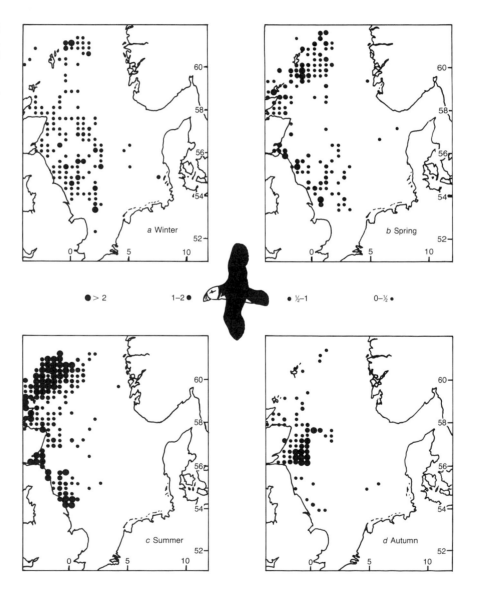

**Fig. 29.** Average densities of Puffins per km² in *c*.30 km squares in the North Sea at different times of year. *a* = winter (October to February), *b* = spring (March to May), *c* = summer (June to July), *d* = autumn (September). Coverage was not complete and the dots only provide a general indication of the distribution of birds. Redrawn from Tasker *et al.* (1987).

**Fig. 30.** Zones of enhanced marine productivity exploited by birds, caused by upwelling, downwelling and the mixing of different types of water. Open arrows = wind directions, filled arrows = water currents. *a* = upwelling caused by offshore or longshore winds blowing water away from the coast. This water is replaced by cold, nutrient rich water from below. *b* = upwelling, and associated downwelling, caused by a submarine obstruction. Based on tidally induced features in the Bay of Fundy, Canada (redrawn from Brown, 1980). In this case, plankton productivity and Red-necked Phalarope distribution in August are closely correlated with the upwelling, and to a lesser extent with the downwelling. Islands and the continental shelf edge produce similar effects. *c* = enhancement of productivity at the region where estuarine waters meet the sea, forming a front e.g. the Skagerrak. Such salinity fronts also occur at river mouths and sewage outfalls. The mixing depth is the depth above which all waters are mixed by tides, winds or currents. The compensation depth is the depth below which there is no net growth of plankton (because of inadequate light levels). *d* = upwelling induced by descending air and divergent surface winds (such as occur in the tropics), and downwelling with associated front induced by convergent surface winds (as at the equator) (redrawn from Vinogradov, 1981). This general pattern of vertical circulation occurs on a much smaller scale in Bénard and

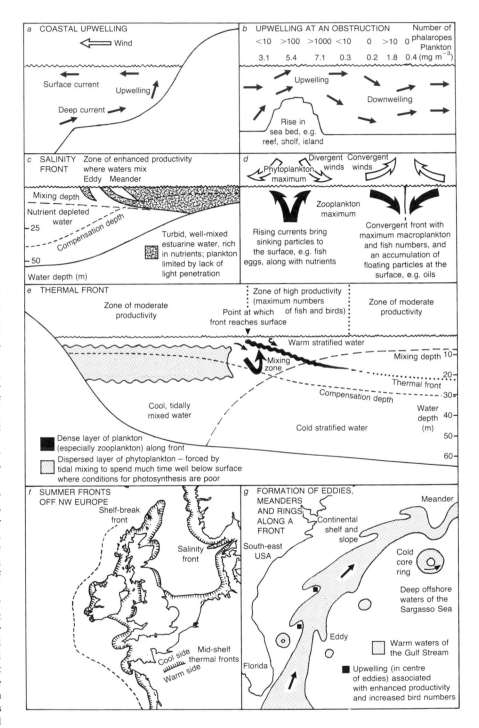

(Fig. 30—cont.)
Langmuir cells (see text). $e$ = enhancement of productivity at the region where tidally mixed inshore waters meet stable, stratified offshore water, forming a thermal front (redrawn from Fogg, 1985) e.g. most of the fronts around the British coast in $f$. Such fronts disappear in winter when the thermal stratification of offshore waters breaks down. $f$ = approximate positions of shelf-break and mid-shelf thermal fronts off the coast of northern Europe during the summer months. Although there is considerable variation in the position of these fronts, most are remarkably persistent features. Plumes and eddies often form along them, enhancing the mixing processes which are so important in stimulating increased productivity. Drawn from satellite photographs (infra-red and Coastal Zone Color Scanner composites) and maps in Pingree & Griffiths (1978), Simpson, Allen & Morris (1978), James (1981), Pingree & Mardell (1981), Simpson (1981) and Holligan, Aarup & Groom (1989). $g$ = formation of productivity enhancing eddies and rings from the Gulf Stream in the South Atlantic Bight of North America. Redrawn from Haney (1986a) and Owen (1981).

Similar studies of the distribution of seabirds at sea off the north-west coast of the British Isles have now been completed (Webb *et al.*, 1990). These have identified a number of areas that support high densities of seabirds throughout the year, including the Minch (between Scotland and the Outer Hebrides). Densities were generally greater over the continental shelf than over the deep ocean, with the shelf break and midshelf fronts (see below) supporting particularly large numbers.

A wreck refers to any unusually large occurrence of dead or moribund seabirds along the coast, or of live ones inland. This quite often occurs during strong winds when birds seek the shelter of inshore waters or simply get blown inland. Young birds are particularly vulnerable during the gales that may occur around the vernal equinox. One of the best examples of a wreck occurred in 1952 when at least 500 Leach's Petrels were recorded inland in Britain following severe westerly gales in October. Many of them were adults in moult (Wynne-Edwards, 1953). Even large albatrosses can occasionally be wrecked during cyclones in the southern hemisphere, juveniles and moulting adults again being most vulnerable (Ryan & Avery, 1987).

Movements of seabirds observed from the land are rather poorly understood. Certain species are observed only under certain conditions. For example, southerly movements of Fulmars, Gannets, Kittiwakes and Sooty Shearwaters off the south Swedish coast, and westerly movements of similar species past Cape Clear in south-west Ireland (Sharrock, 1973; Blomqvist & Peterz, 1984). Such movements generally follow the passage of deep depressions in autumn and winter, and may simply consist of birds returning to their normal feeding areas after they have avoided the high winds and unfavourable feeding conditions that such depressions bring. Thus, in Sweden, the birds would be returning to the Skagerrak and southern North Sea, and in Ireland they would be returning to the Atlantic. The advantage of flying close to the coast during these recovery movements would be to gain some shelter from adverse winds. The fact that more birds are recorded when such winds occur tends to confirm this (Sharrock, 1973).

Large-scale features of the oceanic circulation systems, particularly those causing upwellings, are responsible for major differences in the distribution of food available to seabirds. The largest features are the five circulating gyres which are driven by the prevailing surface winds (North and South Atlantic, North and South Pacific and South Indian Ocean). These gyres contain warm, clear water, rich in oxygen, but depleted of nutrients, so they are relatively unproductive. None the less, these are the areas which are exploited by tropical seabird communities, often by taking advantage of the concentrating effect that predatory fish and cetaceans have on prey (see p. 59).

Upwellings occur along the eastern margins of these oceanic gyres because the prevailing winds in the important summer months (when most light is available for photosynthesis) are either offshore or long-shore (Fig. 30*a*). Longshore winds drive surface water offshore because the earth's rotation deflects southerly winds to the right in the northern hemisphere, and northerly winds to the left in the southern hemisphere (due to the so-called Coriolis effect). These are the prevailing longshore wind directions in the critical marginal areas. Upwellings also occur where there are submarine obstructions, such as reefs, islands or the edges of the continental shelf (Fig. 30*b*), and where winds diverge (Fig. 30*d*). Shuntov (1974) proposed three rules governing the oceanic distribution of birds which stem from these features. These are that the distribution of birds is more closely related to currents and water masses than it is to latitude, that there is a broad increase in bird density with increasing latitude, and that more birds occur near coasts, especially those on the easterly margins of the oceans.

In areas where different water masses meet, fronts often form because of differences in density caused by contrasting water temperature or salinity (Fig. 30*c,d,e* and *f*). They also occur in the vicinity of upwellings when there is a large change in water properties over a short horizontal distance (the definition of a front), as there often is in such circumstances. A variety of effects may cause food to concentrate in such areas. The scale of frontal systems varies from structures less than a metre in size lasting for a few hours, to structures covering many kilometres and lasting for years. The smallest effects are due to evaporation of water from the sea surface causing the upper layers to become denser through cooling and slight increases in salinity. Under light wind conditions, these denser layers sink in a more or less regular pattern known as Bénard cells. These consist of rising water in the middle of an elongated cell a metre or two wide and *c*. 30 m long. The denser water sinks at peripheral convergences extending to a depth of less than a metre. These convergences concentrate plankton and sometimes show up as surface streaks. Langmuir circulation cells, perhaps driven by the Langmuir currents in the air above the sea (Fig. 21), develop under stronger wind conditions. These currents override the Bénard circulation (or may develop from it), forming a set of parallel convergences or windrows which may be anything from a few metres to over 100 m apart and penetrate much deeper than a metre (Owen, 1981). Phalaropes, gulls and terns have been recorded feeding at Langmuir cell convergences, but avoiding the waters between them, off the coast of Peru (Brown, 1980).

On the continental shelf, midshelf fronts may form in the summer months where the stratified offshore water meets the tidally mixed inshore waters. These inshore waters may be less salty, forming a salinity front (Fig. 30*c* and *f*), or colder, forming a thermal front (Fig. 30*e* and *f*). These types of front are important in influencing the summer distribution of seabirds in the North Sea (Fig. 29). Whilst the exact reasons why they are so often zones of enhanced plankton productivity are not fully understood, it seems most likely that the injection of nutrient rich inshore water into the upper layers of warm, stratified water with ideal conditions for photosythesis is the key. This mixing process is enhanced by the formation of meanders and eddies along the front (Fig. 30*g*). These midshelf fronts are also called tidal fronts since it is tidal currents that are responsible for the fact that the phytoplankton in inshore waters spends too little time in the light near the surface to maximise the efficiency of photosynthesis. These fronts break down in winter when conditions are colder and storms prevent the stratification of offshore water. A further tidal front may sometimes occur closer inshore than the midshelf front (Sournia *et al.*, 1990). Plume fronts are merely the plume-shaped fronts associated with a relatively fast moving body of water, such as a river or even a sewer. Plumes of fresh water also burst out of the Baltic periodically (through the Skagerrak), and out of some Norwegian fjords, following periods when westerly winds have held them back (Mork, 1981).

Some of the best studied frontal eddies are those which form near the continental shelf margin off the south-eastern coast of North America. These occur where the Gulf Stream passes north-eastwards along the coast (Fig. 30*g*). Each eddy encloses a tongue of upwelling cold water which is rich in nutrients. Seabirds, including shearwaters, petrels, storm-petrels, gannets, gulls, skuas and terns, are most abundant at the frontal boundaries within these meanders and in the tongues of cold water that have existed for longest (presumably because there has been longer for plankton to develop). For example, storm-petrels occur at densities of 3–16 birds km$^{-2}$ in the cold water parts of the eddies and only 0–4 km$^{-2}$ in the warmer parts (Haney, 1986*a*). Differences are also apparent in the bird communities of fronts in the South Atlantic and the Gulf of Mexico, with Bonaparte's Gulls being commonest at a front rich in zooplankton, and Laughing Gulls at one with abundant small schooling fish (Haney, 1986*b*).

Meanders and eddies may eventually give rise to rings. These often consist of warm Gulf Stream water which in turn contains a central core of cold water. Such cold core rings pass into the Sargasso Sea where they continue to support more productive communities than the surrounding water because of their greater concentration of nutrients. As much as 10% of the Sargasso may actually consist of such rings at any one time, making a significant contribution to its overall productivity. Larger rings may be as much as 300 km in diameter and persist for more than a year (Owen, 1981). Similar eddy and ring forming processes occur along most frontal systems.

The ice edge is an important zone in arctic and antarctic waters, since the underside of the pack-ice supports a community of diatoms, zooplankton and fish, to which birds can only obtain access where the ice is broken. Fulmars, Black Guillemots and Little Auks feed on the crustacean *Apherusa glacialis* which concentrates at the edge of the pack-ice. Leads in the ice, or larger areas of open water in winter (polynias), are also important in providing access to birds. The largest permanent polynia in the Arctic occupies about 15 000 km$^2$ (6000 square miles) at the head of Baffin Bay between Greenland and Ellesmere Island. Polynias generally occur in shallow waters where strong tidal or wind generated currents keep the water flowing and prevent it from remaining at the surface long enough to freeze. The most productive areas in the arctic occur where there is a good deal of mixing, since this ensures that nutrients do not become exhausted, and often where warmer than normal waters occur. The continuous summer daylight ensures that high productivity is achieved in areas such as the Barents Sea, and the Lancaster Sound (off Baffin Bay), where such conditions are met. These regions consequently support enormous concentrations of seabirds during the breeding season.

# 3

## Rocky shores, cliffs and shingle beaches

Cliffs form the most familiar coastal landmarks in the British Isles, from the white cliffs of Dover to the rocky promontories of Land's End in Cornwall and Duncansby Head near John o' Groats. In fact only about 22% of the shoreline of Great Britain is backed by cliffs and 35% is fronted by rocky shores (Fig. 31), but these none the less represent the

**Fig. 31.** Distribution of rocky shores in Great Britain. Some of the areas referred to in the text are indicated. Drawn from information in Doody (1985) and Burd (1989).

Eynhallow

Shiant Islands

Rhum

Murcar
Gourdon

Isle of May
Bass Rock

Farne Islands

North Shields

Grassholm

Skomer

Table 6. *Lengths of the coastline of Great Britain, to the nearest 10 km, occupied by different types of habitat on the landward and seaward sides (percentage in brackets). Simplified from the Nature Conservancy Council's Coastal Resource Survey (P. Doody, personal communication). The totals do not add up exactly because they have been rounded to the nearest ten*

|  | England | Scotland | Wales | Total |
|---|---|---|---|---|
| *Intertidal habitat* | | | | |
| Rocky shore | 580 (10) | 5670 (48) | 460 (29) | 6710 (35) |
| Shingle beach | 540 (9) | 2990 (25) | 100 (6) | 3630 (19) |
| Muddy shore | 1220 (21) | 280 (2) | 290 (18) | 1800 (9) |
| Sandy beach | 980 (17) | 1050 (9) | 320 (20) | 2350 (12) |
| Saltmarsh | 1420 (24) | 360 (3) | 180 (11) | 1970 (10) |
| Other[a] | 1170 (20) | 1450 (12) | 260 (16) | 2880 (15) |
| Total | 5910 | 11 800 | 1620 | 19 340 |
| *Terrestrial habitat* | | | | |
| Cliffs | 1080 (20) | 2460 (21) | 520 (33) | 4060 (22) |
| Vegetated shingle | 300 (5) | 130 (1) | 60 (4) | 480 (3) |
| Sand dune | 330 (6) | 890 (8) | 150 (10) | 1370 (7) |
| Embankment[b] | 1760 (32) | 230 (2) | 180 (12) | 2170 (12) |
| Other[a] | 2030 (37) | 8080 (69) | 650 (42) | 10 750 (57) |
| Total | 5500 | 11 780 | 1560 | 18 840 |

[a]This category includes mostly mixed sediments in the intertidal zone, and mostly grassland inland.
[b]Man-made.

dominant natural coastal habitats (Table 6). These coasts are erosional in nature and are thus continually being eaten away by the sea, except where the shore is protected by man-made structures. Plunging cliffs, with little or no beach at their foot, are rare in Britain. Most are fronted by a wave-cut platform of rock and some also have a shingle beach. Bays form where the rock is softer and these give rise to less steep cliffs. Finer sediments, usually sand, accumulate in the more sheltered conditions in such bays as a result of longshore drift (see Chapter 1). The length of coastline is twice as great in Scotland (*c.* 12 000 km) as it is in England (*c.* 6000 km) and since only a small proportion of the Scottish coast is estuarine, by far the greatest length of rocky shore and cliff is situated there (*c.* 85% of the former and 60% of the latter, see Table 6). The width of the intertidal zone has been measured for only a few areas. In Orkney, rocky shores are about 84 m wide on average, whilst sandy ones are nearly twice as wide (153 m) because of their shallower slope (Tay & Orkney Ringing Groups, 1984).

The birds associated with this type of coastline in Britain belong mainly to the groups described in the previous chapter, notably tubenoses, cormorants and auks during the breeding season, and ducks, gulls, terns and waders throughout the year. There are one or two species, belonging to groups primarily adapted to non-coastal habitats, that are also found on rocky shores. The Order Falconiformes includes all the diurnal birds of prey, and one of these, the White-tailed Eagle, has been successfully reintroduced to the British Isles. This species was once moderately abundant with about 100 known eyries in Scotland, 50 in Ireland and several in England. The last breeding attempt known to have been made by these indigenous birds was in 1916. Attempts were made to reintroduce the species in 1959 and 1968, but it was not realised at that time that only a long-term, persistent programme was likely to be successful. Such a programme was started on Rhum in 1975 (Love, 1983). The first pair bred in 1985 after many birds had been introduced from Norway. In 1991, four pairs managed to raise a total of seven young.

The Passeriformes is the largest order of birds. The majority of species are adapted to life in terrestrial habitats, but some, such as the Rock Pipit, have become adapted to life on rocky shores, and a good many others feed there occasionally. Strand line feeding by Pied Wagtails, Starlings, Jackdaws and Wrens is often opportunistic, though there are some individuals or flocks for which this habitat provides a regular and major source of food. There are quite a few species that nest on cliffs but feed mostly in other habitats, such as House Martins and Choughs. The most specialised of shoreline passerines is the Rock Pipit. It occupies different types of shores in winter, but for breeding it requires suitable rocky shores with well-concealed nest sites, usually deep in vegetation on a cliff or in a rocky recess. The nest is often situated not far above the splash zone and the diet consists mainly of insect larvae, crustaceans and small winkles obtained from the strand line and the upper part of the shore (Gibb, 1956).

## Foraging on rocky shores

On the wave-cut platform at the foot of most cliffs, a typical rocky shore intertidal flora and fauna develops. In many cases, this is below a zone of scree or shingle. The latter is usually too mobile and exposed to support a very diverse community, but the rest of the shore can be extremely productive and thus provides rich sources of food for foraging birds. Not only can the productivity of a kelp bed be higher than that of a tropical rain forest (Table 1, p. 16), but its animal populations can be four times as productive. This is because most plant tissue gets consumed quite quickly rather than being locked up in relatively inedible materials such as wood or being decomposed by microbes in the soil. The relative degree of exposure to wave action is the major factor limiting productivity, since on exposed shores rather few plants and animals are able to find sufficient shelter to survive at all. Drying out and rapid changes in salinity and temperature are other hazards of life in the intertidal zone.

The effects of these factors, and the variation in their intensity at different heights on the shore, produces a characteristic zonation which is most obvious in the plants (Fig. 32). In the splash zone, there are bands of orange and black lichen which can be very conspicuous even from a distance. These bands tend to be widest where greater wave action generates a wider splash zone, e.g. on westerly coasts exposed to Atlantic swell. The intertidal zone itself is dominated by seaweeds in sheltered areas, but these may be replaced by barnacles and mussels

**Fig. 32.** Zonation on the shores of a steeply shelving sea loch in south-western Ireland. In the highest part of the splash zone is a band of grey-green lichens. There then follows a zone of black lichen and Channel Wrack, both of which support winkles. Further down is a zone of Flat and Knotted Wracks, and finally there are kelps just above and below the water level. Few birds except Rock Pipits could feed on a shore as steep as this.

where wave action is stronger. Wracks predominate in the middle part of the shore and kelps in the area around the low tide mark. Many of the commonest animals, which provide the main food of birds that forage on rocky shores, grow on these seaweeds or hide beneath them at low tide e.g., winkles, tube worms and many crustaceans. These animals also show a zonation down the shore as do those which attach themselves directly to bare rock, such as barnacles, limpets, mussels and anemones. The latter groups are reduced in numbers on sheltered shores, but the individuals that are present are often larger in size.

Compared with animals in deeper water, those living on the shore have to be able to cope with physical buffeting by the waves, desiccation, salinity and temperature fluctuations, and exposure to both marine and terrestrial predators. This means that there is a general increase in the density and diversity of animals down the shore, with the kelp zone being the richest, supporting about 2900 invertebrates per 100 g of weed on a typical sheltered shore, compared with 1400 in the lower wrack zone, 300 in the mid-tide region, and 50 or less in the uppermost wrack zone (Colman, 1940). Fortunately for intertidal animals, the same adaptations that enable them to cope with the physical extremes of the environment often protect them from predators as well. For example, the squat shape and surface adhesion shown by limpets (Fig. 33) provides effective protection against many predators. Thick shells, and effective mechanisms for closing them, likewise afford all-round protection. Seeking out sheltered refuges, which may often be associated with good camouflage, is an alternative way of avoiding both environmental extremes and predation. On the lower shore, and in the subtidal area, where the physical difficulties are reduced, many animals do not hide or physically protect themselves, but rely on poisons to deter predators (Fig. 33).

Birds that forage regularly on rocky shores, including Turnstones, Purple Sandpipers, Oystercatchers and gulls, have to find these animals and overcome their defences if they are to survive. Invertebrates that hide in sheltered places beneath stones and weed can be located by moving the cover to one side. Turnstones achieve this by flicking small stones out of the way or by barging growths of weed forward to reveal sand-hoppers, Sea Slaters and crabs. Purple Sandpipers can use their longer bills to probe amongst seaweed, and in patches of soft sediment, in order to locate prey by touch (Summers *et al.*, 1990). On more exposed shores, with less weed cover, Turnstones feed on barnacles by hammering open the terminal plates which guard their openings, on winkles by dislodging them and extracting the flesh through the aperture, and occasionally on sea anemones by probing their oral cavities and extracting the remains of their last meal (Fig. 34).

**Fig. 33.** Adaptations of rocky shore animals to withstand wave action (*a,b,d*), desiccation (*a,b*), temperature extremes (*b*), salinity (*a,b,c*) and predation (*a, b,c,d,e*). Many of the animals illustrated show a combination of different adaptations.

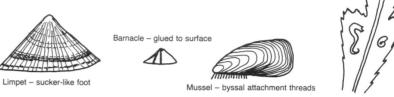

*a* SQUAT SHAPE, SURFACE ADHESION AND CLOSED SHELL OR TUBE

Limpet – sucker-like foot

Barnacle – glued to surface

Mussel – byssal attachment threads

Tube worms – glued to wrack

*b* HIDING IN SHELTERED PLACES

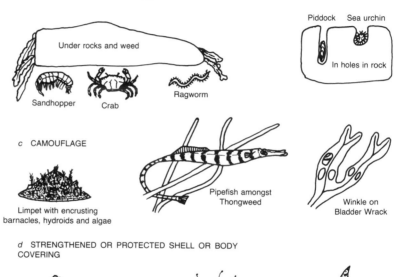

Under rocks and weed

Sandhopper     Crab     Ragworm

Piddock     Sea urchin

In holes in rock

*c* CAMOUFLAGE

Limpet with encrusting barnacles, hydroids and algae

Pipefish amongst Thongweed

Winkle on Bladder Wrack

*d* STRENGTHENED OR PROTECTED SHELL OR BODY COVERING

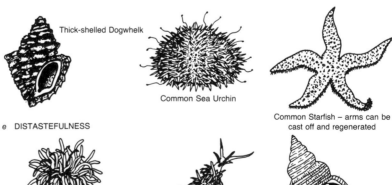

Thick-shelled Dogwhelk

Common Sea Urchin

Common Starfish – arms can be cast off and regenerated

*e* DISTASTEFULNESS

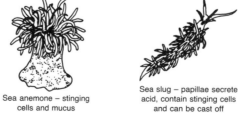

Sea anemone – stinging cells and mucus

Sea slug – papillae secrete acid, contain stinging cells and can be cast off

Spindle Shell – toxic saliva

Oystercatchers and Turnstones use rather different techniques when feeding on limpets. Oystercatchers frequently dislodge the shell by giving it a sudden, sharp blow with the bill, or they may hammer at the shell repeatedly. The majority of these blows are aimed at the right hand side of the head in *Patella aspera* since the attachment to the rock is weakest in the region where the horseshoe-shaped attachment muscle ends. The dotted line in Fig. 34 shows the Oystercatcher's favoured position of attack. Even if the limpet is not dislodged, the shell sometimes breaks and can then be levered off by inserting the bill at an angle into the hole and straightening it. *Patella vulgata* is seldom attacked

**Fig. 34.** Some of the feeding methods employed by birds on rocky shores. For full explanation see text.

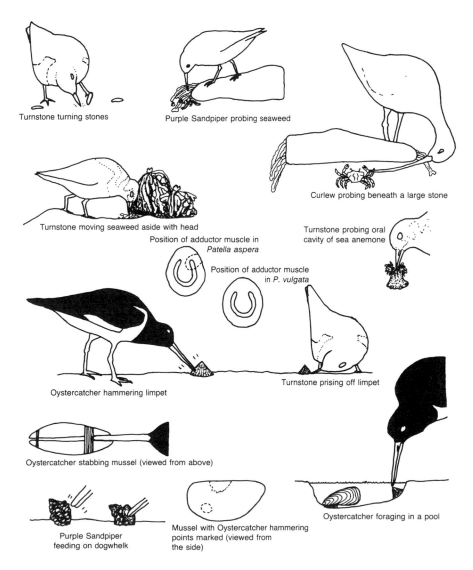

Turnstone turning stones

Purple Sandpiper probing seaweed

Curlew probing beneath a large stone

Turnstone moving seaweed aside with head

Position of adductor muscle in *Patella aspera*

Position of adductor muscle in *P. vulgata*

Turnstone probing oral cavity of sea anemone

Oystercatcher hammering limpet

Turnstone prising off limpet

Oystercatcher stabbing mussel (viewed from above)

Purple Sandpiper feeding on dogwhelk

Mussel with Oystercatcher hammering points marked (viewed from the side)

Oystercatcher foraging in a pool

in this way since it has a thicker shell (relying less on camouflage to escape detection) and has a rounder shape which makes it more difficult for the bird to recognise the head end (Feare, 1971). Turnstones, on the other hand, rely on inserting the tip of the bill under the tiny gap around the rim of the limpets shell and levering it off. Oystercatchers can also do this, but their larger bills cannot get under the edge of small limpets and thus they generally take larger ones.

Oystercatchers use a wide variety of methods for opening mussels. Some individuals stab the bill between the two valves of the shell and cut the larger of the two muscles which hold them together (Fig. 34), while others hammer a hole in the shell at a weak point. The position of the weakest point of the shell depends upon where the mussels are located on the shore and how exposed they are. Ventral hammerers remove a mussel from the mussel bed, wedge it in a firm spot and hammer a hole in the ventral side (usually in the position indicated by the upper dotted line in Fig. 34) so that the bill can be inserted and the adductor muscle cut. These birds select mussels that are brown in colour and have few barnacles on them, probably because this indicates they are fast growing and are therefore thin shelled. Exposed mussels tend to be eroded and thin on the dorsal side. Dorsal hammerers select suitable mussels by tapping them first and then hammering them *in situ* where the shell-closing muscle is attached (lower dotted line in Fig. 34), so that the two valves then fall apart (Durell & Goss-Custard, 1984). Individual Oystercatchers usually specialise in particular methods of opening mussels. In the Exe Estuary, 41% of birds are stabbers, 35% are ventral hammerers and 25% dorsal hammerers, though the proportions vary from one mussel bed to another. The chosen method of feeding influences the shape of the bill tip which becomes worn down if they are hammerers, but remains pointed if they are stabbers. It is the horny outer covering of the bill that wears in this way and not the internal bony part. The horny bill plates grow particularly quickly in Oystercatchers, at over 0.4 mm per day, which is three times faster than human finger nails which in turn grow faster than the bill plates of most birds (Hulscher, 1985). Three main bill shapes can be recognised amongst wintering Oystercatchers in Europe – pointed, chisel-shaped and blunt – together with a wide range of intermediates. A greater proportion of young birds and females tend to have pointed bills. In captivity, birds with all three types of bill are capable of opening 35–62 mm long mussels from the Wadden Sea, but hammerers can take even larger ones than this. If the birds are fed solely on worms such as lugworms, buried in an artificial mudflat, a chisel-shaped bill changes to a pointed one in about 10 days (Swennen *et al.*, 1983). This accords with the fact that when many birds switch from shore feeding to inland feeding (on earthworms, leatherjackets, etc.) during the breeding season their bills become more pointed (Hulscher, 1985).

Stabbers generally have a lower food intake rate than hammerers and often supplement their diet on the shore by feeding on inland fields at high tide. Here their pointed bill is well suited for probing the soil for invertebrates, especially earthworms. Juvenile Oystercatchers are often less successful than adults on the mussel beds, not just because they are usually stabbers, but also because they tend to lose out in competition with adults. This competition may involve 'piping' displays and other activities associated with the ownership of food at a feeding site. Juveniles lose about 30% of their potential food intake, either because it is stolen by other birds, or because they are prevented from feeding by interference (Ens & Goss-Custard, 1984). It takes Oystercatchers a number of years to develop the skills and experience necessary to become completely proficient at feeding under the range of different circumstances that prevail on mussel beds (Goss-Custard & Durell, 1987), not to mention those which operate in the other sorts of feeding areas that they may have to use at other times of year.

Several species of prey animal may be easier to obtain in rock pools or when they are just covered by the tide, because they may become active then. Limpets are easier to dislodge if they are moving and mussels can be taken more easily by stabbing if the shells are gaping slightly as they tend to do when immersed or when they are very cold. Stabbing Oystercatchers spend a lot of time feeding in rock pools and at the tide edge.

The Dogwhelk is a species that feeds mainly on barnacles and mussels by boring through their shells. Dogwhelks often spend the low-tide period in exposed places and have thick shells which even Oystercatchers find difficult to hammer open. However, both Oystercatchers and Purple Sandpipers can feed on them by tipping the shell over quickly and grabbing the body of the mollusc before it can retreat into the shell (Fig. 34). Thus an element of speed and surprise is important for successful foraging, as it often is when capturing limpets and gaping mussels. Dogwhelks have a parasitic worm which castrates its host and causes a change in behaviour. Rather than aggregating with other Dogwhelks, infected individuals tend to lead a solitary life in the open, making them more vulnerable to predation by the birds which are the parasites' final host (Feare, 1971). Some parasites of winkles and *Macoma* similarly manipulate the behaviour of their intermediate molluscan host to their own advantage. Experienced birds probably avoid these parasitised shellfish, for Oystercatchers have been observed rejecting infected *Macoma* after seeing or tasting the flesh (Hulscher, 1982). Young and inexperienced birds may not have learnt to do this, or may have difficulty in finding enough uncontaminated food, and this is one of the reasons they have higher parasite burdens than adults.

Different species of birds commonly take different sizes of the same prey. For example, there is very little overlap in the sizes of mussels taken by Purple Sandpipers, Eiders and Oystercatchers (Fig. 35). Purple Sandpipers take mostly individuals with shells less than 5 mm long, although there is a slight difference between the sexes, with females (which are slightly larger than males) taking slightly larger mussels than males. In the Exe Estuary, Oystercatchers feeding on different mussel beds take a slightly different range of sizes, but only rarely do they take any smaller than 30 mm. Eiders take the widest range of mussel sizes, but they avoid those greater than 40 mm long, presumably because they are too big for them to swallow or to crush in the gizzard. Other species feeding on rocky shores show some overlap with these species, for example Knots take mostly mussels under 10 mm long and thus overlap with Purple Sandpipers (Summers & Smith, 1983).

Some species are able to take even larger mussels than Oystercatchers by dropping the shells to the ground in order to smash them open. This behaviour is widespread amongst gulls and is used to open up a variety

**Fig. 35.** Sizes of mussels taken by Purple Sandpipers on the coast of Angus (Feare & Summers, 1985), Eiders in the Firth of Forth in November–March (Player, 1971), and Oystercatchers on Bed 25 in the Exe Estuary in March (Goss–Custard *et al.*, 1980).

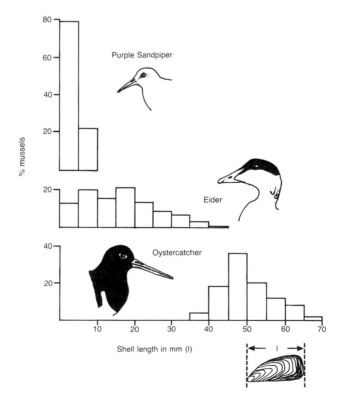

of hard-bodied prey such as crabs, starfish and snails, including hermit crabs in whelk shells as big as 80 mm in length. The dropping behaviour is not always well co-ordinated as a Herring Gull was once observed dropping a shell 39 times in succession onto shallow water, and others have smashed skylights at a London factory by dropping nuts, bolts and pieces of metal onto them (Tinbergen, 1953). Carrion Crows and other corvids appear to be efficient smashers of shore animals, for they often carry them to suitable hard surfaces such as roads and runways to drop them. In one case they even learnt to use mudflats on days when these were frozen. It has been shown that large mussels break more easily when dropped than small ones and dropping them onto the mussel bed from which they came has the advantage of sometimes smashing mussels on the ground as well. Most of the mussels dropped by Carrion Crows are longer than 60 mm and they are dropped from no great height (usually 3 m or less) onto concrete, but from rather higher (5 m) onto sand. They are often dropped from less than the optimal height to achieve breakage, probably because of the risk of other individuals stealing them (Whiteley, Pritchard & Slater, 1990). Birds often drop them after they have started to dive, presumably because they then stand less chance of losing sight of them or having them stolen by other birds. Such behaviour might also increase the force of the impact, especially when the height of the drop is small.

Gulls and crows regularly forage on the shore in a more sedate fashion. Gulls take smaller mussels (under 20 mm long) by pulling them off the rocks and swallowing them whole, as they also do with small limpets. Crows sometimes take small fish from rockpools by immersing their heads into the water, or even executing crude plunge-dives.

The strand line at the top of the shore represents a very important and regularly renewing source of food for a number of species. Many of the smaller passerines, including pipits, wagtails, finches, buntings and Starlings forage there for crustaceans, insects and seeds. Larger species such as gulls and crows compete for items of offal that may have been washed ashore, such as dead fish and birds. Since the strand line is available at a time when the rest of the intertidal area is covered, it provides an opportunity for many intertidal feeders such as Turnstones (Fig. 36) to extend their foraging time. Even species normally thought of as being entirely terrestrial, such as Song Thrushes and Blackbirds, will exploit the strand line occasionally, for example when the ground elsewhere is frozen or snow covered. Likewise, Kingfishers and Dippers may be found on the shore during extreme conditions when fresh waters freeze or dry up. Cliffs too can be important for foraging when other ground is snow covered. Skylarks, Greenfinches, Linnets and Meadow Pipits have all been observed feeding on cliff ledges under such conditions (Alder, 1982). In addition, some species such as Black Redstarts utilise this habitat regularly.

The activities of such a wide range of rocky shore predators can have a significant impact on some invertebrate populations. For example, Purple Sandpipers are responsible for nearly all of the 89% winter mortality experienced by one-year-old Dogwhelks in north-east England (Feare & Summers, 1985). An individual Rock Pipit may consume as many as 14 300 very small winkles and 3500 chironomid larvae (the latter mostly from *Entermorpha* near the high tide mark) in a day. This preference for small winkles explains why the Rock Pipit is one of the few birds that is more abundant on exposed rocky shores than on sheltered ones where shore invertebrates tend to be larger (Gibb, 1956). It has been suggested that differential selection by birds may be responsible for maintaining

**Fig. 36.** Turnstones and Purple Sandpipers moulting into summer plumage. The former is one of the most abundant waders on British rocky shores and the latter is almost entirely dependent on this habitat. Chris Rose.

the marked colour polymorphisms that occur in winkle populations (Pettitt, 1975).

The Eider referred to above is the only species so far considered that feeds on the intertidal area at high tide. As well as mussels, it takes winkles, crabs, whelks and starfish. Mussels also predominate in the diet of Long-tailed Ducks, although they take a somewhat narrower range of sizes than Eiders, and usually avoid individuals more than 20 mm long. They also take whelks, winkles, crustaceans and gobies (Owen, Atkinson-Willes & Salmon, 1986). Common Scoters take a similar range of species when feeding over rocky shores, but, like Long-tailed Ducks, considerable numbers feed from soft sediments (see Chapter 4). All three species often feed on areas well below the extreme low water mark, especially at low tide when the bottom is easier to reach. The Velvet Scoter tends to feed closer inshore, mainly in the wrack zone of rocky shores, and consequently takes a higher proportion of winkles and whelks than other species. The sawbills (mergansers, Goosander and Smew) generally have a higher proportion of fish in the diet than other ducks and will follow the fish into the intertidal zone as the tide rises. Red-breasted Mergansers feeding on rocky shores take gobies and sticklebacks as well as midwater fish such as Herrings. All three species of divers (Great Northern, Black-throated and Red-throated) tend to feed further offshore than most species of ducks and take a variety of midwater fish such as Sprats, Herrings and whitings as well as gobies and pipefish from the sea bed.

**Abundance of wintering birds**

The first comprehensive survey of the birds occurring along the coasts of Britain and Northern Ireland (the Winter Shorebird Count) was organised by the British Trust for Ornithology and the Wader Study Group in the winter of 1984–85 (Moser & Summers, 1987). Observers walked along the shore between the tide marks, within a few hours of low tide, counting all the shorebirds they passed, or that flew over them. Estuaries were excluded from the survey because they are regularly counted by the Birds of Estuaries Enquiry. Some of the more remote areas were surveyed by specially organised expeditions. Total coverage amounted to 83% of the 13 525 km of non-estuarine coast and nearly 300 000 waders were counted (Table 7).

Oystercatchers were the most abundant species, followed by Curlews and Turnstones. The presence of Curlews in such large numbers on non-estuarine shores is a little surprising. The majority of them were found on the Scottish coast, with 17 700 on Orkney (Tay & Orkney Ringing Groups, 1984). More were associated with rocky shores than

Table 7. *Numbers of waders counted on the non-estuarine coasts of the United Kingdom during the 1984–85 Winter Shorebird Count (from Moser & Summers, 1987). Counts greater than 1000 rounded to the nearest 100*

| Species | Number |
| --- | --- |
| Oystercatcher | 70 100 |
| Ringed Plover | 18 300 |
| Golden Plover | 11 600 |
| Grey Plover | 2600 |
| Lapwing | 18 900 |
| Knot | 4800 |
| Sanderling | 9300 |
| Purple Sandpiper | 15 800 |
| Dunlin | 30 100 |
| Jack Snipe | 10 |
| Snipe | 1800 |
| Bar-tailed Godwit | 5200 |
| Whimbrel | 3 |
| Curlew | 48 600 |
| Redshank | 22 900 |
| Greenshank | 130 |
| Common Sandpiper | 7 |
| Turnstone | 38 500 |

with sandy ones (the latter form only 11% of the shoreline in Orkney). These birds probably fed on grass fields inland as well as on the shore itself and it is not known what proportion of their food they obtained in each habitat. However, their curved bill is well adapted to obtaining food on rocky shores (Fig. 34).

The next most abundant species after the Turnstone were Dunlin, Redshank, Lapwing, Ringed Plover and Purple Sandpiper. Most of the Lapwings probably fed inland and most of the Dunlins and Ringed Plovers were associated with sandy shores (see Chapter 4). The same association with sandy areas was true of most Sanderlings, Bar-tailed Godwits and Grey Plovers on non-estuarine coasts. In Britain, rocky shores are therefore of greatest importance to Oystercatchers, Turnstones, Redshanks, Purple Sandpipers and possibly Curlews. Of these, the species that is most dependent on this type of shore is the Purple Sandpiper, with over 95% of the population occurring there. However, the British population is tiny compared with the hundreds of thousands of Purple Sandpipers that winter along the 55 000 km of Norway's coast. They are one of the few waders able to winter inside the Arctic Circle where they appear to fill a Dunlin-like niche for some of the time, moving between rocky shores and muddy ones. Both habitats are perhaps necessary for survival given the very short winter days and the fact that the majority of feeding must therefore take place at night (Strann & Summers, 1990; Summers, 1990). The Purple Sandpipers wintering in eastern Scotland and north-east England originate from breeding areas in Norway, whilst those in the north and west of Britain probably breed in Greenland or Canada (Nicoll *et al.*, 1988; Summers Ellis & Johnstone, 1988).

Birds other than waders were also counted in some areas during the Winter Shorebird Count. Coverage of the west and north-west coast of Scotland was particularly thorough in this respect, enabling population estimates to be made for several species in an area that represents about a third of the total coastline of Britain (Moser *et al.*, 1986). Divers were well distributed throughout the area, albeit at low density. Great Northern Divers were commoner (464) than Red-throated Divers (216), which contrasts with the situation off the east coast of Britain where Red-throated Divers are more numerous. A few grebes were recorded, with Little Grebes (135) being commonest and occurring almost exclusively in the sheltered sea lochs. Nearly 2000 Grey Herons were counted, with the highest densities on shores sheltered from westerly winds. In sea lochs they reached a density of one bird per kilometre (Moser *et al.*, 1986). Grey Herons were regularly spaced and were probably defending feeding territories. Although some may have been Scandinavian immigrants, the majority probably bred locally, a few of them on cliffs (Fig. 37). Eiders (*c.* 9000), Goldeneyes (*c.* 1300) and Red-breasted Mergansers (*c.* 2400) were also common throughout the study area, though like the grebes they were most abundant in sheltered sea lochs in the south. Significant proportions of the British wintering populations of these three species occurred in this region (19%, 9% and 24% respectively).

On British rocky coasts as a whole, Great Northern Divers are more strongly associated with rocky shores than the other two species of divers and also tend to feed further offshore (up to 10 km from land). Significant numbers are also found off Orkney and Shetland. As many as 4000 may winter in Britain, of which about 1000 are part of the local breeding population, the rest originating from Iceland, Scandinavia and perhaps Greenland. Divers feed on a variety of small fish, such as gobies, and crustaceans.

The midwinter population of Shags in the British Isles probably exceeds 100 000, which is about four times the number of Cormorants. In contrast to Cormorants, which take mainly bottom-dwelling fish, Shags take many midwater species, and also blennies and crustaceans from the bottom in rocky intertidal areas. The wintering distribution of Shags largely reflects their distribution in the breeding season and they are seen much less often than Cormorants in estuaries and inland. Neither species strays far from land because of the need to dry the plumage after bouts of feeding and this cannot be satisfactorily achieved at sea.

**Fig. 37.** Grey Heron in nesting tree. A surprisingly large number of Grey Herons feed on rocky shores throughout the year, particularly in Scotland. One great advantage of this habitat is that it very rarely freezes, even in severe winters. Keri Williams.

The wintering and breeding distribution patterns of Eiders also coincide quite closely, though there is a tendency for the birds to spread further south in winter. Of the 70 000 or so that winter in Britain, nearly half occur in eastern Scotland, and in turn half of these occur in estuaries, though it is mainly the estuarine mussel beds to which they are attracted. The Firth of Tay supports the greatest numbers. The main moulting areas are at Murcar and Gourdon (see Fig. 31) and in the Firth of Forth. The European wintering population numbers about two million, with most birds occurring in Scandinavia and Iceland, so the British total represents only about 4% of this (Owen, Atkinson-Willes & Salmon, 1986). Velvet Scoters also occur mainly in north-east Scotland, with the British population of 5000 or so representing only a small proportion of the European wintering total of about 200 000.

Gulls are particularly efficient scavengers on rocky shores backed by cliffs. This is because they exploit the updrafts generated by both onshore and offshore winds to glide relatively effortlessly along the coast surveying the shore and concentrating wherever food becomes available. These updrafts are similar to those which form in the vicinity of waves (see Fig. 21, p. 47), but are more powerful because cliffs are usually taller than waves and, unlike waves, do not move in the direction of the wind. Many species exploit these updrafts including Cormorants, Shags, Fulmars, corvids and birds of prey. Others exploit the shelter afforded by cliffs when they are migrating against adverse winds.

In the nineteenth century, littoral and inshore areas probably represented the most important feeding sites for all of the larger species of gulls. This has changed during the present century as many individuals switched to feeding inland on pastures and at refuse tips, but considerable numbers still remain on the shore, including some that feed at sewage outfalls. A survey of wintering gulls in Britain in 1983 (Bowes, Lack & Fletcher, 1984) showed that the following numbers of each species were present on the coast (percentage of total wintering population in brackets) – Herring Gull 202 000 (73%), Great Black-backed Gull 26 000 (65%), Common Gull 214 000 (46%), Black-headed Gull 838 000 (45%) and Lesser Black-backed Gull 14 000 (24%). However, these are roost counts and, since a great many gulls that feed inland continue to roost on the coast, this exaggerates the importance of the coast. It does, none the less, indicate that Herring Gulls and Great Black-backed Gulls are more maritime than Black-headed Gulls, Common Gulls and especially Lesser Black-backed Gulls. The latter species used to be almost wholly migratory, wintering in Iberia and North Africa (less than 200 were recorded in Britain in the winter of 1959), but has steadily increased as a winter resident, especially in the south. A few Scandinavian Lesser

Black-backed Gulls also spend the winter in Britain now. The southern half of England and Wales supports about 80% of the coastal roosting population of Black-headed Gulls, 98% of Lesser Black-backed Gulls, 65% of Common Gulls, 55% of Great Black-backed Gulls, but only 40% of Herring Gulls.

In the Bristol Channel, only about 5300 Black-headed Gulls (12% of the total wintering population), 1000 Herring Gulls (5%) and fewer than 50 (5%) Lesser Black-backed Gulls actually feed on the shore during daylight hours and some of these show evidence of having fed on pastures as well. Of the shore feeders, 90% of the Black-headed Gulls exploit soft sediments, such as mudflats, whereas 58% of the Herring Gulls occur on rocky shores (Mudge & Ferns, 1982*a*). Since rocky shores are not as extensive as soft sediments in the Bristol Channel this suggests quite a strong preference for rocky shores amongst Herring Gulls. This helps to explain why quite a high proportion of wintering Herring Gulls tend to occur in the north of Britain where there are more extensive rocky shores. Quite a few of our wintering gulls are immigrants, particularly those on the east coast. For example, Coulson *et al.* (1984) found that Scandinavian Herring Gulls began to arrive on the north-east coast in September, reached a peak (representing about 25% of the wintering population) in December and January, and then disappeared in late January or February. Fewer of the immigrants were juveniles (11%) compared with the local birds (45%), and about 70% were female.

Many Peregrines haunt rocky coasts in winter, where they feed on a variety of species. In Scotland, for example, their main prey includes feral pigeons, Rock Doves, passerines, gulls, auks and occasionally waders (Mearns, 1982). Rock Doves, and the coastal populations of feral pigeons that are derived from them and with which they freely interbreed, frequently roost on sea cliffs or in caves, but for the most part feed on cultivated fields. Rock Pipits winter along the same rocky coasts where many also breed, though they spread to a wide variety of other coastal sites too.

Black Guillemots show a winter distribution pattern covering north and west Scotland and Ireland, corresponding with the rocky coastlines where they breed. The population of some 70 000 birds is thus considered to be relatively sedentary, staying within 5 km of the breeding sites, although birds that nest in the more exposed colonies may move up to 100 km to winter in sheltered areas. Their winter diet consists of many species of fish taken from rocky shores, including Butterfish, blennies and pipefish, together with crustaceans such as crabs, lobsters, hermit crabs (which occasionally form as much as 75% of the diet) and even whelks, winkles and mussels. The availability of fish varies at different times of year and on different types of coast with rocklings commonest in exposed areas and eelpouts where weed is denser. Both of these fish are nocturnal in their activity, and so do not figure as prominently in the diet of Black Guillemots as do diurnal species such as Butterfish and Sea Scorpions (Kruuk, Nolet & French, 1988).

### Breeding sites

Some of the largest concentrations of breeding coastal birds are found along Britains' rocky coasts, mainly where these form part of offshore islands or cliffs. Most birds that forage in the open sea are forced to nest at some distance from their feeding areas. Moreover, since many are such specialised flying or swimming machines, they tend to be clumsy on land and, not surprisingly, tend to nest in areas that minimise their vulnerability to ground predators. Thus they choose cliffs, remote islands or underground burrows. Colonial nesting may itself offer advantages by more efficient communal defence or by swamping predators with a super-abundance of food (see p. 198).

The ideal cliff, from the point of view of supporting the greatest diversity of breeding coastal birds, has a mixture of large and small ledges, together with an accumulation of fallen material of different sizes at the base (Fig. 38). Needless to say, there are few sites that

**Fig. 38.** Diagrammatic section of a cliff showing typical nesting oportunities for different species of coastal birds in Britain. The location of these different types of habitat on a cliff is mainly dependent on geological factors and weathering.

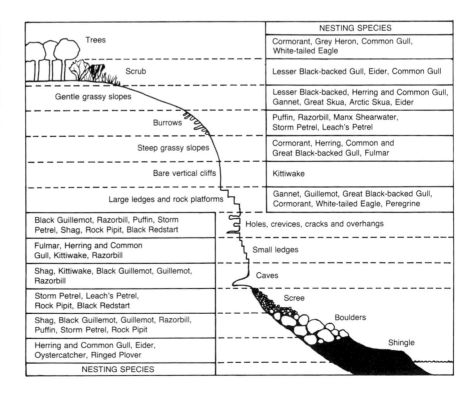

| | NESTING SPECIES |
|---|---|
| Trees | Cormorant, Grey Heron, Common Gull, White-tailed Eagle |
| Scrub | Lesser Black-backed Gull, Eider, Common Gull |
| Gentle grassy slopes | Lesser Black-backed, Herring and Common Gull, Gannet, Great Skua, Arctic Skua, Eider |
| Burrows | Puffin, Razorbill, Manx Shearwater, Storm Petrel, Leach's Petrel |
| Steep grassy slopes | Cormorant, Herring, Common and Great Black-backed Gull, Fulmar |
| Bare vertical cliffs | Kittiwake |
| Large ledges and rock platforms | Gannet, Guillemot, Great Black-backed Gull, Cormorant, White-tailed Eagle, Peregrine |

| NESTING SPECIES | |
|---|---|
| Black Guillemot, Razorbill, Puffin, Storm Petrel, Shag, Rock Pipit, Black Redstart | Holes, crevices, cracks and overhangs |
| Fulmar, Herring and Common Gull, Kittiwake, Razorbill | Small ledges |
| Shag, Kittiwake, Black Guillemot, Guillemot, Razorbill | Caves |
| Storm Petrel, Leach's Petrel, Rock Pipit, Black Redstart | Scree |
| Shag, Black Guillemot, Guillemot, Razorbill, Puffin, Storm Petrel, Rock Pipit | Boulders |
| Herring and Common Gull, Eider, Oystercatcher, Ringed Plover | Shingle |
| NESTING SPECIES | |

provide all of the best features (Fig. 39), and since different birds use different nest sites, the number of species and their population densities in particular areas depend on what sites are available as well as on what food supplies are situated within flying range. Gannets and Cormorants need broad ledges or flat ground, and since these are in short supply, the largest colonies of these species tend to be on grassy slopes on offshore islands.

Cliff eyries of Peregrines and White-tailed Eagles are often used for many years. Some Peregrine sites have a history that can be traced back to Medieval times (Sharrock, 1976). Sites can only persist this long if they are in relatively sheltered positions with little erosion. Peregrines require ledges at least half a metre deep, whilst White-tailed Eagles usually build on massive ledges, several metres wide, where large accumulations of nesting material may build up over several years. Eighty per cent of the traditional White-tailed Eagle eyries in Scotland during the last century were on cliffs, though in some parts of Europe nests in trees are commoner. The oldest known eyrie in Scotland is on the Shiant Islands, which can be traced back to 1716 (Love, 1983).

Fulmars and Herring Gulls need medium sized ledges on which to nest simply because they are relatively large birds. Flying ability also influences the choice of nest site to some extent. Little Auks and Black Guillemots have little difficulty in landing at sites that are difficult to approach, but the heavier guillemots are less proficient, and hence prefer easily accessible ledges. Even so, they may sometimes have to make several attempts to land. Guillemots prefer to nest at high density on large ledges and rock platforms. Razorbills nest in crevices, often in boulder scree, or in the tight corners of open ledges.

Kittiwakes need only a suggestion of a ledge since they can cement their nest to the tiniest notch or projection. They achieve this by collecting wet mud, seaweed or other algae and sticking it to the cliff to form a platform. Plant material or mud is then trampled into this to form the nest cup itself, the whole thing being strengthened to some extent by guano, which is deposited around the outside of the structure.

Shags nest on ledges, in large sheltered crevices, or amongst scree. Other species that exploit any fallen material at the foot of the cliff are indicated in Fig. 38. Obviously, species such as Storm Petrels and Leach's Petrels require a smaller crevice size than most of the others. Only sites sheltered from large amounts of spray, which may be driven ashore by summer storms, are usually successful. Plunging cliffs that have no rock platform to catch falling debris obviously lack any nesting sites of this sort. Sea defence works around the coast are designed to reduce the rate of cliff erosion and this greatly reduces the amount of scree available. This type of seabird nesting habitat is absent from virtually the whole of southern England.

**Fig. 39.** A real cliff showing the more limited range of nesting opportunities that usually exist at any one site. Chris Rose.

Eiders prefer to nest amongst vegetation on offshore islands, but they will sometimes nest in open rocky areas at the top of the shore and amongst sand dunes. The female builds up large fat reserves prior to the breeding season which allow her to incubate almost continuously. She relies on camouflage to avoid being found by predators in the more exposed sites and remains perfectly still on the nest when danger threatens.

Several cliff-nesting species have adapted to using a variety of man-made structures. For example, Fulmars nest on ruined buildings and occasionally on occupied ones. Cormorants nest on old sea forts, Mulberry harbours and even electricity pylons. One of the world's best studied seabird colonies is the group of Kittiwakes nesting on the window ledges of an old warehouse in North Shields on the River Tyne. In this case, the adults and young can be ringed and measured readily because they are accessible through the warehouse windows. The most widespread users of buildings for nesting are Herring and Lesser Black-backed Gulls (p. 221).

The dry stone walls of old field systems, ruined houses and barns provide conditions similar to scree slopes and are often used for nesting by Storm Petrels. Black Guillemots sometimes nest in holes in break-waters, harbour walls, occupied houses and even amongst timber debris washed up on the shore. The availability of the latter type of site has allowed Black Guillemots to colonise parts of Alaska where there are no suitable natural nest sites (Harris & Birkhead, 1985). Swallows, Swifts and House Martins, which now nest almost exclusively on and in man-made structures, must originally have been restricted to coastal and inland cliffs, and some tree-holes. A number of House Martin colonies still remain along our coast, but apparently there are fewer than there were at the beginning of this century (Sharrock, 1976). Two other species that nest on cliffs have partially transferred to buildings – Jackdaws and Rock Doves.

Shingle and pebble beaches offer nesting opportunities for a quite different range of species. Crevices are usually too small for any hole-nesting species to exploit, but large expanses of stable shingle, which gradually become vegetated, attract species that are associated with sand and saltmarsh. For example, Oystercatchers nest on all these types of shore. They are distinctly coastal in their breeding distribution in Ireland, Wales and southern England, but are also widely distributed inland in northern England and Scotland. Ringed Plovers are less regular inland breeders than Oystercatchers. Of 1811 nests found during a BTO survey in 1973/4, 91% were coastal; 24% of these were on shingle, with a further 44% on a mixture of shingle and sand. Only 4% were on sandy beaches and less than 0.5% on rocky shores (Prater, 1976). A survey in 1984 showed substantially the same pattern, though the great importance of machair for Ringed Plovers had by this time been more fully recognised (see p. 203) (Prater, 1989).

All five British breeding species of coastal terns (Sandwich, Roseate, Common, Arctic and Little) are colonial, with nests situated on rocky, sandy or grassy islets or on shingle or sand bars just above the high water mark. Such sites are often short-lived because of wave action and consequently breeding colonies move quite frequently. They are also vulnerable to disturbance by humans and predators such as Foxes, Peregrines and gulls. Little Terns nest most frequently in shingle habitats (Fuller, 1982). In the 1960s, over 60% of their colonies were situated within 20 m of the strand line and no more than $1\frac{1}{2}$ m above the high water mark (Norman & Saunders, 1969).

## Population sizes and trends

As with almost all other animal populations, serious attempts to estimate the population sizes of breeding seabirds were not made until the first half of the present century. At that time, the numbers of most species were still at a low ebb following the slaughter that took place during the late nineteenth and early twentieth centuries (see Chapter 6). One of the first species for which an attempt was made was the Gannet (Gurney, 1913). A more systematic census was made by Wynne-Edwards, Lockley & Salmon (1936). Gannets and Fulmars were counted in both 1939 and 1949 (Fisher & Vevers, 1951; Fisher, 1952), and although it had been hoped to repeat the counts in 1959, only partial coverage was actually achieved. By this time, however, a great many people were counting seabirds on a more regular basis.

A major breakthrough occurred in 1969 when, following an initiative by W.R.P. Bourne, the first attempt at a complete census of Britain's breeding seabird populations was made. Operation Seafarer, as it was called, extended through two breeding seasons – 1969 and 1970. The published results (Cramp, Bourne & Saunders, 1974) also collated a great many of the earlier counts made at a number of the more important colonies. It did not prove possible to organise a complete repeat survey in 1979, though a number of long-term census plots for selected species were started up by the NCC and RSPB. The most recent nation-wide survey was the Seabird Colony Register (Table 8), based on field work from 1985 to 1988 (Lloyd, Tasker & Partridge 1991). The register is designed to be capable of constant and regular updating whenever an area is counted in the future, even if only a single species is counted. This will hopefully provide a permanent source of data for determining future population trends.

Table 8. *Sizes of the breeding seabird populations of Britain and Ireland in 1985–87 (from Lloyd, Tasker & Partridge, 1991). Estimates greater than 10 000 rounded to nearest 1000, 1000–10 000 to nearest 100*

|  | Number of pairs |
|---|---|
| *Mainly cliff-nesting species* |  |
| Fulmar | 571 000 |
| Storm Petrel | 160 000 |
| Leach's Petrel | 55 000 |
| Cormorant | 12 000 |
| Shag | 47 000 |
| Herring Gull | 206 000 |
| Great Black-backed Gull | 23 000 |
| Kittiwake | 544 000 |
| Guillemot | 806 000 |
| Razorbill | 22 000 |
| Black Guillemot | 20 000 |
| *Species nesting mainly in other habitats* |  |
| Manx Shearwater | 275 000 |
| Gannet | 188 000 |
| Arctic Skua | 3400 |
| Great Skua | 7900 |
| Black-headed Gull | 233 000 |
| Common Gull | 71 000 |
| Lesser Black-backed Gull | 89 000 |
| Sandwich Tern | 19 000 |
| Roseate Tern | 650 |
| Common Tern | 16 000 |
| Arctic Tern | 81 000 |
| Little Tern | 2800 |
| Puffin | 467 000 |

Accurate counts of breeding seabirds are surprisingly difficult to achieve, even in species where the adults are conspicuous when they are incubating. The Gannet is perhaps the most conspicuous of all and, not surprisingly, was one of the first to be counted using photography (Salmon & Lockley, 1933). Harris & Lloyd (1977) found that different observers counting nests (i.e. incubating adults) on an aerial photograph of a portion of the Grassholm gannetry, obtained values ranging from 2823 to 3362 (mean 3170). With species that are less obvious and the nests less regularly spaced, photography can still be valuable even though greater variations in accuracy may occur. For example, Harris & Lloyd found that different observers underestimated the known number of Guillemot and Kittiwake nests at a colony by an average of 11% and

20% respectively when using photographs to assess numbers. Estimating flock sizes on projected slides is a slightly different problem, but in this case flocks of 100–400 birds tend to be overestimated in size by 10–20% and larger flocks underestimated. The average underestimation is about 24% for flocks of 3000 (Prater, 1979; Erwin, 1982).

The flight-line method (i.e. counting the number of birds entering or leaving the colony) can be useful in cases where the nests are difficult to see because of thick vegetation and where there is a relatively simple pattern of colony attendance. Both these conditions often occur in heron and egret colonies in trees or mangroves. The relationship between the number of birds arriving and leaving colonies of known size usually has to be established first (since one or both parents can be involved). There are differences in the frequency of flights at different stages of the nesting cycle, and between different species, so the method has to be used with great care (Erwin, 1981).

Even when it is possible to count individual nests accurately on the ground (e.g. many gull and tern colonies) it can be difficult to avoid missing some and double-counting others. In such cases, the use of a small spot of dye to mark each nest as it is counted, followed by repeat counts to determine the proportion of marked and unmarked nests, can be used to advantage, provided this does not attract predators to the nests. For example, in the case of Lesser Black-backed and Herring Gulls in quite dense vegetation, observer efficiency at finding nests varied from 66% to 95% (average 83%) (Ferns & Mudge, 1981). Care needs to be taken in deciding what constitutes a true breeding attempt e.g. should it just be the construction of a nest, or should at least one egg be laid? Care also needs to be taken in deciding at what time of year to undertake counts, since new nests are often added until quite late in the season. Where this is the case, e.g. in both Lesser Black-backed and Herring Gulls, counts should be made as late as possible, i.e. when laying has been completed for the colony as a whole (Wanless & Harris, 1984). It is also important that a consistent methodology be used from year to year if it is hoped to assess short-term population changes at a colony.

A particular problem arises in the case of burrow-nesters and others where the nest itself cannot be seen. The Seabird Group recommends using the apparently occupied nest site in Black-Guillemots and Puffins, and a head count of individuals in Guillemots and Razorbills. In fact it is surprising how often a head count of adults at a breeding colony is quite close to the number of nests. For several auks, gulls and terns, the relationship is often accurate to ±20% during the incubation period (Erwin, 1980), though numbers tend to rise during the evening when both members of the pair are likely to be at the nest, and during bad weather that may prevent the off-duty bird from feeding. In Puffins, where one member of the pair is underground incubating the egg most of the time, it is the peak evening count that is most likely to come closest to the number of pairs. In many species, the presence of non-breeding birds at the colony can complicate the situation.

The most difficult species of all to census are the burrow-nesters that only visit the colonies at night, notably the petrels and shearwaters. Trying to determine which burrows are occupied can be extremely difficult and the size of colonies can often be estimated only to the nearest order of magnitude. In some intensively studied colonies, it is possible to improve on this. For example, the Manx Shearwater colony on Skokholm has been estimated at 35 000 pairs as a result of ringing chicks in burrows (only one chick is produced by each pair) and counting the proportion of ringed ones amongst fledglings found later (Harris, 1966; Perrins, 1967). The fertilising effect of Manx Shearwater droppings on the vegetation at the colony on Rhum has been used as a means of mapping the extent of the occupied area. Combined with burrow counts of sample plots within this area, an estimate of 116 000 pairs was reached for the 1965–69 period (Wormell, 1976).

Estimates of breeding success are subject to as many difficulties and errors as are estimates of colony size. One of the major problems is making an allowance for breeding attempts that may start and fail between census visits (Mayfield, 1975; Hensler & Nichols, 1981). If young birds tend to hide, and therefore may miss being counted, the use of small, wire-netting enclosures to prevent them from moving too far can be a useful way of measuring the productivity of small samples of nests. This is particularly valuable with gulls, though care has to be taken that the enclosures neither increase nor decrease mortality (Erwin & Custer, 1982).

The published results of the Seabird Colony Register (Lloyd, Tasker & Partridge, 1991) show that the majority of British breeding seabirds increased in numbers between 1969/70 and 1985–87. This is the case in the Fulmar (Fig. 40), Gannet, Cormorant, Shag, Common Gull, Lesser Black-backed Gull, Sandwich Tern, Little Tern, Razorbill and Black Guillemot. The difficulties of assessing the breeding populations of the Manx Shearwater, Storm Petrel and Leach's Petrel mean that it is impossible to detect any significant overall change in their numbers. Several species have increased, but show signs of stabilisation or decline during the late 1980s. These include the Kittiwake and Guillemot, in both of which breeding success in some areas and seasons has declined (e.g. Pienkowski, 1991), apparently as a consequence of reduced food supplies. In the Black-headed Gull, some quite large colonies, such as that at Ravenglass, have disappeared (possibly as a result of Fox predation in this particular case). Even though some new colonies have formed, there has been a net overall reduction in the number of Black-headed Gulls breeding in coastal colonies since 1973. Several of the species that had otherwise been increasing during the 1970s and 1980s, have recently declined as a result of the breeding failures in the Shetlands (see p. 263). These include the Arctic Tern, Arctic Skua and Great Skua.

Three species have not changed significantly in numbers – the Great

**Fig. 40.** Pair of Fulmars at the nest. This species has shown a steady increase in both range and numbers for more than a century. The current rate of population increase is 3% per year. Yet we still do not fully understand the reasons for this. Harold E. Grenfell.

Black-backed Gull, Common Tern and Puffin. The Common Tern does not have such a significant portion of its population on the Shetlands as does the Arctic Tern, and is less dependent on sandeels (Uttley, Monaghan & White, 1989). Only two species have shown significant decreases between 1969/70 and 1985–87 – the Herring Gull and the Roseate Tern. The former has suffered widespread mortality during the breeding season as a consequence of botulism and the latter during the winter as a result of hunting along the African coast (see Chapter 6).

### Adaptations to cliff nesting

Most of the adaptations to cliff nesting are closely related to the exact type of breeding site chosen. For example, Kittiwakes have longer and sharper claws than most other gulls to provide a secure grip at the often precarious nest site. The nest has a deep cup to prevent the eggs rolling out. Since the nests are not accessible to ground predators, the nestlings do not need to be camouflaged like other gulls. Several other aspects of the Kittiwake's antipredator behaviour, such as alarm-calling, mobbing of predators and egg-shell removal, are greatly reduced (Cullen, 1957). Even bird predators find it difficult to land at the nest and steal eggs or young (Tinbergen, 1974). The food delivered to the female by the male, or to the young by both parents, is regurgitated carefully into the mouth of the recipient, rather than being dropped to the ground as in other gulls, since there would otherwise be a much greater risk of it being lost. It is sometimes claimed that the reduced clutch size of the Kittiwake (the commonest clutch size is two, compared with three in most other gulls) is an adaptation to the restricted size of the nest, but there is evidence that birds which have a clutch of three are more efficient feeders (Coulson & Porter, 1985).

Despite the fact that Kittiwakes do not need to compete with other species for their nest sites, there is fierce fighting between the Kittiwakes themselves for nesting ledges, especially those at the centre of the colony. In fact, birds breeding at the centre of the North Shields warehouse colony raise 11% more chicks than those at the edge (Coulson & Thomas, 1985). In most species this effect is due to higher predation at the edges (p. 197), but predation does not occur at this colony and the effect seems to be due to the fact that high quality birds obtain the central sites. These birds refrain from breeding less often than others, though what advantage these sites offer them is not yet clear.

The chicks of several cliff-nesting gulls have an aversion to approaching the edge of a ledge, which is obviously advantageous in preventing them from falling off. This is particularly well marked in Kittiwakes, which use both sight and touch to avoid the edge (McLannahan, 1973).

Newly hatched Herring Gull chicks avoid the cliff edge more strongly than do Lesser Black-backed Gulls, which corresponds with the former's more frequent use of cliff ledges for nesting. Moreover, Herring Gull chicks reared on ledges avoid the cliff edge more strongly than do those reared on flat ground showing that experience (short of a serious fall causing death or injury) plays a part in this behaviour (Tinbergen, 1974). The tendency of Kittiwake chicks to face inwards on the nest rather than outwards may be learnt as a result of adults tending to brood them in this position. The adults do this partly because their tails are too long to allow them to brood the other way round (Hodges, 1974). The black band across the back of the neck of the young is very conspicuous in this position and may be a signal to prevent any risk of fighting at the nest. Herring and Black-headed Gull chicks reared in Kittiwake nests (achieved experimentally by egg swapping) soon perish because they are too active and fall out of the nest.

Even the calls of the Kittiwake have been influenced by their choice of nest site. Cliff-nesters have 14 different vocalisations discernible to the human ear, whereas individuals that nest on flat ground at some island colonies only have nine. While we still have much to learn about these calls, several of them are concerned with ensuring efficient and safe arrival at, and departure from, the nest site (Daniels, Heath & Rawson, 1984).

While cliff nesting generally provides a good deal of protection from mammalian predators (Fig. 41), avian predators remain a major threat, and this exerts a strong influence on behaviour. For example, Razorbills nesting in burrows or crevices on Skomer raised 0.70 chicks per pair on average, whereas those nesting on ledges raised only 0.55 (Hudson, 1982). This difference was due mostly to egg predation by Herring Gulls, Great Black-backed Gulls, Jackdaws and Carrion Crows. Only Jackdaws were able to reach burrow and crevice nests to remove eggs, but they could not pull the adult away from the nest (despite trying) and thus could not rob occupied sites. Despite this, 70% of the Razorbill population on Skomer does nest on ledges, presumably because of a shortage of more protected sites. They do avoid the largest and most exposed ledges, however. Guillemots, on the other hand, favour some of these sites where they nest at very high densities. When Great Black-backed Gulls attempt to seize Guillemot eggs, they first try to remove the shielding adults, but are usually rebuffed by the high density of sharp, lunging beaks at a well-packed colony. Late nests in a colony are much more likely to be predated, even on densely occupied ledges, since the adults leave once their chicks fledge and any late-developing chicks are thus more exposed (Hatchwell, 1991). On Skomer, there are marked differences in breeding success on ledges with different densities of birds.

Dense colonies on wide ledges lose only 13% of eggs and young, but dense colonies on narrow ledges lose 26%, and sparse colonies on wide ledges 69% (Birkhead, 1977). Curiously, the majority of Guillemots nest at high density on narrow ledges, whilst many large ledges remain unoccupied, even though they have been occupied successfully at times of higher population density in the past. Presumably this is because it is better to remain on a narrow ledge than to risk being stranded at low density as a pioneer on a potentially better site. Needless to say, if Guillemots or any other species leave their ledges as a result of human or other disturbance, gulls and corvids quickly take advantage of the situation to seize eggs and young. The latter may also be accidentally knocked off the ledges by panicking adults.

Given the risk of predation, several species have adopted crevice- or hole-nesting habits, even on cliffs (Fig. 38). The eggs of hole-nesters tend to be very pale or white with few markings, presumably so that they show up clearly in the reduced light of the nest site. The eggs of ledge-nesters, on the other hand, are conspicuously marked and very variable. In species such as Guillemots, which nest at high density, such variation assists the ability of individuals to recognise their own eggs.

**Fig. 41.** The Gower peninsula in South Wales, with the Worm's Head in the background. Nesting on cliffs and islands affords some protection against mammalian predators. Harold E. Grenfell.

The fact that clusters of similarly coloured eggs tend to occur together in the colony may be because Guillemots tend to breed near close relatives. This would also help to explain why they are prepared to attack predators at some personal risk. Those individuals which enhance the survival of their relatives donate more of their genes to future generations (through the copies that they share with their relatives) than they would by behaving entirely selfishly. Ledge-nesters without nests also have thick-shelled eggs, presumably to help prevent breakage.

The number of pores, through which the egg breathes, tends to be larger in those species which develop more quickly, or in which the egg becomes covered in faeces. For example, Fulmar's eggs have c. 5200 pores, whereas those of Brünnich's Guillemots, with a much shorter incubation period, have c. 7600 (Rahn *et al.*, 1984).

It has been suggested that the pear-shaped egg of ledge-nesting species has a reduced rolling circle and hence is less likely to roll off the ledge. In fact, when rounder Razorbill eggs were substituted for Guillemot ones they proved no more likely to roll away. However, Guillemots do change incubating positions less frequently than Razorbills and do not leave the egg unattended for so long, both of which are presumably adaptations to reduce the likelihood of their eggs moving (Ingold, 1980). The pear shape could instead be important in maximising the area of contact with the brood patch in a species that, because of the density at which it breeds, sometimes incubates in a half upright position (Harris & Birkhead, 1985).

**Are nest sites limiting?**

The answer to this question appears to differ in different species. In Gannets, which Nelson (1978*a*) argues are adapted to cliff nesting, ledges do appear to be lacking in many areas, and consequently there has been an overflow onto the tops of suitably remote islands such as the Bass Rock and Grassholm. Lack of nest sites does not at present limit the abundance of this species. None the less, there may be some species in which there is an abundance of food, but an inadequate number of nest sites. For example, in the productive Humboldt and Benguela Currents, off the coasts of Peru and Namibia, there are areas with abundant food resources exploited by relatively small numbers of breeding birds because they are situated too far from suitable breeding sites. In some species, notably burrow-nesters, the birds themselves may be the cause of the loss of nest sites. Puffins are the best known example of this. Several thriving colonies became so honeycombed with burrows that the soil eventually collapsed and eroded, and became too shallow to contain new burrows, e.g. Grassholm (Harris, 1984).

Within colonies there is some evidence that good nest sites may be a limiting factor. For example, most of the Shags on the Farne Islands nest in faulted or eroded gullies in the relatively low cliffs (Potts, Coulson & Deans, 1980). Some nests are situated on more exposed sections of cliffs and a few (about 26%) are so low down on the cliffs that they are washed away during rough seas. The colony as a whole is unusual in that about 92% of nests are not protected by overhanging or projecting rocks, which is normally the case with Shags. About 13% of the occupied ledges are too small to accommodate three fully grown chicks and 32% have escape routes that are blocked by other nests. All in all, only 4% of sites can be regarded as optimal. Good quality sites consistently fledge more young (average 1.6 per pair) than poor ones (average 0.6).

In May 1968, there was a 'red tide' caused by a bloom of toxic marine dinoflagellates (single-celled planktonic organisms). As a result, 82% of the breeding Shags on the Farnes died from consuming fish contaminated with this toxin (Armstrong *et al.*, 1978) (see Fig. 87, p. 258). Some humans suffered paralytic shellfish poisoning as a result of eating contaminated mussels during the same period. A year later, in 1969, the average quality of Shag nest sites rose as the majority of the greatly reduced population was able to be accommodated in the better sites. Even young birds, which normally obtained poor sites because they arrived at the colony too late to claim good ones, increased their breeding success by 70% following the red tide. This study shows that even where the absolute number of nest sites may not be limiting, their quality could be low enough at high population levels to provide a potential limit to population growth. However, this is not very common, since many places clearly offer abundant unused sites (e.g. Olsthoorn & Nelson, 1990), and even in the case of the Farne Shags the breeding population has continued to increase during the 1980s (see p. 258).

A similar situation has been found in hole-nesting Black Guillemots on Shetland. Nests that were visible from the burrow entrance, with only a single exit, or that were too close together, did not produce as many young as sites without these deficiencies. This species usually performs best on offshore islands with an absence of ground predators such as rats, Hedgehogs and Mink (Ewins, 1989). Fledging success in Herring Gulls is also reduced at high density, but in this case it is due to increased intraspecific fighting and predation (Kilpi, 1989). However, Birkhead & Furness (1985) have argued that all these circumstances are exceptional and that other factors are usually more important in limiting seabird numbers. For example, it can be inferred that species for which colonies on islands are larger than those on promontories (or for which those on promontories are larger than those on straight sections of coastline) are likely to be limited in numbers by the availability of food within range of the colony. This is because islands are surrounded by potential feeding sites on all sides, whereas straight coasts have the sea on one side only. However, this argument is greatly complicated by the reduced numbers of predators on islands and perhaps promontories. It is also clear that a great many species are currently recovering from low population levels imposed by human predation (p. 227).

## Clutch size

Seabirds in general are long lived, slow to reach sexual maturity, and have rather small clutch sizes. It is usually argued that natural selection favours the clutch size that gives rise to the maximum number of surviving offspring, without compromising the survival rate of the parents. This is apparently contradicted by the fact that the Gannets on Bass Rock lay only one egg, but are capable of rearing two youngsters when an extra chick of the same size (to prevent it being killed by its foster sibling) is added to the nest (Nelson, 1964). One explanation for this may be that feeding conditions for Gannets are unusually favourable in the North Atlantic at present, and they have not yet evolved a larger clutch size (Nelson, 1978*a*). Cape and Australian Gannets are less successful at rearing artificial broods of two. Alternatively, the extra stress of rearing two young may compromise the parents' chances of breeding in future years. However, there is no simple trade-off between clutch size and adult survival in the Kittiwake, in which individuals with larger clutch sizes (and larger eggs) tend to breed more consistently from year to year, rather than less so (Coulson & Thomas, 1985).

Some boobies do have a natural clutch size of two, but when the eggs hatch the larger chick always kills the smaller one. This siblicide occurs in a number of species, especially birds of prey, where it is usually a result of the parents' inability to provide enough food for both chicks, rather than a matter of course. It does constitute a mechanism whereby the brood size can be adjusted to the food supply. The same end can be achieved if the largest chick monopolises all the food and the smaller ones simply starve to death. Siblicide requires that the chicks become more aggressive when they are hungrier and that there is sufficient disparity in their sizes that the fighting is over quickly without damage to the larger individuals. The parents have evolved a very simple way to achieve the latter, and that is by commencing incubation immediately the first egg is laid, thus ensuring that this egg begins development earlier and hatches first. Most species want to ensure synchronous hatching and thus do not start incubating until the last egg is laid. In the case of two-egg clutches in which the first chick to hatch always kills the second, regardless of the amount of food available, the second egg seems merely to represent an insurance in case the first is infertile or is lost to predators.

Most tubenoses and auks, and many Pelecaniformes, have a clutch size of one, whilst the gulls fairly consistently have a clutch size of three. Terns are intermediate and the clutch size varies from one to three in different species. For example, in Caspian Terns the commonest clutch size is two, and the first egg weighs 5% more than the second. Hatching success is about 79% for both eggs, but only 8% of nests fledge two young, whilst 70% fledge just one. Of the nests that rear only a single chick, this chick hatches from the first egg laid in 68% of cases. This chick hatches two days earlier than the second. Thus Caspian Terns show the classic method of brood reduction. In this case the smaller young does not usually starve and it is not usually killed by its sibling, but instead suffers a higher predation rate from Ring-billed Gulls. This is because the smaller young are fed less by the parents and thus remain at a vulnerable size (the size that is capable of being swallowed by a Ring-billed Gull) for longer (Quinn & Morris, 1986).

In gulls, the survival of three young is sufficiently uncommon to suggest that in this case the third egg is only an insurance. The fact that the third egg is smaller and sometimes less well camouflaged than the others, with fewer spots and streaks, has even led to the suggestion that it is a kind of bait to attract a predator's attention away from the other two eggs if the nest is discovered (Verbeek, 1988). Most of the waders that breed in temperate regions have a clutch size of four. The Oystercatcher is an exception, with a clutch size of three, probably because it is one of the few waders that actually brings food to the young and thus could not adequately provision four chicks.

The extent to which incubation is shared by the parents differs between species. In Kittiwakes, Herring Gulls and Black-headed Gulls, for example, males are present at the nest during the egg stage for 46%, 50% and 53% of the time on average, respectively (Drent, 1970; Coulson & Wooller, 1984). Both parents were missing for 0.3% of the time in Kittiwakes and 3% in Herring Gulls. There may, however, be considerable variation between pairs with the range of attendance times in Kittiwakes being 45–65% in females and 35–56% in males (see also next section).

**Growth and fledging of the young**

Newly hatched chicks require a considerable amount of brooding since they are so small that their rate of heat loss is very high. They also lack the ability to control their own body temperature during the first few days of life and are difficult to feed because of the small size of their gape. For these reasons, the growth rate of chicks is rather slow at first, but usually picks up to a steady value in about a week (Fig. 42). However, in larger species such as the Gannet, it takes about two weeks before the steady rate of growth is reached. During this early period the adults of species such as Guillemots, which normally consume quite large fish, have to switch to small fish.

In general, the growth curve is S-shaped or sigmoid (Fig. 42). The upper part of the S is caused by a reduction in growth rate which takes place as the young approach fledging. The latter may occur because of a reduced rate of feeding by the adults, or because the young expend much more energy on exercise in order to prepare themselves for a life of normal activity away from the nest.

The slopes of the straight regions of the growth curves in Fig. 42 do not differ all that greatly, despite the large differences in body size and clutch size between the species, and despite the fact that the absolute growth rates (in grams per day, as opposed to percentage increase per day) are higher in the larger species. This might be due to some kind of physiological limit to the rate at which the growth and development of the organs can occur. There is certainly a lot of evidence that growth in many cases is not limited by the ability of the parents to provide food. For example, Leach's Petrel chicks in the field given supplements of either oil or protein grow no faster than unsupplemented control chicks. This implies that the chicks fail to absorb the excess calories or burn them off (Ricklefs, Place & Anderson, 1987). Quite a few species have

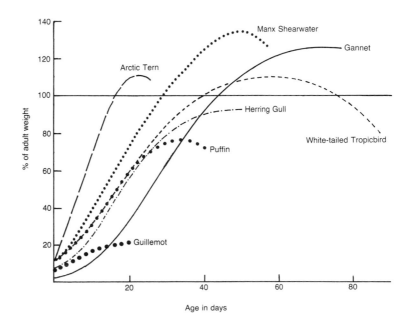

**Fig. 42.** Growth curves of individual seabird chicks of different species. Based on information in Nelson (1978*a*, from Poulin, 1968), Mudge (1978), O'Connor (1984), Ewins (1985), Gaston (1985) and Phillips (1987).

been found to be capable of rearing extra young without a reduction in the growth rate of the individual chicks, and there are often no significant differences in the growth rates of chicks in different sized broods. Pearson (1968) found the latter to be the case in Arctic Terns, Kittiwakes, Lesser Black-backed Gulls (broods of one and two) and Shags (broods of one, two and three). Manx Shearwaters can rear an extra chick but both members of the brood then fledge at a reduced weight and with reduced

**Fig. 43.** Kittiwake at the nest with chicks. Guano helps bind the outer parts of the nest together. With one parent in attendance the chicks can be protected from predators such as Herring Gulls and Jackdaws. Harold E. Grenfell.

survival prospects (Perrins, Harris & Britton, 1973). The average number of surviving young left by such pairs is still greater than those that rear a single chick, but the possible cost to the parents, in terms of reduced survival chances, is not known.

One way in which adults can cope with larger broods is simply to spend longer foraging. For example, in many species, including terns and Kittiwakes, one parent normally remains at the nest whilst the other forages (Fig. 43). If the chick's hunger is not being satisfied it is always possible for both parents to forage, though this increases the risk that the chicks will be predated. Kittiwakes on the Isle of May in 1988 left 50% of their chicks unattended at the nest due to a dearth of sandeels and the consequent need for both adults to forage. Breeding success was reduced by 20% that year, possibly because more chicks became chilled, soiled by droppings from other nests, or attacked by other Kittiwakes (Wanless & Harris, 1989). High rates of predation of chicks at Kittiwake colonies by Ravens, large gulls and Great Skuas has been observed in recent years on Shetland. This again has been a consequence of a lack of sandeels (p. 263) causing both parents to leave the nest to forage. Many predators are more inclined to feed Kittiwake chicks to their young in the absence of sandeels (Heubeck, 1989). The fact that particular colonies may have difficulty in provisioning their young in some years helps to explain why there are many seabirds incapable of rearing extra young. For example, Herring Gulls in the Netherlands were unable to do so, and when chicks were hand reared artificially, they did grow faster than in the field (Spaans, 1971).

There is a broad tendency for offshore feeders to have a clutch size of one and to have slow-growing young fed at 2–3 day intervals, whilst inshore feeders have a clutch size of two or three and faster growing chicks which are fed several times a day. Overall, it therefore seems likely that clutch size is adapted to the adults' ability to provide food under average conditions, but that there is considerable flexibility in the rate at which adults can do this, provided adequate supplies are available. A compromise may have to be made between the growth rate of the young and their survival if food supplies are low or if the clutch size is too large.

In some species, such as the Gannet, the young fledge at a weight exceeding that of their parents (see Fig. 42). They fly, or partially glide to the sea, but then are unable to take off again until their weight has dropped from 4 kg or so to the 3 kg more typical of adults (Nelson, 1978a). They swim out to sea, apparently unaccompanied by their parents, and the large amount of fat they carry is presumably essential to tide them over the period when they are learning to forage for themselves. How they learn to feed is not known because it is difficult to follow them as they swim southwards. Presumably they feed at the surface initially, as the adults occasionally do in the presence of shoals of small fish.

Adult Gannets clearly have few problems in providing enough food to enable the chick to achieve a large body size. For many of the auks, however, this seems to be difficult, perhaps because they have had to compromise their flying ability in order to forage efficiently underwater and consequently are not suited to commuting long distances to fetch food. Instead, they have evolved a system in which the young fledge at an early stage of their development when they are only about 20% of adult weight (Fig. 42). The advantage of this is that they can swim out to the feeding grounds with the male parent and can then continue to be provisioned by him at a more rapid rate for several further weeks. The chicks at sea can thus gain weight at least as fast as they can during the most rapid period of growth at the colony (Harris, Webb & Tasker, 1991).

When they leave the cliffs, young Guillemots and Razorbills use their rudimentary wings to exercise some control over their otherwise head-long plunge to the sea, but none the less they may strike obstacles on the way down. The young of both species emit a piercing 'piu piu' call once they reach the sea and this enables the male to locate them. This call is first made on the nesting ledge, usually on the day of fledging, presumably so that the male can learn to recognise his own chick. Most fledging takes place in the hour or two before dusk to minimise the considerable risk posed to the young by gulls and other predators. If the nest is on a cliff with a beach beneath it, the young may have to land on the beach first and then run to the sea. In the presence of scree, the chicks have to scramble through it on foot as best they can.

Some auks fledge at a relatively higher body weight. Puffins are able to achieve this (Fig. 42) because of the adult's ability to carry many food items at once. Both Puffin and Black Guillemot chicks apparently fledge on their own and are independent of their parents thereafter. Little Auks, however, continue to be accompanied by their parents (Harris & Birkhead, 1985). The auks that do fledge early have a relatively slow growth rate (Fig. 42), but even they show a slightly sigmoid growth curve. This may be because in the last few days of growth they spend time strengthening the legs and feet ready for swimming and developing a plumage suited to the more rapid heat loss they will encounter in the water.

For all cliff-nesting species, fledging necessitates a sudden and dramatic change in behaviour on the part of the chicks. The previous strong attachment to the nest site and avoidance of the cliff edge has to be replaced by a sudden leap into space. Those which fly at fledging undergo a long period of transition during which they exercise their wings, but for the early fledgers it is a more sudden process. In Brünnich's Guillemots, the adults brood the chicks less and the chicks show bursts of wing flapping. The chick approaches the cliff edge several times but does not leave unless the male is beside it. The male jumps with the chick and glides a short distance behind its tail all the way to the ground, presumably so as not to lose it. If they do become separated, they can find one another again by their calls (Gaston & Nettleship, 1981).

Many gulls fledge a little short of adult weight, but they continue to be fed by one or both parents for some time, occasionally right up until the following breeding season (Holley, 1982). The fact that chicks reared from Herring Gull eggs transferred to Lesser Black-backed Gull nests migrate to Iberia and North Africa during their first winter, unlike normal Herring Gull chicks (Harris, 1970*b*), strongly suggests that the young generally accompany their parents (in this case their foster parents) for some time after fledging. This prolonged parental care occurs in a number of species of gulls, terns, pelicans, cormorants, shags and frigatebirds (Burger, 1980).

## Diet and feeding range during the breeding season

The diet of all nine species of breeding seabirds studied by Pearson (1968) on the Farne Islands consisted of fish. The largest species (Cormorant) brought back the largest prey to its chicks, on average, but it also brought back the largest range of prey sizes. The smallest species (Arctic Tern) did not bring back the smallest prey, instead the Puffin did. This was because Puffins were more heavily dependent on sandeels and these were smaller in size. The next most frequent prey taken by Puffins was Sprats and young Herrings which arrive in inshore waters a little later in the breeding season. They form the prey most frequently taken by Arctic and Common Terns. There was a considerable overlap between all species in the size of sandeels that they took, with Guillemots taking the largest. Puffins were able to feed efficiently on sandeels because of their ability to carry several at once, whereas the terns, and other auks, can only carry a single fish at a time. Other fish of importance were Cod, in the case of the Lesser Black-backed Gull, and flatfish in the case of the Cormorant. The Cod may have been taken as rejects from trawling activities in the area. The gulls and terns as a whole were largely limited to surface dwelling fish because of their inability to dive. The Shag and Cormorant took both bottom-dwelling and mid-water fish with the Cormorant showing a greater preference for flatfish from areas of soft sediment. Recent studies of foraging of Shags from the

Isle of May show that they feed on average 7 km from the colony and range over quite widely differing types of seabed (Wanless, Harris & Morris, 1991). The remaining diving species in Pearson's study took mostly mid-water fish.

The five species of sandeels found off the north-east coast of Britain formed one of the most important groups of prey for these nine species of seabirds. Sandeels spend the early winter in offshore waters, spawning in late winter. The young tend to move inshore in spring and summer and are found close to the surface either when feeding or when chased by predators such as Pollack. Herrings spawn offshore, but the young do not move into shallow water until rather later in the year – about July near the Farne Islands.

There have been a great many other studies of the diet of seabirds during the breeding season but few have simultaneously studied as many species as Pearson. Gannets show little overlap with other species since they take quite large Mackerel, Herrings and sandeels. Very few other seabirds are able to catch large Mackerel and Herrings. It has been suggested that they might range as far as 480 km (300 miles) from the colony to reach favourable feeding areas. Early in the breeding season, Herrings, sandeels and Sprats are the most important foods, but by the time the young have hatched, Mackerel have moved inshore to spawn and become an important food at nearly all colonies (Nelson, 1978a). The Fulmar is another species which it has been suggested can forage at a considerable distance from the colony during the breeding season. One ringed bird from Eynhallow in Orkney was recorded 466 km (290 miles) away in the middle of the North Sea during the breeding season, though most individuals are seen or caught closer to the colony than this. This particular individual did fledge a chick in the year in question (Dunnet & Ollason, 1982). However, it has been suggested that this bird may not have returned to the colony that season and it would certainly have been possible for the other parent to have reared the chick on its own (Furness & Todd, 1984). Fulmars take a wide variety of planktonic crustaceans, annelids, squid, cuttlefish, pelagic molluscs, small fish and even jellyfish. They also take a considerable amount of fish offal (especially fish livers, oil and fat) by following trawlers and other fishing vessels (Fisher, 1952). In the breeding season, 72% of the diet on Foula in Shetland consists of sandeels, with fish offal constituting 14% and pelagic zooplankton 11%. By contrast, 71% of the diet on St Kilda in the Outer Hebrides consists of zooplankton, with the commonest species being the widespread, large (4 cm), and shrimp-like crustacean *Meganictiphanes norvegica* (Furness & Todd, 1984), sometimes referred to as the krill of the northern hemisphere. The fact that there are such large differences in breeding colonies only 445 km apart clearly indicates that

Fulmars do not regularly forage at large distances from their colonies.

Presumably, the St Kilda Fulmars caught most of their *Meganictiphanes* at night when they were at the surface. In West Greenland, Little Auks showed a marked increase in the frequency with which they fed their chicks at night, corresponding to vertical migrations of the zooplankton which occur even in the continuous daylight of the arctic summer. In the middle of the breeding season, 94% of the diet consisted of *Calanus finmarchicus*. Most feeding was done within 2.5 km of the colony, although birds occasionally fed 32 km away (Evans, 1981). Whilst Storm Petrels also take invertebrate zooplankton, a surprisingly high proportion of the diet in the breeding season consists of fish fry, with Sprats being particularly common. Larger Sprats are an important source of food for Manx Shearwaters, together with young Herrings, Sardines and sandeels (Cramp & Simmons, 1977).

Gulls that forage in the intertidal area show some changes in diet during the course of the breeding season. For example, on the island of Terschelling in the Netherlands in 1966, starfishes, cockles and other soft shore bivalves dominated the diet of Herring Gulls in April and May. These were largely replaced by mussels and crabs in June and July. On the island of Vlieland, on the other hand, mussels were consumed more consistently throughout the year. This contrasts with the situation in winter when cockles formed the bulk of the diet, though a lot of food was also obtained at refuse tips – 30% most of the time, but 80% during bad weather when cockle feeding was difficult or impossible. There was also a clear tendency for the adults to bring more fish (mostly Sprats, young Herrings, whiting and sandeels) to their chicks than they would normally consume themselves. There are several possible explanations for this. It could simply relate to differences in food availability, but Spaans also points out that fish contain less indigestible material in the form of shells and exoskeletons than intertidal invertebrates, and are thus a more profitable food to carry back to the young (Spaans, 1971). Also, adults need to forage regularly throughout the day in order to satisfy hungry young, and at high tide fish may be the only food available to them. Western Gulls in North America switch from feeding on refuse and take small fish to their young instead (Annett & Pierotti, 1989).

Studies on Herring Gulls in Newfoundland have demonstrated that predominantly mussel-feeding birds lay earlier, have larger, heavier clutches and sometimes rear more young than those feeding on refuse, fish, squid or on the chicks of other birds. This is despite the fact that mussels are the least efficient diet, at least in terms of the energy yield per meal delivered to the chicks. The reasons for this are not known but it may relate to the higher calcium content of the diet of mussel feeders.

All adults tend to take more fish to their young as the breeding season progresses (Pierotti & Annett, 1987, 1991).

It is important that both members of a breeding pair should be co-ordinated and compatible in their feeding strategies. For example, if they feed in different ways or in different areas, they will be less vulnerable to local food shortages, or if one feeds by day and the other by night, one parent can always guard the chicks (Coulson & Wooller, 1984). In Kittiwakes these compatible birds are the ones that nest at the centre of the colony. The females also call more intensively at the beginning of the season and hence receive more courtship feeding from the males, which helps them form eggs (Wooller, 1979). They also divorce less frequently than other birds. Divorce offers the opportunity to find a more compatible mate, but it may take several years to achieve optimum breeding performance, since a lot of time is spent in courtship behaviour at first, and rather less in foraging (Chardine, 1987).

Nearly all species are sufficiently adaptable that they can take advantage of sudden gluts of food, such as whitebait or *Meganictiphanes*, during the breeding season. Amongst the specialist feeders, it is those with the narrowest range of feeding depths at the surface of the sea, i.e. the terns, that are most vulnerable to food shortages or weather conditions that make foraging more difficult, such as thick fog or high winds (Sagar & Sagar, 1989). They generally have the fastest growing young, provided they can provision them with enough food, but they are the first to show the adverse effects of a general decrease in food abundance (see Chapter 6). The generalists, on the other hand, can switch to alternative sources of food and hence are more resistant to such changes. The best examples of this are the gulls, which, as indicated above, can switch between the open sea, the intertidal area and terrestrial feeding according to circumstances. However, it may take a long time for the young to develop the diverse skills needed to feed in such contrasting circumstances and it may therefore be five years or more before they are able to breed.

# 4

# Mudflats and sandy shores

In England, the majority of the 47% of the British coastline that is not made up of rock or shingle, consists of soft sedimentary shores inside estuaries (see Fig. 44). In Scotland, however, sandy beaches predominate (Table 6). The way humans value these two types of habitat could

**Fig. 44.** Distribution of muddy and sandy shores in Great Britain. Some of the areas referred to in the text are indicated. Drawn from information in Doody (1985) and Burd (1989).

125

hardly be more different. Everyone loves a sandy beach, but many people seem to regard mudflats with distaste. This probably stems from the fact that a sandy beach is a good place to walk, swim and play, whilst a mudflat is a potentially dangerous place where there is a risk of getting bogged down or caked in mud. Such superficial reactions are misleading, since mudflats are much more productive for both fish and birds (Fig. 45).

The transition from coastline to river bank at the mouth of an estuary is often arbitrary and this is one of the factors that makes it difficult to estimate the length of our coastline. In practice, it is often more useful to define estuaries as discrete units and measure subtidal and intertidal areas rather than the length of shore. When this is done for the British Isles, it emerges that there are about 155 estuaries, containing a total of 3100 km$^2$ of intertidal mudflat, sandflat, rock, gravel and saltmarsh. The total subtidal and intertidal area of 5300 km$^2$ compares with a European total of about 19 000 km$^2$ (Davidson *et al.*, 1991). The fact that the British Isles has such a significant proportion of this total (28%), means that our

**Fig. 45.** Mudflats of the Beaulieu River, Hampshire, with glasswort saltmarsh in the foreground.

responsibility to conserve it is that much greater. Yet this is a time when our estuaries are threatened by a wider number and range of developments than ever before. The fundamental problem is that developers, engineers and politicians still regard tidal mudflats as wastelands.

The birds associated with mudflats and sandy shores are dominated by the wildfowl, waders, gulls and terns. The penetration of seabirds into estuaries is not usually very marked (except sometimes in adverse weather conditions), since they mostly require relatively clear water in which to feed. The Cormorant is the main exception to this. Several species are also able to exploit the least turbid estuaries, and the clearer regions of some others, such as the area of transition from salt to brackish water near the mouth. These species include Red-breasted Mergansers and several grebes (Little, Great Crested, Slavonian and Black-necked). Estuaries are also of major importance as spawning grounds and nursery areas for young fish, e.g. Sprats and whiting, which later in life may be important foods for birds in the open sea.

The Order Ciconiiformes contains a number of species of herons, bitterns, egrets, storks and ibises which are frequently associated with sand and mudflats on the coast as well as with freshwater areas. They are mostly medium to large wading birds with long legs and dagger-shaped bills. They can feed in many types of aquatic habitat, with the coast being particularly important in countries where inland waters freeze in winter, because brackish and salt waters remain ice-free for longer. The herons, egrets and bitterns have a characteristically kinked neck due to the elongated 6th neck vertebra. This acts as the main fulcrum for the quick lunge with which they typically capture their food. These species also have patches of powder down feathers on the breast and rump. These feathers gradually break up into small, powdery particles during preening and are very effective in blotting up the fish slime with which they often become coated, especially when handling struggling prey such as Eels. A comb-like claw on the middle toe is then used to remove accumulations of powder and slime from the plumage. Ibises and spoonbills feed quite differently. Ibises probe in vegetated shallows or mud with their curlew-like bill and spoonbills sweep their spoon-shaped bill from side to side through the water column. Both feed with the mouth slightly open and snap it closed to catch any small animals that they touch.

The Order Phoenicopteriformes contains the flamingos, of which there are only five species world wide. They are closely associated with saline or alkaline waters in the tropics and subtropics and consequently are also found in some coastal areas. Apart from their extraordinarily long legs, they are best known for their large, downwardly curved beaks. When feeding, the head is held upside down in the water, sometimes touching the bottom, and the bill is used to filter out small organisms, some of which are stirred up by movements of the feet. The tongue acts like a piston, sucking water in through the front part of the open bill and pushing it out sideways past a series of lamellae equipped with filtering projections which trap the food particles. These particles are periodically removed with the aid of the tongue, and then swallowed.

Most of the remaining birds associated with sandy and muddy shores belong to the Order Charadriiformes. Seven families of waders are usually included in this order. The families Rostratulidae and Dromadidae each contain only a single western Palearctic species, neither of which occurs in Britain – the Painted Snipe and Crab Plover respectively.

The Family Haematopodidae includes 11 species of oystercatchers. These birds are commonly found on sandy shores as well as rocky ones, because they specialise in feeding on bivalve molluscs which are often abundant in both types of habitat. Their adaptations for feeding on shellfish include a stout, fast-growing bill and powerful neck and jaw muscles. The Family Recurvirostridae contains three species of stilts and four of avocets. These are slim, graceful waders and are unusual (for this group) in that the males are usually larger than the females. The same is true of the Family Burhinidae, which contains nine species including the Stone Curlew, thick-knees and dikkops. These species are associated with a wide variety of habitats from the coast and fresh waters to semi-arid regions and deserts. Their large eyes and powerful legs are adaptations to a nocturnal and terrestrial way of life. The Family Glareolidae includes the Egyptian Plover, eight species of coursers and eight pratincoles, none of which occur in Britain.

The Family Charadriidae includes a large number of species of lapwings, plovers, sand plovers and dotterels – about 65 world wide. They are characterised by large eyes, short bills and a method of progression on the ground which involves taking a few steps at speed interspersed with short periods when they stand very still. The Family Scolopacidae is the largest group of waders and includes about 88 species – 25 sandpipers, 4 stints, 2 knots, 3 dowitchers, 4 godwits, 6 curlews, 2 whimbrels, 18 snipes, 6 woodcocks, 2 redshanks, 2 greenshanks, 2 tattlers, 2 yellowlegs, 2 turnstones, 3 phalaropes, as well as the Sanderling, Dunlin and Ruff. Most of these birds have long legs and long bills and are well adapted to feeding in water and soft sediments. The bill tip is particularly sensitive and the upper jaw has a movable connection to the skull allowing flexibility and movement of the bill which is useful when grasping prey hidden in sand or mud. The eyes are smaller than in the Charadriidae, since sight is less important in feeding, and the head in consequence has a slimmer shape. All of the Scolopacidae are primarily associated with wet habitats and all, except woodcocks, occur regularly on the coast at some stage of the year even if only during migration. Most species are migratory and a great many breed at high latitudes.

The Family Rynchopidae contains three species of skimmers, which are related to the gulls and terns, but occur only in Africa, America and South-East Asia.

### Abundance of wintering birds

If the measures we take to conserve shorebird populations, or to allow them to be exploited by shooting, are to be soundly and rationally based, it is essential that we have reliable information on their population sizes. The most useful data we have on the numbers of shorebirds wintering in British estuaries is that collected by the Birds of Estuaries Enquiry (BoEE). This enquiry was started in pilot form by the BTO in the winter of 1969/70 following a suggestion in 1968 by W.R.P. Bourne. In the following season a full time organiser was appointed (A.J. Prater) and the BoEE became jointly sponsored by the BTO, RSPB and the NCC. The Irish Wildbird Conservancy launched a 'Wetlands Enquiry' in the following year and thus information soon became available for the whole of the British Isles. In 1972, the Wildfowl Trust, which had organised wildfowl counts on inland waters since 1954, joined the BoEE.

The organisation of the counts themselves has changed little over the years. Three key winter counts are conducted, corresponding wherever possible with the National Wildfowl Counts and International Waterfowl Census dates in December, January and February. However, regular monthly counts are encouraged and this is achieved for many sites. The counts are generally made at spring tide roosts. The results up to 1974/75 have been published by Prater (1981) and there have been regular yearly updates. For assessing the national and international importance of sites it has become the practice to use the average peak population during the last five years. For wintering shorebirds this means that, for each site, the average is taken of the highest count obtained each year between November and March for the five-year period. Spring and autumn passage populations are assessed separately.

Total numbers for most species have not changed greatly during the course of the BoEE (Table 9), except for the Scaup which has declined dramatically (see p. 255). Long-term trends can be examined by deriving indices based on matched pairs of sites counted in January in consecutive years (coverage varies sufficiently for the total counts each year not to provide a reliable trend). The indices, of which some are shown in Fig. 46, show that most of the common species have changed rather little

Table 9. *Changes in the total numbers of waders counted in the British Isles in January during the Birds of Estuaries Enquiry and the National Wildfowl Counts. Counts greater than 1000 rounded to nearest 100, 100–1000 rounded to nearest 10. The counts are not directly comparable because of differences in coverage (e.g. some counts for Eire are included in the 1972–75 totals, but not in those for 1986–89), but do give a general indication of population changes*

| | Mean January count | |
|---|---|---|
| Species | 1972–75 | 1987–90 |
| Shelduck | 62 200 | 67 600 |
| Pintail | 16 500 | 21 500 |
| Scaup | 20 800 | 4600 |
| Oystercatcher | 177 200 | 261 600 |
| Avocet | 95 | 670 |
| Ringed Plover | 7500 | 10 600 |
| Golden Plover | 46 200 | 44 800 |
| Grey Plover | 10 000 | 31 400 |
| Lapwing | 126 500 | 143 200 |
| Knot | 290 000 | 257 200 |
| Sanderling | 5400 | 5900 |
| Dunlin | 538 200 | 388 500 |
| Ruff | 200 | 81 |
| Jack Snipe | 58 | 24 |
| Snipe | 3000 | 2500 |
| Black-tailed Godwit | 3800 | 5300 |
| Bar-tailed Godwit | 43 800 | 53 200 |
| Curlew | 56 900 | 68 900 |
| Spotted Redshank | 58 | 61 |
| Redshank | 69 300 | 76 700 |
| Greenshank | 190 | 250 |
| Turnstone | 10 500 | 19 200 |

in abundance (Shelduck, Ringed Plover, Knot, Sanderling, Curlew and Redshank), or have increased markedly (Pintail, Oystercatcher, Grey Plover and Turnstone). The Pintail is included because it is the one regularly counted species of duck, apart from Shelduck, in which more than 80% of the wintering population occurs on the coast. Most of the seaducks are not counted regularly enough to enable indices to be calculated. Two other ducks in which a significant proportion of the population is coastal (about 5%) are the Teal and Wigeon.

Some species, such as the Bar-tailed Godwit, tend to show wider fluctuations in numbers than others (Fig. 46), but these are not linked in

**Fig. 46.** Indices of population change in some selected coastal birds in Britain counted by the National Wildfowl Counts and the Birds of Estuaries Enquiry. The anchor year (index score = 100, shown by the dashed line) is 1971 for wildfowl and 1973 for waders. Hard winters (those with mean November to February temperatures more than 4°C below the 1960–88 average, i.e. 1977, 1979, 1982, 1986) are indicated by the vertical dotted lines. The arrowheads indicate winters following periods of inferred high predation in Siberia. Based on information in Summers & Underhill (1987), Salmon, Moser & Kirby (1987), Salmon, Prys-Jones & Kirby (1988, 1989), Marchant *et al.* (1990) and Kirby, Waters & Prys-Jones (1991).

any obvious or simple way with hard winters. Grey Plovers, Knots, Sanderlings, Little Stints, Curlew Sandpipers, Bar-tailed Godwits and Turnstones nesting on the Taimyr Peninsula and wintering along the coasts of southern Africa tend to show increased wintering populations after good lemming years (Martin & Baird, 1988). This is because predators on the breeding grounds tend to ignore bird prey in such years (see p. 186). Moreover, Mason (1988) showed evidence of 2–7 year cycles in the numbers of Ringed Plovers, Curlew Sandpipers, Dunlins and Bartailed Godwits on autumn passage at Eyebrook Reservoir in Leicestershire (July–October). In the cases of Curlew Sandpiper and Dunlin, these were significantly correlated with good breeding seasons for Brent Geese nesting in the same area. No clear trends are apparent in the wintering indices in Fig. 46, though there is a slight tendency for the population to rise in years of inferred low predation in the three Siberian breeding species – Grey Plover, Dunlin and Bar-tailed Godwit.

The species that shows the clearest evidence of the effect of hard winters is the Shelduck (Fig. 46), which had distinct population peaks in 1979, 1982 and 1986. Influxes of Shelducks were noted in at least the first two of these years (Owen, Atkinson-Willes & Salmon, 1986). These presumably consisted of birds displaced from the continent, especially the Wadden Sea, when the mudflats froze. Influxes of other species

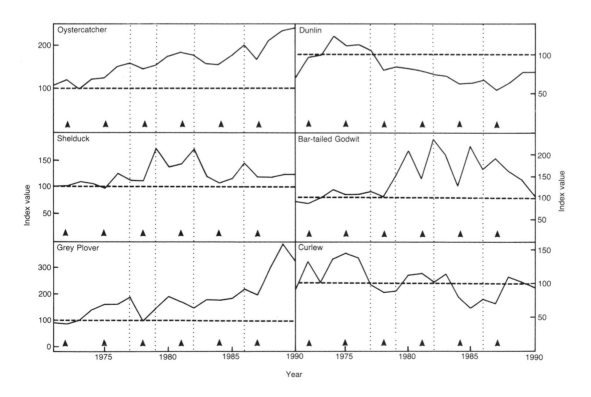

Table 10. *Densities of shorebirds (per km$^2$) on rocky and sandy shores in Scotland*

| Species | Orkney[a] | | Lewis and Harris[b] | |
|---|---|---|---|---|
| | Rocky | Sandy | Rocky | Sandy |
| Oystercatcher | 46 | 23 | 67 | 50 |
| Ringed Plover | 16 | 57 | 22 | 16 |
| Golden Plover | 36 | 47 | | |
| Grey Plover | | 1 | | |
| Lapwing | 59 | 42 | | |
| Knot | | | | 7 |
| Sanderling | 6 | 43 | | |
| Purple Sandpiper | 101 | 23 | 42 | |
| Dunlin | 14 | 98 | 4 | 14 |
| Snipe | 11 | 2 | | |
| Woodcock | 1 | | | |
| Bar-tailed Godwit | 4 | 43 | | 6 |
| Curlew | 271 | 236 | 34 | 5 |
| Redshank | 117 | 51 | 40 | 4 |
| Turnstone | 102 | 45 | 81 | 9 |
| Total | 784 | 711 | 290 | 111 |

[a] Tay & Orkney Ringing Groups (1984).
[b] Buxton (1982).

tended to occur in these years too, e.g. Knot in 1979. Cold winters do not always increase the numbers of shorebirds in Britain. Very severe conditions, when the mudflats freeze for prolonged periods, such as occurred in 1962–3, can result in many birds leaving the country or dying (Boyd, 1964). Curlews and Redshanks sometimes occur in reduced numbers on the coast if it is cold in January. This is because birds sometimes move inland to feed on fields at high tide, and hence are missed by the BoEE counts. In years of low winter rainfall, the soil is too dry to enable waders to feed on fields, but water levels in lakes and reservoirs are usually low enough to make shallow water and mud available as alternative feeding sites.

The birds on sandy shores outside estuaries in Britain are not regularly counted but were included in the 1984–85 Winter Shorebird Count. In the results of this survey the numbers on rocky and sandy coasts have not been separated, but information is available for Orkney and parts of the Outer Hebrides (Table 10). This shows that sandy shores usually support higher densities of Ringed Plovers, Sanderlings,

Dunlins and Bar-tailed Godwits than rocky shores, and that the reverse is true for Oystercatchers, Purple Sandpipers, Redshanks and Turnstones. The numbers of Golden Plovers, Lapwings and Curlews are more variable, probably because in both types of habitat the shore is used mainly as a roost site for birds which feed on fields.

Overall, the numbers of waders in British estuaries are holding up well and this is reflected in the increasing indices and increased total counts. In the case of the Oystercatcher, the winter increase has been accompanied by a rise in the size of the British breeding population which represents a significant proportion of the wintering population (Marchant *et al.*, 1990). The breeding areas of most other increasing species are too dispersed and difficult of access for us to know whether these populations have increased in size as well, but it seems likely that they have. One of the main reasons for increases in the wintering populations of Britain may be a reduction in mortality caused by hunting. For example, Redshank and Curlew have both been removed from the list of quarry species during the lifetime of the BoEE. However, the populations of these two species have not changed dramatically, possibly because their breeding areas in damp, lowland grassland and upland moors are disappearing and this has cancelled out any benefit of increased survival. There has probably been reduced hunting pressure on all shore waders and this may have allowed the populations of some species to increase. In the case of Grey Plovers there may even have been a shift in winter distribution, with fewer birds wintering in parts of the range such as France, where hunting pressures remain high.

One species that showed a steady decline during much of the 1980s is the Dunlin (Fig. 46). One possible reason for the permanent declines that have occurred at some sites is the increasing area of mudflats which have become colonised by *Spartina anglica*. This spread has been partly natural and partly a consequence of human agency. The species is deliberately planted because of its ability to trap silt and hence protect sea walls from wave action. Dunlin numbers have decreased most at sites where the spread of *Spartina* has been greatest (Davis & Moss, 1984; Goss-Custard & Moser, 1988). Although this plant occupies only a small zone at the top of the shore, the loss of high-level mudflats does have the effect of reducing the length of time for which shorebirds can feed and this may be critical during cold and windy conditions in winter.

Some species have greater difficulty than others in surviving adverse weather and poor feeding conditions in winter, and this partly reflects the ease with which they are able to find their daily food requirements under ordinary conditions. For example, Oystercatchers, Grey Plovers, Bar-tailed Godwits and Curlews did not use all of the available feeding time on either spring or neap tides during two mild winters on the Wash

(Fig. 47). Knots, Dunlins and Redshanks, on the other hand, came close to doing so in December, January and February, especially during spring tides when the enforced fast at high tide is more prolonged. During severe winters, the latter are the species that are most adversely affected. For example, during a cold spell in the south in January 1985, Redshanks comprised half of the waders found dead during a special survey, and Dunlins 15% (Davidson & Clark, 1985). During even worse conditions the following winter, nearly 600 waders were found dead on the Wash and Stour estuaries, comprising 55% Redshanks, 36% Grey Plovers and 3% or less of other species (Clark & Davidson, 1986). The reason Grey Plovers were badly affected on this occasion was probably that high winds were associated with the cold weather. Wind increases the chilling effect of cold for all species, but it is particularly serious for plovers because it also hides the signs of invertebrate activity the birds depend on to locate food (Dugan *et al.*, 1981). Juveniles (Fig. 48)

**Fig. 47.** Amount of time spent feeding by different species of waders on the Wash during spring and neap tides. Redrawn from Goss–Custard *et al.* (1977).

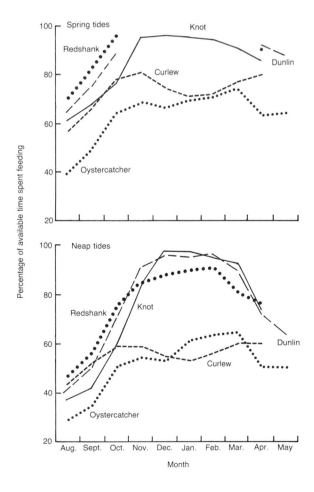

generally spend longer feeding than adults because of their lack of skill, and they are often the first to die in unfavourable conditions. The most recent cold snap, in February 1991, killed over 850 of the 4000 Redshanks wintering on the Wash (Mead, 1991).

The numbers of casualties were even greater during the most severe winter of the last 30 years – 1962/63. On that occasion Redshanks formed 47% of the wader corpses found, Knots 34% and Dunlins 11% (Table 11). We cannot judge how long it took populations to recover from this winter because the BoEE indices do not go back that far, but the National Wildfowl Counts showed no clear depression in numbers in 1963/64. However, it is possible that wildfowl were able to escape the worst of the weather by migrating elsewhere, and some of those found dead may themselves have been continental immigrants. On the other hand, it took Grey Herons about seven years to recover to their pre-1962/63 breeding population levels (Reynolds, 1979).

Oystercatchers and some other species feeding on mussels may actually be able to feed quite well in short cold snaps because bivalve adductor muscles weaken at low temperatures and the slightly gaping shells become much easier to open. Prolonged cold has an adverse effect, however, because it eventually kills the mussels and whole beds may then be destroyed, as happened in 1962/63.

**Fig. 48.** Dunlin (right) and Little Stint feeding by stitching in submerged vegetation. Both are juveniles since the Dunlin has a spotted belly and broad, pale-fringed coverts (though some have been replaced by more uniform grey winter feathers), and the Little Stint has the characteristic whitish eye stripe and 'braces'. Harold E. Grenfell.

Table 11. *Number of shorebirds found dead on a 14 km (9 mile) stretch of the north shore of the Wash on 26/2/63, 2/3/63 and 3/3/63. Some of the wildfowl had been shot and were therefore in good condition (especially the Wigeon), whereas most of the rest of the birds were emaciated. Even the Wigeon may have been made more vulnerable to shooting because the hard weather concentrated them on a smaller number of feeding sites. From Pilcher (1964)*

| Species | Number |
|---|---|
| Red-throated Diver | 12 |
| Black-throated Diver | 12 |
| Great Crested Grebe | 9 |
| Red-necked Grebe | 2 |
| Fulmar | 1 |
| Gannet | 1 |
| Grey Heron | 2 |
| Mute Swan | 1 |
| Pink-footed Goose | 2 |
| Brent Goose | 12 |
| Shelduck | 58 |
| Wigeon | 35 |
| Teal | 1 |
| Mallard | 7 |
| Tufted Duck | 2 |
| Common Scoter | 10 |
| Red-breasted Merganser | 1 |
| Water Rail | 1 |
| Moorhen | 2 |
| Oystercatcher | 1 |
| Grey Plover | 7 |
| Lapwing | 2 |
| Knot | 104 |
| Dunlin | 33 |
| Bar-tailed Godwit | 2 |
| Curlew | 15 |
| Redshank | 144 |
| Black-headed Gull | 26 |
| Common Gull | 51 |
| Herring Gull | 8 |
| Great Black-backed Gull | 6 |
| Kittiwake | 9 |
| Guillemot | 19 |
| Razorbill | 3 |
| Total | 601 |

### Distribution

Not surprisingly, larger estuaries support greater numbers of birds, but the relationship is not a simple one. The overall density of birds is usually lower in the larger sites (Prater, 1981; Davidson, 1991), perhaps because they contain a greater proportion of mobile and unproductive central sand banks and a relatively lower number of feeding sites at high tide (see p. 193). The distribution of waders and wildfowl that exploit sandy and muddy shores is very closely related to the distribution of our major estuaries. There are clusters of large estuaries on the east coast in Kent and Essex, on the south coast in Hampshire and West Sussex and in the Irish Sea from Cheshire to Cumbria. The distribution of birds is strongly associated with these clusters. For example, Shelducks (Figs. 49 and 50) occur in greatest numbers in these areas, and in the Wash, although the species also occurs in smaller numbers at a wide variety of other sites. The Pintail is only about one third as numerous as the Shelduck, and tends to be concentrated at major sites in north-east Scotland. There is also a substantial population on the sandy areas of the west coast of the Outer Hebrides but these birds remain very difficult to count adequately. Considerable numbers of Long-tailed Ducks, Velvet Scoters and Common Scoters occur in the Moray, Cromarty and Dornoch Firths associated with subtidal sandy sediments. There is also a large concentration of Common Scoters in Carmarthen Bay off the South Wales coast.

Other wildfowl such as Wigeon, Teal, Mallard and Shoveler are widespread and numerous on the coast, but are at least equally dependent on inland feeding areas. In recent years increasing numbers of Pochard and Tufted Duck have been recorded in estuaries such as the Thames, Firth of Forth and Severn, but these still comprise only a very small proportion of the total population. The Scaup's main habitat consist of shallow coastal waters and it is equally common feeding on cockles and *Macoma* in sandy and muddy areas as it is on mussels in rocky ones. It is mostly restricted to sites in Scotland and around the Irish Sea at present. A large concentration associated with sewage outfalls in the Firth of Forth has disappeared in recent years (see p. 255). Goldeneyes and Red-breasted Mergansers are both much more widespread and occur along rocky shores as well as along sandy ones and in estuaries. In fact, Red-breasted Mergansers avoid the high tidal range estuaries, such as the Severn and Mersey, presumably because of the difficulty of locating fish in turbid waters. Of the three species of divers wintering in Britain, the Red-throated Diver is most frequently associated with sandy shores and estuaries.

Waders that winter in low numbers in Britain, often have a rather restricted distribution. Thus Avocets occur regularly only on the Alde,

Exe and Tamar estuaries (Fig. 49). Numbers had increased to over 500, 300 and 100 individuals at each of these sites respectively by 1988/89 (Reay, 1989). Although this amounts to approximately the same total as the number of birds breeding in Britain, plus their offspring, there is likely to be some exchange of individuals with the continental population.

It is not just the uncommon species that are restricted to a relatively small number of sites. Knots (Fig. 49) and Bar-tailed Godwits are

**Fig. 49.** Distribution of some wintering wildfowl and waders in the British Isles. Redrawn from Prater (1981).

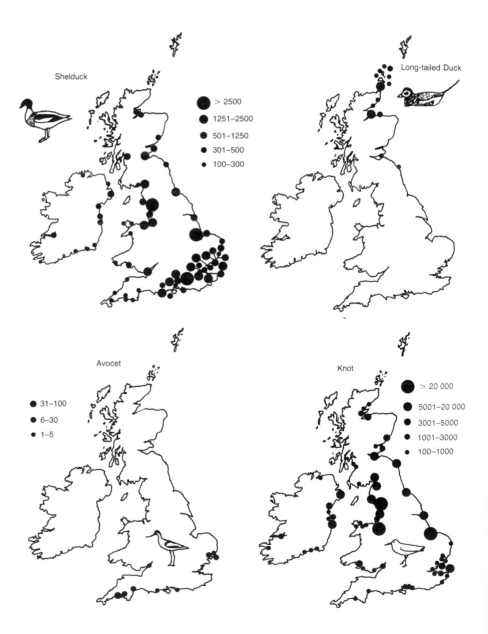

abundant, but are concentrated mainly on a small number of large estuaries such as the Wash, Dee, Ribble and Morecambe Bay. There are differences between these two species, however. Only a small proportion of the Knot population occurs in southern Britain and Ireland (except in hard winters), whereas the Bar-tailed Godwit is fairly common in these areas. The Sanderling is another rather concentrated wader with large numbers of birds in the relatively sandy estuaries in the southern Irish Sea and on the western coast of the Uists. This is even more marked during spring when nearly six times as many birds are present en route for breeding areas in Greenland and Canada (wintering birds are thought to breed in Siberia). Over 90% of these *c.* 40 000 passage birds occur in north-west England.

Black-tailed Godwits have a very southerly distribution and are particularly common in southern Ireland. This species increased rapidly in numbers during the middle years of the present century (1930–1970), possibly because larger numbers bred in Iceland as a result of climatic

**Fig. 50.** Shelduck drakes displaying at one another, while one partner feeds and the other looks on. The males usually defend an area around their mates, rather than a fixed territory, but they often stay in the same general area of a mudflat. Chris Rose.

amelioration. A decline in numbers was predicted by Prater (1975) as a consequence of cooler springs and this had indeed occurred by the late 1970s. The population has since increased again. The Dunlin is both the most abundant wader in our estuaries and one of the most widely dispersed. Curlews and Redshanks are also widespread. Species that only winter in small numbers, because Britain is on the northern edge of their range, tend to have a very southerly distribution. Thus the 200 or so Spotted Redshanks, the 100 Common Sandpipers and 1500 Ruffs are concentrated along the south coasts of England and Ireland. The 1200 or so Greenshanks are concentrated in Ireland and south-west Britain. Only about 30 Green Sandpipers winter on the coast in Britain (though as many as 700 do so on inland freshwaters) and these are mostly in the south-east.

Many waders, like the Sanderlings already mentioned, occur in larger numbers during spring or autumn migration than they do in winter, and populations often originate from different breeding areas. This applies to Ringed Plovers, Black-tailed Godwits, Curlews and Redshanks, amongst the commoner species, and to nearly all of those which winter in small numbers. Peak counts underestimate passage numbers because there is often a continuous turnover of birds, whereas populations are more stable in winter. For example, Moser & Carrier (1983) dyed the plumage of Turnstones and Ringed Plovers in the Solway Firth in April and May, so that they could detect departures of marked birds and influxes of new ones. For Turnstones there was a single influx beginning at the end of April, and no turnover amongst these birds once they had all arrived. All Turnstones left for the breeding grounds in mid May. There was also an influx of Ringed Plovers in late April, but of the 45 birds dyed on 30 April, all had left by 14 May, despite the fact that consistent counts of 400 were made from 30 April to 25 May. Thus at least 800 Ringed Plovers probably passed through the Solway in spring 1983. For species such as Dunlin, there are indications that the rates of turnover during passage may be even greater.

Some species are more abundant in one migratory season than another, with numbers usually being higher in autumn. The Curlew Sandpiper is a classic autumn migrant, which has had occasional bumper seasons when breeding success is good in the Siberian Arctic and weather conditions force birds further west than normal during the migration (Stanley & Minton, 1972). Sanderlings and Whimbrels are the only regularly occurring waders that are more abundant in spring than in autumn or winter.

**Food resources of mud and sand**

Large plants find it difficult to establish a foothold on fine shifting
sediments and consequently most of the plants that occur on mud and
sand in the intertidal area are microscopic in size. They are restricted to
the surface film of water on top of the sediment. The most abundant of
them are the mobile, unattached diatoms. They may impart a greenish
or brownish tinge to the surface of the mud, especially in the summer
months. Few birds are able to exploit such plants as a source of food, but
one group that can are the flamingos. They also take small sediment
dwelling invertebrates when these are present and can even feed on the
organic debris associated with the surface layers of mud. The Lesser
Flamingo is a specialist feeder on diatoms and other algae, but feeds
mostly on high-altitude salt lakes where these plants occur at very high
density. The more generalist feeders, such as the Greater Flamingo, feed
regularly in coastal waters and open mudflats. Escapes from captive
collections in Britain also manage to feed on intertidal mudflats, and two
survived on the Fleet at Abbotsbury for about six years.

Diatoms produce copious mucus which traps and binds fine sediments. Indeed if their invertebrate grazers (such as *Hydrobia* and *Corophium*) are removed, the resultant diatom bloom traps sediment at an
even faster rate (Coles, 1979). In the stablest areas of estuarine mudflats
and sandflats, plants such as *Zostera* (a flowering plant) and *Enteromorpha* (an alga) become established, producing 'sea-grass meadows',
extensive enough to support grazing birds such as Brent Geese and
Wigeon (see Chapter 5).

Life for invertebrate animals in sand and mud is far from easy. It is a
rather unstable environment, not just because of the mobility of the
sand and mud itself, but because of the wide changes in salinity that
occur in estuaries. This is one of the reasons that the number of
invertebrate species is small compared with more stable habitats. Many
filter feeders, such as mussels and cockles, are absent from the turbid
parts of estuaries because their gills, which are used to filter small
organisms and organic matter from the water, rapidly become clogged
with sediment. Their place is taken by bivalves such as *Macoma* and
*Scrobicularia* (Fig. 51), which have only a limited ability to filter water in
this way. Instead, they feed mainly from within the relative safety of the
mud via long siphons. Deposits on the surface layers of the mud (which
are rich in both settled organic materials and microscopic plants and
animals) are sucked down the siphon and passed directly to the gut for
digestion. Deposit-feeding may be supplemented by some suspension
feeding or by a mixed strategy in which the siphon is used to stir up the
surface layer and resuspend it in water so that the slower settling

organic materials can be sucked in. By contrast, the bivalves that occur on sandy shores, such as cockles, clams and Razor Shells (Fig. 51), are conventional filter feeders, which suck in water through their burrow entrances and filter out food from it using their gills. Cockles have short siphons since they burrow just beneath the surface, but clams burrow more deeply and consequently have longer ones. Razor Shells have short siphons since they feed near the surface, but can retreat very rapidly to a greater depth using their muscular food to pull themselves down, as anyone who has tried to dig them up can testify.

The tiny snail *Hydrobia* is one of the few gastropod molluscs that is common in soft sediments (Fig. 51). It feeds on both algae and detritus in the surface layers of mud, especially diatoms on particles of 2–250 micrometres diameter. Extremely high densities of these snails may be present, over 30 000 individuals per square metre being recorded on occasions. The worms that inhabit coastal sediments feed in a variety of

**Fig. 51.** Some of the most characteristic invertebrates of muddy and sand shores (approximately to the same scale – half life size). Redrawn from various sources, including Barrett & Yonge (1958), Prater (1981), McLusky (1981) and Hulscher (1982).

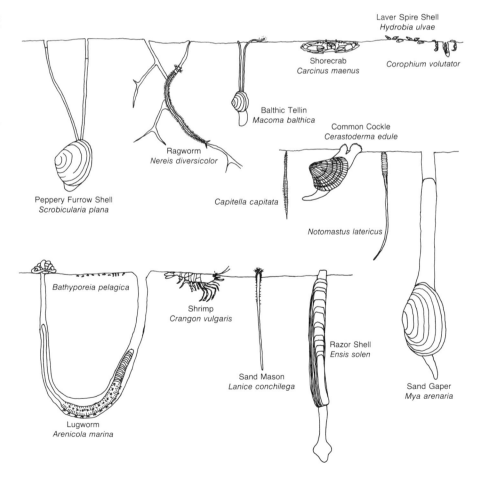

ways. The Ragworm ingests the rich surface layers of sediment, as well as being an active predator, using its jaws to capture small worms or scavenging on dead animals. It is also capable of using a mucus net which it secretes in its burrow and irrigates in order to trap fine suspended materials. The Lugworm can only maintain its large burrow in relatively coarse sandy sediments and is commonest where these are rich in organic materials. It ingests such materials directly as they fall, or are wafted, into its burrow. The Sand Mason feeds both on surface deposits and by filtering the water column. *Notomastus* is a bristle-worm which ingests sediment rich in organic matter directly and sometimes excavates a tight spiral burrow through suitable horizontal strata in the sand or mud (Fauchald & Jumars, 1979).

*Corophium* is a small crustacean which lives in a U-shaped burrow in the upper layers of mudflats (Fig. 51). It is a deposit feeder consuming mainly bacteria on particles of 4–63 micrometres diameter. Larger crustaceans, such as crabs and shrimps, are amongst the most important predators of other invertebrates. Both tend to move into shallower intertidal waters during the summer months to take advantage of seasonal abundances of food.

All of these invertebrates have a considerable advantage compared with their relatives on rocky shores, in that they can bury themselves completely in the sediment and thus escape detection by surface predators such as fish and birds. Burrowing also enables them to escape from extremes of temperature, drying out of the surface layers of the sediment and short-term changes in salinity. Thus they need none of the protective armour, sealing mechanisms, camouflage and other defences which are so characteristic of rocky-shore animals. Birds feeding on them are consequently faced with quite a different set of problems to those feeding on rocky shores. Detecting the presence of invertebrates, and catching them, are the major difficulties. Opening them up is not usually a problem.

In winter, most intertidal invertebrates (like all cold blooded animals) are less active than in summer, because of the lower temperatures. This poses few problems for bird predators on rocky shores, since most prey remain just as detectable, though a few species such as crabs may remain in subtidal waters during the winter months. However, this is a time when birds in estuaries and on sandy coasts usually have the greatest difficulty in finding sufficient food, because their prey tends to burrow deeper in cold weather and become less active. For example, *Macoma balthica* is found nearest the surface in June and is buried most deeply in December. Large individuals are always buried more deeply than small ones. In June, over 90% of the total biomass of this species is located in the top 3 cm of the mud and is available to birds such as Knots (though very few of these birds are present in our estuaries in June). In December, only 5% is available at a similar depth (Reading & McGrorty, 1978). This clearly indicates that it must be of great advantage to waders to exploit any behaviour of *Macoma* that brings individuals closer to the surface, such as feeding, burrow maintenance, defaecation, etc. Signs of such activity are often invisible to the human eye so we are often not sure exactly what clues the birds are using to detect their prey, but there are other cases where the evidence is clearer (see below).

## Foraging on soft sediments during the day

Two main methods have been used to study the diet of shorebirds. The first is direct observation and the second is the analysis of pellets or droppings. Both suffer from some disadvantages. With direct observation it is not always possible to identify the prey species with certainty. Even measuring its size by comparing it with the bird's bill length or height from the ground can be difficult. Pellets are the hard parts of prey (such as mollusc shell, worm jaws and bristles, and crustacean and insect exoskeletons), which are regurgitated from the gizzard. The latter does most of the 'chewing' of food in birds, and retains some grit to help in this process. By getting rid of most of the indigestible material, birds can achieve quite high absorption efficiencies, e.g. 64% for Oyster-catchers on mussels (Speakman, 1987) and 76% for Sanderlings on mealworms (Castro, Stoyan & Myers, 1989). The problem with pellets as a guide to diet, is that the remains of small prey may be ground up completely or pass on through the rest of the gut (Pienkowski *et al.*, 1984).

Shorebirds employ a variety of senses to detect prey, but by far the most important are touch and sight (Fig. 52). Several species take advantage of the fact that invertebrates tend to become more active as they are covered by the advancing tide, and to remain active until they are exposed to the air by the retreating tide. Thus there is often a concentration of birds feeding along the tide edge (Fig. 52*a*). In the case of Knots, such birds are believed to forage by touch, with the water merely bringing more prey within reach of the bill. For lugworms, greater activity at the tide edge has been directly observed, with larger numbers appearing at the rear entrance of their burrows to defaecate when covered by the tide (Smith, 1975). Bar-tailed Godwits take advantage of these defae-catory movements and probe in the burrows when they see such activity. Probing at the burrows of inactive lugworms is seldom success-ful since they are too deep for even the godwit's long bill. Occasionally, godwits only succeed in breaking off a few tail segments from the worm, perhaps because they choose the wrong moment to probe or pull. In these circumstances, the worm is able to regenerate the missing segments. The ability to break off tails or siphons and eventually regenerate them is quite widespread amongst those invertebrates which may lose them to birds. Waders in turn show great care, once they have grasped prey, to try and extract it from the sediment without breaking it. For example, they may pull at it slowly and evenly, or may let it go and attempt to grasp it further down the body by probing again.

Not only do many waders feed at the tide edge, but many also prefer to forage in wet areas of mud for the same reasons (Fig. 52b). Though there is as yet no proof that prey are more active in wetter areas, the density of some species such as crabs is higher. More subtle differences in microtopography may also influence the way in which mudflats drain

**Fig. 52.** Main dispersion patterns and feeding techniques of shorebirds foraging on soft sediments, viewed diagrammatically from above. Dashed lines = previous track. 'Stitching' (*a*) is often done in a dense flock since birds feeding by touch do not interfere with one another. Many sandpipers feed by night in this way. Besides feeding at the tideline itself, many birds feed in damper mud or in running water (*b*), because prey activity is higher there, or because the sediment is softer and easier to probe. Redshanks (*c*), and many other sandpipers, feed by visually scanning a strip of mud (indicated by the lines on either side of the dashed one) as they walk along. When potential prey is spotted an attempt may be made to take it by pecking if it is small, e.g. *Corophium* (closed circles) or by probing if it is larger, e.g. *Nereis* (stars). Plovers (*d*) stand still on one spot (large filled circles) and scan an area in front of them (indicated by the circle sectors). After a while they move rapidly to a new scanning position, or to grasp prey if they see a suitable sign of activity (stars). Redrawn from Pienkowski (1983*b*). The Shelduck (*e*) filters the surface layers of wet mud through its bill by moving it from side to side, leaving characteristic marks on the surface. Patches of high density prey, such as capitellid worms in buried kelp on sandy shores, may support dense patches of birds (*f*).

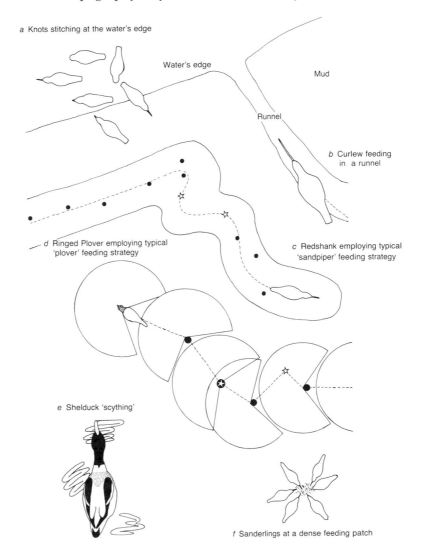

*a* Knots stitching at the water's edge

Water's edge

Mud

Runnel

*b* Curlew feeding in a runnel

*d* Ringed Plover employing typical 'plover' feeding strategy

*c* Redshank employing typical 'sandpiper' feeding strategy

*e* Shelduck 'scything'

*f* Sanderlings at a dense feeding patch

and hence in how quickly areas dry out. These effects may help to explain why the use of a superficially uniform mudflat by shorebirds is uneven, as an examination of the distribution of their footprints often shows (Wilson, 1988).

Waders feeding on open mudflats during the day face a major difficulty in that prey usually respond to unfavourable conditions by retreating more deeply into their burrows. Such unfavourable conditions include the presence of birds and other sources of disturbance at the mud surface (but probably not the approaching tide). For example, the frequency with which *Corophium* protrudes its anterior end from its burrow varies with a number of environmental factors, including temperature (it declines to zero at 4 °C) and rainfall (which halves such activity). When a Redshank walks across the mud, the number of *Corophium* protruding from their burrows, in an 8 cm wide strip around the birds path, declines to zero and only returns to 50% of normal levels five minutes later (Goss-Custard, 1970). A Redshank feeding in an area just visited by another individual thus takes about three times as long to find and capture a *Corophium* as when feeding in an area that has not been visited for some time (Selman & Goss-Custard, 1988). This phenomenon is sometimes known as resource depression, to distinguish it from the situation in which the availability of food declines because it is actually consumed by other predators (resource depletion) (Charnov, Orians & Hyatt, 1976). When the area around individual foraging Bar-tailed Godwits (unoccupied by other individuals) declines from ten to five square metres the number of lugworms appearing at the surface to defaecate falls by a half, from six to three per minute. Surface activity is also temperature dependent, though lugworms are not as sensitive as *Corophium*, with activity being reduced to zero only at −2 °C (Smith, 1975). The activity of *Notomastus* worms at the surface, which can be detected because of small outflows of water from their burrows, is also depressed by low temperatures and high winds (Pienkowski, 1981). In this case an attempt was made by Pienkowski to 'peck' at the outflows with a pair of forceps. Worms were successfully captured by this method at 68% of holes, compared with only 22% at worm holes when no outflows were present (Pienkowski, 1983a). This provides a strong indication that outflows are the clue used by foraging waders (in this case Ringed Plovers) to detect *Notomastus*.

The majority of sandpipers, feeding on open mudflats during the day, appear to adopt a strategy that exploits a mixture of visual and tactile clues to detect prey. They move across the mud, avoiding close contact with other individuals, pecking at the surface to catch small prey, such as *Corophium*, and probing more deeply in those spots where they see signs of the activity of larger prey, such as worms, near the surface (Fig.

52c). Even if prey have retreated to some extent by the time they are reached, they may still be located by touch with a few well directed bill probes. Small plovers, on the other hand, have to rely on visual clues more exclusively, since their shorter bills preclude them from probing very deeply. Grey Plovers can probe to at least 3 cm, which is as far as most small sandpipers. All plovers adopt a strategy in which they stand still on one spot scanning for signs of prey activity. When they see a prey movement judged worthy of an attempt at capture, they run quickly to the spot and make a grab for it (Fig. 52d). Since a plover is standing still when scanning, it probably disturbs invertebrates less than a walking sandpiper, but it requires good eyesight to spot prey under poor light conditions and the capacity to accelerate fast to be able to catch it. Lapwings foraging on mudflats capture *Corophium* within a relatively short distance of their scanning position – usually two paces or less. Even large ragworms are usually taken in three paces or less, though the greatest distance moved to capture prey is eight steps (Metcalfe, 1985). By increasing the number of paces taken when moving to a new scanning position from a previously unsuccessful one, and reducing the number of paces when successful, both Ringed Plovers and Grey Plovers tend to concentrate their activity in areas with a high density of active prey (Pienkowski, 1983b).

When prey activity is low, for example at low temperatures, plovers cannot simply switch to feeding by touch, which is at least an option for most sandpipers. Instead, they may use another foraging tactic, known as foot trembling or foot vibration. This involves placing the foot on the sediment, slightly forward of the other leg and rapidly vibrating it up and down. There has been much argument about why this technique works, but the fact that it does has been demonstrated by Pienkowski (1981, 1983a) who was able to increase the surface activity of the small crustaceans *Bathyporeia* and *Eurydice* by simulating plover foot movements with a pair of forceps. This was in the same upper-shore areas where Ringed Plovers employed foot trembling, and was caused either by the liquefaction of the sand or by simulating the approaching tide. Both of these conditions are likely to stimulate swimming activity in these species. The overall level of activity induced by these vibrations (at least those visible to a human observer) represented an increase of 30% over activity levels without vibration.

Some, but not all, of the clues used by waders may be visible to the human eye. When feeding on *Macoma*, Oystercatchers probe in the centre of the clearly visible star-shaped feeding depressions these bivalves make in the mud surface (Fig. 51). When feeding on cockles they use clues undetectable to us since the sand looks uniform, but they successfully locate prey more often than random probing can achieve

(Hulscher, 1976, 1982). Shelducks rely on touch alone to detect prey. They utilise a number of different feeding techniques including scything, digging, dabbling, head dipping and upending (Bryant & Leng, 1975). The last three are used mainly at the tide edge or on the water itself, the most frequent method employed on open mudflats being scything (Fig. 52*e*). This involves side to side movements of the bill through the surface layers of the sediment. The mud is filtered in the mouth and organisms such as *Hydrobia*, *Nereis* and small oligochaete worms are removed and swallowed. The filtering mechanism is highly efficient and allows one of the largest European shorebirds to thrive by eating some of the smallest prey organisms.

One of the most familiar shorebirds on sandy shores outside estuaries is the Sanderling, which is usually seen scampering back and forth between the breaking waves. In fact we know rather little about why they feed in this way, but it is probably a quite different situation from that in which waders exploit the tide edge on muddy shores. It seems unlikely that invertebrates would voluntarily become more active near the surface in the presence of breaking waves, but rather more plausible that the swash and backwash disturb the upper layers of the sand to such an extent that buried animals are involuntarily brought to the surface. As they hurriedly bury themselves once more, their movements may then betray their presence to Sanderlings. *Bathyporeia* is a small crustacean living in the surface layers of beach sediments and that could well be taken in this way since it forms a major part of the Sanderling's diet. The ability to feed on such small prey requires the ability to peck fast and move rapidly. Although Sanderlings are sandpipers, they lack the hind toe which is present in nearly all the other sandpipers. The hind toe is also absent in the plovers, which are likewise capable of rapid movement across the sediment.

Work on captive shorebirds has revealed other possible methods of prey detection. Gerritsen & Meiboom (1986) suggested that Sanderlings are able to detect the worm *Nerine* with the bill tip in the sediment at a distance of about 2 cm by means of 'remote touch'. In other words, they detect movements of the worm transmitted through the sediment. It has also been shown that both Sanderlings and Dunlins spend more time foraging in jars of sand in which *Nereis* have been kept, than in clean sand, suggesting that they use the sense of smell or taste (van Heezik, Gerritsen & Swennen, 1983). Taste sensors have been found towards the tip of the lower mandible and inside the mouth in these two species, and in Purple Sandpipers and Knots (Gerritsen & Sevenster, 1985). How useful a sense of taste might be in locating individual prey items remains to be determined, but it certainly could help in locating high density patches of prey. For example, Sanderlings sometimes feed on the swarms of worms that surround decomposing kelp buried in the sand. These patches are invisible at the surface, but once found become surrounded by dense groups of Sanderlings all probing rapidly in a small area (Fig. 52*f*).

Birds frequently seem to feed in particular ways and places by observing the behaviour of others and copying it, or by observing the behaviour of others and stealing the food that they obtain! For example, Sanderlings run to join groups of birds feeding at the food patches described above. Food robbery is particularly likely to occur when food items are large or difficult to handle. Sanderlings, for example, rob Ringed Plovers feeding on polychaete worms on the sandy beaches of the Outer Hebrides where the plovers take some time extracting the worms in order to avoid breaking them. The Sanderlings steal the worms during or after extraction, or sometimes usurp Ringed Plovers and probe at the spots from which they were just about to extract worms.

There are a number of waders specialised for feeding in shallow standing water (Fig. 53). The upturned bill of the Avocet enables it to keep the maximum length underwater while 'swishing' it from side to side, thus searching a greater area of water, or liquid mud. They locate a wide range of estuarine and freshwater invertebrates in this way, particularly those that occur at high densities in standing brackish water, such as the crustacean *Palaemonetes*. At one of their main British wintering sites on the Tamar Estuary, they spend about 51% of their time feeding on mud, 37% in water and 12% swimming (Reay, 1988). They swallow items at a rate of 23–42 per minute so these must consist of very small prey. Spotted Redshanks feed on average in deeper water and swim more often than other sandpipers (Cramp & Simmons, 1983). They will also upend in deep water to feed on prey such as fish fry.

Gulls are also abundant predators of invertebrates on muddy and sandy shores (Fig. 54), but much less attention has been paid to their foraging in estuaries than to that of waders. In the Severn Estuary, Black-headed Gulls constitute about 90% of the gulls counted on muddy shores in winter, though these individuals comprise only about 12% of the total Black-headed Gull population of the area, with most of the rest feeding on fields (Mudge & Ferns, 1982*a*). In summer, few Black-headed Gulls remain in the Severn Estuary, but in the Clyde Estuary, where a detailed study of feeding behaviour has been made (Curtis *et al.*, 1985), more birds were present in summer than winter. In winter, the rate of food intake per unit time was only 30% of the rate in summer, indicating the much greater difficulty of foraging when invertebrates are less active and buried more deeply in the mud. Their main prey in summer was the Ragworm, which occurs nearer the surface at this time of year, and which they obtained either by walking quickly and pecking at the mud surface, or more frequently by walking slowly and 'crouching' (leaning forward and downwards with the head close to the sediment) before taking prey. Crouching also occurs in Lapwings and Golden Plovers feeding on fields, where it allows them to locate larger earthworms (Barnard & Thompson, 1985), possibly by means of sound, though the evidence for the use of auditory clues in these and other plovers is equivocal (Lange, 1968; Hale, 1980; Pienkowski, 1983*b*). One further difference in the way gulls feed in winter is the use of an additional feeding technique known as 'paddling'. This consists of standing in one spot, mostly in water greater than 5 cm deep, and treading repeatedly. This probably stimulates animals into greater activity by agitating the sediment. Gulls also feed extensively on sandy shores outside estuaries

**Fig. 53.** Greenshank grasping an Elver in shallow water. Many waders will feed on fish provided they are small enough to swallow. They will also take pieces of flesh from the carcasses of larger fish if these have been opened by larger predators. Keri Williams.

where they take a range of prey, including Edible Crabs, Shrimps, cockles and Sand Gapers (e.g. Sibly & McCleery, 1983).

For most species of intertidal foragers, on sandy, muddy and rocky shores, there are usually peaks of feeding activity on the ebbing and flowing tides with something of a lull around low tide, when many birds roost for a while. This is true of Oystercatchers feeding on cockles by stabbing, partly because cockles gape slightly when the sand is wet at the beginning and end of the tidal cycle, and are therefore easy to open at this time. Hammering Oystercatchers require dry sand to support the cockles whilst smashing them open and, hence, in the laboratory at least, do most of their feeding on dry sand in the latter half of the simulated tidal cycle (Swennen *et al.*, 1983).

The sizes of prey taken by different species on sandy and muddy shores are often different, just as they are on rocky shores. Thus Greenshanks in the Wadden Sea take far smaller crabs than Curlews, which in

**Fig. 54.** Mixed group of gulls on a sandy shore in winter plumage, comprising (starting foreground left and proceeding clockwise) Great Black-backed Gull, Lesser Black-backed Gull, juvenile Herring Gull, two Black-headed Gulls squabbling over a piece of seaweed, Herring Gull, Common Gull and juvenile Black-headed Gull (asleep). Chris Rose.

turn take smaller ones than Herring Gulls (Fig. 55). Similarly, Little Stints, Ringed Plovers, Dunlins, Curlew Sandpipers, Reeves and Ruffs take successively larger sizes of fly larvae when feeding on piles of wrack along the Norwegian coast. In this case, the mean length of the maggots taken by each species was correlated with the birds' body weights, but not their bill lengths. This may be because the main factor determining the size of food items that can be swallowed is gape size, and this in turn may be better correlated with body size than bill length (Lifjeld, 1984). Wrack piles can form the most important sources of food in areas where the tidal range is small, such as the Baltic. They form on both rocky and soft sedimentary shores since dead seaweed can be transported for some distance before being stranded.

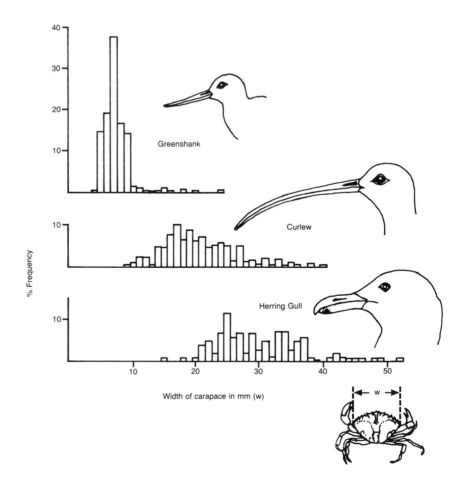

**Fig. 55.** Sizes of Shorecrabs taken by three species of shorebirds in the Wadden Sea. Based on remains in pellets, and redrawn from Zwarts (1981).

### Foraging at night

Sandpipers, which normally feed by day using a combination of sight and touch, seem to switch at night to touch alone (perhaps aided by other factors such as remote touch and taste). This can be inferred because species such as Dunlins employ 'stitching' at night. Stitching involves rapid up and down movements of the bill in a restricted area. Dunlins also employ stitching under conditions of low temperature, rainfall, high winds, high bird density, and when a layer of fresh mud has been deposited on the surface by the previous tide. All these are factors that depress or hide prey activity and hence preclude the use of the normal sandpiper search strategy. Their diet then changes from one showing positive selection for *Nereis*, to one in which invertebrates are taken at more or less the same frequencies as they occur in the mud (Pienkowski *et al.*, 1984).

Oystercatchers feeding at night on both cockles and *Macoma* also revert to a feeding method dependent entirely on touch since their rate of prey capture becomes no greater than that which would be predicted by the random collision of their bills with the bivalves' shells (Hulscher, 1982). This suggests that Oystercatchers at least are unable to use remote touch, taste or sound to help locate prey.

There is evidence that many invertebrates are more active near the surface during the night than during the day. This is well known in the case of terrestrial earthworms which take advantage of the damper condition of the vegetation and soil surface at night. Dugan (1981) and Evans (1987) measured prey activity on the surface of mud and sandflats at night by suddenly switching on a torch and counting the number of animals visible. They found that *Nereis*, *Macoma*, *Hydrobia* and *Corophium* were all more active than they were in the same areas by day, indeed only *Hydrobia* was seen at the surface at all during the day. It has also been found that more *Corophium volutator* and other crustaceans swim actively in the water column during the night (Hughes, 1988; Robert & McNeil, 1989). This all suggests that waders might be able to feed quite well at night if they were able to locate prey.

There is some evidence that sandpipers take advantage of any available light at night to exploit greater prey activity. For example, Curlews roost inland at night during the breeding season, but these roosts are deserted during full moon periods. The birds are presumed to fly to the coast to feed on *Scrobicularia* at such times because shell fragments of this genus appear in their pellets (Hale, 1980). Waders may also take advantage of street lighting and other artificial sources of illumination if it is available near feeding areas. The stomachs of Knots, however, contain only about a quarter as much food after nocturnal feeding as they do after daytime feeding (Prater, 1972). This seems curious for a species that routinely spends so much time feeding by touch and therefore might have been expected to do as well by night as by day. One possible explanation for this is that they select the best general areas in which to feed by means of visual clues, such as the density of *Macoma* feeding marks, but are unable to see these at night.

Given their reliance on feeding by means of visual clues, it might be expected that plovers rather than sandpipers would encounter the greatest difficulty in foraging at night. In fact, the reverse seems to be the case, with plovers somehow being able to exploit the higher levels of prey activity that occur then. In the case of the inland feeders, foraging is so efficient at night (especially during the full moon period) that species such as Lapwings may scarcely feed at all by day (Hale, 1980; Milsom, 1984). Grey Plovers may also obtain a major part of their food intake at night. For example, radio-tagged birds in the Tees Estuary were found to feed at night during spring tides on an area where large *Nereis virens* were abundant, but which was never visited during the day (Townshend, Dugan & Pienkowski, 1984). On very dark nights, however, Pienkowski (1982) found that the feeding rate of both Ringed and Grey Plovers was lower than during the day and that it was depressed further by strong winds. He also noted that invertebrates were more active at night and considered that plovers normally took advantage of this. Some invertebrates were luminescent, probably as a consequence of feeding on luminescent dinoflagellates, and plovers might have been able to locate some individuals in this way. This all suggests that plovers are visual foragers, even at night.

Turnstones may not feed as much by night as do other waders. They are not seen or heard on the shore as often as during the day, except perhaps in some tropical habitats (Schneider, 1985). This may be because they are dependent on visual foraging techniques and because they are one of the most efficient waders when feeding along the strand line at high tide. This means they can take advantage of all the available daylight hours, including the high tide period when most other waders are forced to stop feeding. They tend to use remote headlands and islands as nocturnal roosts rather than roosting nearer to the feeding areas, perhaps because of the greater security such sites offer during a prolonged roosting period.

### Spacing, flocking and roosting behaviour

Given the depression in prey activity which can be caused by the birds themselves, it is not surprising that they spread out when feeding. For example, Dunlins split up into smaller flocks as the winter progresses and invertebrate activity declines (because of lower temperatures). The distance between members of these feeding flocks reaches a maximum (a density of about one bird per 50 m$^2$ of mud) in February. This compares with about one bird per 15 m$^2$ in October and one per 10 m$^2$ in April, even though fewer birds are usually present in autumn and spring (Worrall, 1981). This increase in spacing in midwinter is likely to be needed because of the greater undisturbed area within which each bird needs to forage in order to find enough active prey. Spacing is achieved without much overt aggression between the birds, but it is not surprising that in some cases waders find it worth while to actively defend a winter feeding territory against intruders. These territories may occur in patches of moderately high prey density such as those of Horseshoe Crab eggs in North America which may be defended by Sanderlings (Recher & Recher, 1969) or Short-billed Dowitchers (Mallory & Schneider, 1979). In patches of very high density food, the number of intruders may be so great that defence becomes uneconomical since too little time is left for feeding (Myers, Connors & Pitelka, 1979).

Longer term feeding territories are usually larger than transient ones. For example, of the 350 Grey Plovers that winter at Teesmouth, the majority feed in a large dispersed flock with no fixed territories. Some territories are defended, however, by a few birds at mid tide level in an area with deep creeks. Townshend, Dugan & Pienkowski (1984) argued that the benefits of these territories are a lower rate of prey depletion throughout the winter and the opportunity to feed (in the creeks) during high winds more effectively than birds on the open mudflats. The costs of territoriality include the extra effort required to chase off intruders and a greater risk of predation by Foxes at night. The balance of advantage between the two strategies is very close, depending on factors such as the severity and frequency of high winds and predator activity, and is one reason why a mixed strategy occurs in the population as a whole. Myers (1984) found that Sanderling territoriality broke down in years when Merlins were more abundant, showing that the balance of advantage may change from year to year and that birds adjust their behaviour accordingly. Curlews behave similarly to Grey Plovers, with some individuals defending territories. In the Taff Estuary, the small number of Curlew territories present do not include creeks, but instead a section of river channel, which also offers shelter (Siman, 1989). Males and females often occupy adjacent territories in winter, but in spring a pair may share the defence of a single territory at the top of the shore.

Even more complex feeding territories are defended by Ringed Plovers on the sandy shores of the Outer Hebrides. Some birds feed in flocks, whilst others defend short term territories, including separate ones at the top and bottom of the shore (Fig. 56). In the latter case, they do not feed in the intervening area and maintain the same positions relative to one another. In some cases, long term territories are defended by a pair (male and female). Some of these territories last throughout the whole winter, but others only as long as beach topography remains the same. A major influence on the location of territories is the deposition of kelp on the shore by storms. Before and after being buried in the sand, kelp attracts the invertebrates that make territoriality worth while. These include flies and their larvae early in the cycle of decomposition and sediment-dwelling worms in the later stages (Wells, 1991). Most waders show a degree of territoriality on the wintering grounds if the appropriate feeding conditions that make it worth while are present. In coastal California, for example, 60% of species showed it to some extent, mostly on narrow sandy beaches and saltmarshes, rather than on more extensive mudflats.

Occasionally, wading birds will combine together to feed in a co-ordinated group. This is most clearly developed in pelicans which feed in lines, arcs or circles; all the birds simultaneously open their pouches underwater thereby herding and capturing fish in shoals more efficiently. Avocets, Greenshanks and Spotted Redshanks may co-ordinate their movements through the water in a similar way, but this is much less obvious and may arise as much from the need to avoid previously disturbed areas as to gain any advantage from herding invertebrate prey.

**Fig. 56.** Ringed Plover feeding territories occupied at high tide (MHWS = Mean High Water Springs) and low tide (MLWS = Mean Low Water Springs) at Howbeg Rocks on South Uist in the Outer Hebrides. Redrawn from Wells (1991).

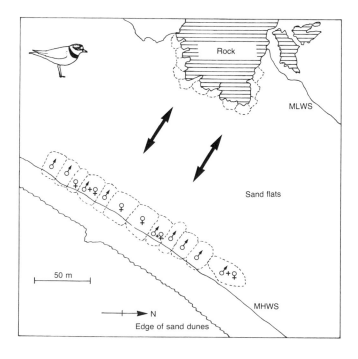

Given that it is frequently disadvantageous for birds to feed close to one another, mainly because of prey depression, why do shorebirds so frequently associate together in flocks? One advantage may be the greater efficiency with which patchily distributed food may be located. This has been demonstrated for small groups of Great Tits in the laboratory (Krebs, MacRoberts & Cullen, 1972), but it remains to be seen whether it is the case in the generally much larger flocks of shorebirds. One problem is that we know very little about the degree and scale of invertebrate patchiness on mudflats.

One of the most tangible benefits of flocking is greater security from predation (Fig. 57). Larger flocks tend to be subjected to a lower rate of predation either because they see predators earlier and can therefore take more effective evasive action, or because it is more difficult or dangerous for a predator to attack a large flock without risking injury (Bertram, 1978; Hale, 1980). Earlier predator detection cannot be the sole benefit, since flocking may be initiated after a predator has been spotted (Myers, 1984). The presence of a Sparrowhawk causes Redshanks to

**Fig. 57.** Flock of Sanderlings landing. There are several benefits that may be gained from group living, including greater efficiency of feeding and protection from predators. Harold E. Grenfell.

desert their feeding territories in favour of flocking for 15–60 minutes and causes Turnstones to bunch together more tightly – at a spacing of about one bird length, as opposed to the more normal three bird lengths (Whitfield, 1988).

In the presence of aerial predators, many waders adopt communal aerial manoeuvres to evade capture, including flashing (alternately exposing the darker upperparts and light underparts of the body), rippling (waves of flashing passing gradually through the flock instead of through the whole flock virtually simultaneously), and towering (vertical rotating) flight. Merlins attempting to catch Dunlins were successful, i.e. achieved a kill, on 9% of attacks on rippling flocks, 15% on flashing flocks and 29% on towering flocks. Not surprisingly, Dunlin flocks towered much less frequently when being attacked by Merlins (Buchanan *et al.*, 1988). Flocks often split up when a predator stoops and this provides an opportunity for the hunter to single out one bird and separate it from the rest.

Solitary Dunlins avoid further stoops by lateral dodges or by landing. Flocks may 'retrieve' individual birds from the ground or from the water by flying over them and thus giving them the opportunity to rejoin (Boyce, 1985). The aerial manoeuvres occur so quickly in such flocks that birds cannot simply be following the movements of their immediate neighbours, but instead must visually monitor the movements of individuals some distance away in order to prepare themselves to move at the appropriate moment. This has been termed the 'chorus-line hypothesis' of co-ordination because of its similarity to the methods used by human dance groups to achieve co-ordinated movement.

Flocking also enables more time to be spent feeding because the task of keeping a lookout for predators can be shared. This has been demonstrated in Bar-tailed Godwits feeding on the shore (Smith & Evans, 1973) and in Curlews feeding on inland fields. The latter spend less than 5% of their time with their heads up in an alert, scanning position when in flocks of more than ten. When on their own, they spend over 20% of the time with their heads up (Abramson, 1979). As a consequence, the rate of food intake is higher in larger flocks, provided the density does not become so high that mutual interference through prey depression begins to occur. The density at which most wader flocks feed presumably represents the optimum balance between these two factors – increased feeding time through shared vigilance and reduced feeding efficiency through mutual interference (Goss-Custard, 1970).

Turnstones and Purple Sandpipers feeding on rocky shores share vigilance with one another. Both species have well camouflaged, dark upperparts, and by remaining still (and in the case of Purple Sandpipers hiding behind rocks) they are able to remain undetected on rocky shores for far longer than on open mudflats. Turnstones also gain anti-predator vigilance benefits when foraging with Oystercatchers and Redshanks, whilst avoiding the competition and aggression that often occurs when feeding close to other Turnstones. However, when other species are not available they are forced to flock closer together. Purple Sandpipers do not gain the same vigilance benefits from these larger species (Metcalfe, 1989), and indeed they show a slight tendency to avoid gulls (perhaps because of the risk of food robbery). However, both species increase the time they spend scanning for danger, from less than 10% to more than 25%, as their view becomes more restricted by rocks and other obstacles, suggesting that there is a significantly higher risk of predation in the latter type of habitat (Metcalfe, 1984).

Flocking may also bring advantages by allowing individuals to copy the feeding techniques of more successful neighbours, or because some individuals benefit from the prey that may be flushed by others. The opportunity for efficient team work certainly exists in waders. Turnstones, for example, tend to remain in the same small foraging range each winter and to associate with the same small group of 50 or so other individuals whose behaviour they must get to know quite well (Metcalfe, 1986).

Waders feeding on muddy and sandy shores generally choose roost sites very carefully. If the tide is low when they stop feeding, they usually remain as close to the feeding area as possible, but in much denser groups than when feeding. At high tide, they are forced to find other sites, and prefer areas that combine good all-round visibility with freedom from disturbance and shelter from the wind. There may be a conflict between these qualities, for example the most sheltered sites usually have poor visibility, so the choice of roost site usually represents a compromise. Ringed Plovers prefer to roost on beaches where the pebbles are of a certain size, so that they can see over them but most of their body remains sheltered. For larger species, such as Curlews, sites with taller obstacles such as saltmarsh vegetation are preferred. When Curlews are present, smaller species may be able to roost in long vegetation too, taking advantage of the better shelter and relying on the Curlews to act as sentinels and react when danger threatens. Roosting waders usually emit a continuous cacophony of sound, perhaps to attract other individuals into the roost. This may in turn be advantageous for the callers because it increases the number of pairs of eyes on the lookout for danger. It could also increase the amount of shelter for the birds at the centre of the roost and reduce their risk of predation because late joining birds, settling at the edges of the roost, are likely to be in the most vulnerable positions. Roost calling usually stops at once if danger threatens.

**Fig. 58.** Roosting flock of Redshanks and Oystercatchers. Some individuals (mostly on the outside of the flock) are alert, others are standing on one leg. Some are asleep with their bills tucked in, and at least one is peeking and one

Roosting birds adopt postures that minimise heat loss. They face into the wind in order to avoid disruption to the feathers, shorten the neck, tuck the bill under the feathers and often stand on one leg (Fig. 58). All these behaviours result in significant reductions in heat loss to the environment, and therefore reduce the amount of food that the birds have to consume during the rest of the day. If danger threatens, birds adopt an alert posture with neck extended and the carpal joints of the

(Fig. 58—cont.)
is preening. Most, but not all,
are facing into the wind.
Several of these postures
minimise heat loss and thus
save energy. Chris Rose.

wings held away from the body (normally these are tucked into the body feathers) ready for take off. Birds roost close enough together to gain some wind shelter from one another, but not so close that they impede each other's wing movements if they have to take off quickly. Under exceptionally cold conditions, however, they may choose to risk difficulty in taking off, and huddle together much closer than normal to gain additional shelter.

Sleeping may represent a further adaptation to reduce energy expenditure. For example, Glaucous-winged Gulls spend 54% of the day inactive. True sleep (33% of the day), as opposed to resting wakefulness (21% of the day), has a lower energy expenditure and results in an overall saving of about 15% in the daily energy budget (Opp & Ball, 1985; McFarland, 1989). The fact that these gulls do not spend all of their resting time asleep suggests that other factors, such as readiness to respond to predators and other sources of disturbance are also important. In roosting waders, the flock is more difficult to see if all the individuals in it are completely still because they are asleep. Thus sleep may decrease the likelihood of predation as well as reducing energy expenditure. This stillness is in marked contrast to the situation after arrival at the roost, or before departure, when the birds generally indulge in bouts of preening and are much more conspicuous.

Wakefulness is not necessary when keeping a look-out for predators since 'peeking', i.e. opening one or both eyes occurs whilst birds show the brain waves characteristic of true sleep. Birds with their bills tucked in, usually tuck them in on the side of the body facing the greatest danger, e.g. the outside of the flock (Cutts, 1985), and the whole body may be periodically rotated through a few degrees in order to widen the visual field of the single peeking eye. Herring Gulls and Mallards peek when roosting, at a rate of about ten times per minute, though this rises to as much as 25 peeks per minute in circumstances of greater danger (Amlaner & McFarland, 1981; Lendrem, 1983). Drakes are also more vigilant than ducks. This may be because they are less well camouflaged. The fact that their peeking rates drop close to female levels when they are in their dull, eclipse plumage tends to support this (Lendrem, 1983).

Many species of waders prefer to roost in shallow water at night. This must increase heat loss through the legs, but means that terrestrial predators can be heard approaching. Wildfowl and gulls form floating roosts on deeper water, and some waders such as Avocets and Redshanks may do this too if subjected to extreme disturbance. Of the various types of disturbance that may affect waders at roosts, that by humans is often the most frequent (e.g. Furness, 1973b; see also p. 266), though Carrion Crows may be a more frequent source of disturbance for the smaller species. Crows appear to test the flock for the presence of sick or injured birds. Fit waders do not appear unduly disturbed by this, merely keeping a few metres between themselves and the crows. The presence of an aerial predator such as a Peregrine causes much more severe disruption, with birds flying around for a considerable period and often deserting the roost altogether.

Communal roosts provide an opportunity for birds to follow others to new feeding areas and this explanation of their function has been termed the 'information centre' hypothesis (Ward & Zahavi, 1973). However, flocks at feeding areas are just as likely to be centres of information about new roosting sites, as roosting sites are about feeding areas. In other words, all aggregations of birds may be information centres, and the opportunity to learn by following and copying may be an important driving force behind flocking in all species. This is likely to be particularly true for young birds whose experience of the best feeding and roosting areas under different environmental conditions is limited.

### Plumage of wading birds

On soft sedimentary shores, camouflage is much more difficult to achieve than on rocky shores. Generally, waders that feed in such areas have grey backs which blend in moderately well with a wide variety of such backgrounds. However, the birds do not allow the close approach (for example by humans) which is permitted by the much darker backed and better hidden waders such as Turnstones, Purple Sandpipers and Oystercatchers on rocky shores.

White underparts actually appear much darker under normal lighting conditions because they are in the shadow of the rest of the body. This phenomenon of underparts being lighter than upperparts is widespread in the animal kingdom and is known as countershading. Most waders exhibit countershading, though the undersides may also be white in order to make them as difficult as possible to see from beneath, i.e. by potential prey animals.

In open habitats, with a uniform background, plumage coloration is only of limited value in achieving camouflage since the body stands out as a silhouette under most lighting conditions. There is nothing that can be done to avoid this, apart from responding early to potential sources of danger, and this is why waders on muddy and sandy shores are generally rather shy and flighty. Some species, such as Oystercatchers and Ringed Plovers, have bold and conspicuous markings which help to hide the outline or silhouette of the body against certain backgrounds, notably rocks and stones. These patterns are known as disruptive camouflage. In Ringed Plovers, the neck bands serve to make the head and the body appear as two separate, round (and therefore not bird-like) objects. Another feature, which would often reveal the presence of a hiding bird, is the characteristic shape of the eye. Hence the eye itself is often hidden by disruptive dark bands or blotches, as in all the *Charadrius* plovers and in the many sandpipers which have a darkish streak through the eye and lighter bands above it (the latter also help to disrupt the head shape).

On mudflats and sandflats, Ringed Plovers are less conspicuous than many other waders from behind, because of their grey-brown backs, though they are more visible from the front. Oystercatchers are extremely conspicuous on mudflats, as are stilts and Avocets. In the former species, this might be because they are primarily adapted to feed on rocky shores where they are quite well camouflaged. Alternatively, it might be that one or more of these species are actually distasteful to predators, and thus be exhibiting warning coloration. It is certainly true that the shorebirds most regularly consumed by humans, and classified in Britain and Europe as game species (Table 18, p. 232), e.g. Golden Plover and Snipe, are very cryptically coloured, whereas both Oystercatchers and Shelducks are said by some to be unpalatable (e.g. Cott, 1947).

Snipe are amongst the best camouflaged of all waders. The complex, dark brown and yellow blotches and stripes which cover their upperparts blend perfectly with the vegetation of the areas these species normally frequent (Fig. 61, p. 175). They are consequently the most approachable of all waders in these habitats, Jack Snipe almost allowing themselves to be trodden on before taking flight. Careful observation of the latter species has shown that it does not freeze completely when approached, but turns its back so as to present the best camouflage and least helpful silhouette (Pedersen, 1988). Not surprisingly, such species are rather difficult to census adequately!

In many species, quite a substantial change in plumage coloration takes place in the breeding season, associated with the major change in habitat which also usually occurs. Many of our wintering sandpipers breed on arctic and subarctic tundra, and the need for camouflage whilst incubating eggs is a major factor determining the coloration of their upperparts. Thus Dunlins have a dorsal plumage that blends in well with the grasses and sedges amongst which they nest. The conspicuous black belly patches, which are strongly displayed during territorial song flights, are invisible when they are incubating. Knots and several other species have apparently conspicuous salmon-pink breeding plumage, but this is actually very difficult to spot amongst the mixture of dif-ferently coloured rocks, stones and sparse vegetation in which they nest.

Some parts of the plumage do not show any variation between winter and summer and these include the wing bars and rump patches which many species possess. These are normally hidden when the bird is on the ground, but they are very conspicuous in flight. The range of different markings is quite wide, though several species often share the same patterns. It is usually assumed that the main function of these marks is to facilitate species recognition, so that individuals can more easily join flocks of their own kind. However, there is no evidence that species sharing similar marks are more likely to form mixed flocks than those which differ. For example, Dunlins, Curlew Sandpipers and Ringed Plovers are quite often seen in mixed flocks but have different rump patterns, whereas Dunlins, Little Stints, Purple Sandpipers and Ruffs share similar rump and wing patterns and yet are not seen together all that often. On the other hand, the preferred habitats of the latter group differ more than the former, so they do not usually occur together in the same areas anyway. It has been suggested that species preferring similar feeding sites tend to have similar rump patterns and, since rumps are often displayed during aggressive encounters over food, that they form part of an interspecific communication system designed to lessen fighting between these species. This in turn reduces the likelihood of attracting predators and increases the time available for feeding (Stawarczyk, 1984). There does appear to be less aggression between species with similar rump patterns, but more work is required to establish that this is the reason why these patterns evolved.

It is possible that wing bars fulfil a much more important function during flight, namely helping to co-ordinate birds during the aerial manoeuvres they undergo when evading predators (as described on p. 158). Wing bars serve to make the wings more conspicuous so that changes of direction within the flock can be monitored more easily by other birds and appropriate actions taken to time movements and avoid collisions. Wing bars are not needed on the undersides of the wings, since body and wings can be seen clearly in outline against the light sky from below. This idea seems to be broadly supported by the fact that species without wing bars tend not to exhibit such spectacular flock movements, either because of their large body size (e.g. Curlew) or because of their more solitary habits (e.g. Greenshank, Green Sandpiper).

Many of the snipes and woodcocks, despite their well-camouflaged plumage, usually have prominent display marks in the form of white or pink spots or patches hidden on the upper- or underside of the tail. These can be exposed during ground displays concerned with the defence of food or territories by cocking the tail. Thus Snipe cock the tail to expose the white tips of the tail feathers, whilst Woodcock arch the fanned tail over their backs to expose silvery white, light reflecting feather tips on the underside.

### Body size, weight cycle and migration

Most coastal birds have a 'normal' sexual dimorphism in which the male is larger than the female. This includes the divers, grebes, tubenoses, gannets, cormorants, wildfowl, gulls, terns and auks. Most of the waders, on the other hand, have a reversed dimorphism in which the female is slightly larger than the male. The reasons for this remain a puzzle. An exception is the Ruff, which is quite a lot bigger than the Reeve. In this case, the males indulge in a good deal of display and fighting at leks during the breeding season. Presumably therefore, large size is an advantage in this competition. It enables a male to gain a good site on the lek and hence plenty of females with which to copulate. Competition between males, or sexual selection by females, is usually considered to be the reason why males are larger than females in most birds and mammals.

Plovers on the whole show less sexual dimorphism than sandpipers, but the females are clearly the larger sex in species, such as the Dotterel, in which the female does most of the territorial defence. She is also the more brightly coloured of the sexes. This 'role reversal' occurs in several groups of waders, including jacanas and phalaropes. Even in species where the males defend the territories there is a general tendency for them to play the major role in caring for the young. While the parents take an approximately equal share of incubation, the male usually remains with the young for longer than the female. It has also been suggested that females might prefer those males with the most acrobatic display flights and that small size is an advantage in this (Jehl & Murray, 1986). In the case of bill length, there has been a good deal of argument about whether the most important selective forces operate on the breeding or wintering grounds. Hale (1980) argues that the critical period for both bill and tarsus length is towards the end of the breeding season when food supplies are dwindling. Dunlin pairs that consist of larger than average females and smaller than average males have been found to nest earlier and to have larger eggs than more evenly matched pairs, and so probably have a greater breeding success. However, the reasons for this are not known (Jönsson, 1987). Large female Redshanks nesting in Iceland seem to be at an advantage in being able to breed earlier (Summers & Underhill, 1991), though there must of course be counter-selection at some stage, for if this were the whole story, Redshanks would be steadily increasing in size!

Regardless of when the critical selection occurs, there is no doubt that the size dimorphism influences the behaviour of waders in winter. In some cases, there is a latitudinal difference in wintering range, e.g. Ruffs winter further north (mostly in Europe) than Reeves (mostly in North Africa). Consistent differences have been found in the sex ratios of Dunlins wintering at sites only a few kilometres apart in the Severn Estuary. Males are common in areas of liquid mud, whilst females are commoner in areas of firm mud where most large ragworms are out of reach of the males' bills. There are also disproportionately more juvenile birds at one sandy site where they continue to feed for much longer than the adults as the tide comes in (Clark, 1983).

During the course of the year, waders undergo a number of changes in body weight similar to those which occur in a wide range of temperate birds, including many wildfowl and passerines. This involves a peak in the middle of winter and a further peak before migration in the spring. By contrast, species that spend the year in relatively uniform climatic and feeding conditions, such as in the tropics, tend to be rather constant in weight. Most of the extra weight that birds in temperate regions carry in winter and spring is fat, but because of the need to carry the extra load, the leg and flight muscles are also larger.

Throughout the breeding season, most species remain at relatively low weights. This is partly because food supplies are plentiful and there is no need to carry any reserves, and partly because it is more efficient to avoid carrying unnecessary weight. Being lighter means that birds expend less energy in carrying out their daily activities and can therefore spend less time feeding themselves and more time feeding their young and performing other activities associated with breeding. One other important factor that helps to keep body weight low is the risk of predation. Heavy individuals are likely to accelerate more slowly and be less manoeuvrable in flight than light ones and may therefore be relatively easier for an aerial predator to catch. Since it probably takes heavier birds longer to get airborne, they may also be more vulnerable to terrestrial predators. In a flock of waders being chased by a Peregrine, it is not unreasonable to expect that heavier birds lag behind and thus become more vulnerable (Dick & Pienkowski, 1979; Lima, 1986). The available evidence on the weight of prey taken by predators, both avian (Kenward, 1978) and mammalian (Temple, 1987), does not appear to support this idea since the average prey weight is less than the average weight of birds in the population as a whole. However, the average prey weight is biased by the fact that Kenward's Goshawks, for example, took many sick, weak and solitary Woodpigeons which were light in weight and easy to capture. They also took some individuals that had a higher than average body weight (inferred from an index of their body condition).

When the risk of starvation is higher, it is obviously advantageous for greater fat reserves to be carried, either to enable the bird to survive whilst waiting for conditions to improve, or to enable it to migrate to a more favourable feeding area. Oystercatchers in the Netherlands adopt the former strategy, remaining at the roost all day if the mudflats freeze or if high winds prevent the tide from falling (Swennen, & Duiven, 1983; Swennen, 1984). Some birds actually starve to death at the roost if the mudflats remain frozen for several days. Overall, the best strategy for Oystercatchers in the Wadden Sea when weather conditions worsen seems to be to stay put as long as possible, but to leave while they still have sufficient reserves to fly elsewhere, such as to France. Lapwings and Golden Plovers have gone one stage further. Due to their reliance on earthworm activity near the surface (which may be reduced by ground frosts), they build up large enough fat reserves in winter to enable them to undertake movements in search of frost free ground. For example, they may desert Britain completely when the ground is snow covered and move west to Ireland or south to France, Iberia and even North Africa (Elkins, 1983). Estuarine Grey Plovers do not need to be as mobile as this, but individuals which probably spend the whole of mild winters in the Wadden Sea or Wash will none the less move west or even north to such sites as Teesmouth if the local mudflats freeze (Townshend, 1982). The winter peak fat reserves of plovers (22–25% of body weight) are greater than those of sandpipers (12–15%) (Davidson, 1981). This enables them to perform hard weather movements, but they also avoid areas of more extreme climate by not wintering as far north as sandpipers (Pienkowski, 1979).

The typical winter weight cycle can thus be understood as a trade-off between the risk of starvation and the risk of predation, though many aspects of this explanation remain to be proven (Rogers, 1987). During breeding, and during the annual moult when the wing area is reduced by the gradual replacement of the flight feathers, the weight is low. It then increases throughout the winter as the risk of starvation increases. Predators ought to find it easier to catch these heavier birds (assuming that they themselves do not have to carry increased reserves), but none the less this increased risk of predation must on average be less than the risk of starvation. In spring, feeding conditions improve and thus body weight can drop considerably prior to the period at which large reserves have to be accumulated ready for migration to the breeding grounds.

The cycle differs from species to species. For example, Oystercatchers tend to rise in weight throughout the winter with a late spring reduction and no premigratory increase. This is probably because their food supplies continue to diminish in late winter and because they breed not far from the wintering areas. Many juvenile birds of all species do not show the premigratory peak, because they do not breed in their first year. They also tend to have slightly larger fat reserves than adults in midwinter. This is despite the fact that their body weights are often lower because they are not fully grown. The extra reserves are needed to compensate for the lack of foraging skill which makes them less able to cope during poor feeding conditions. Both adult and juvenile waders have been captured following bad weather conditions at low weights, indicating that they had used their body reserves. This was the case for Grey Plovers at Teesmouth in January 1979 (Dugan *et al.*, 1981) and for Redshanks in eastern Britain even in some mild winters (Davidson, 1982).

Waders wintering in tropical areas, such as West Africa, do not exhibit this type of winter weight cycle (Dick & Pienkowski, 1979; Puttick, 1980). This is not because food supplies are more abundant (in some cases the productivity of the shore is less than in temperate areas), but rather that they are more predictable. Birds may actually consume a very high proportion of what is available (Piersma & Engelmoer, 1982; Wolff & Smit, 1990). This means that although waders have to spend a large proportion of the day feeding, there is little risk of the food suddenly becoming unavailable through cold weather or other adverse conditions. One disadvantage for birds wintering along tropical coasts is that they have to cope with heat stress and salt loading, which results from feeding all day at high environmental temperatures with only salt water to drink. Thus they have well-developed salt glands to allow them to drink sea water, but heat stress remains a problem (Klaassen, 1990; Klaassen & Ens, 1990).

Shorebirds that undergo long migratory flights naturally enough put on more fat before departing than do those which fly only a short distance or which travel a long distance in a number of steps with feeding areas along the route. Fat is by far the most cost-effective fuel for such journeys since it yields about 39 kJ (J = Joule, 1 kJ ≈ 4 kilocalories) of energy per gram, compared with about 18 kJ for a carbohydrate such as glucose or a protein. However, as mentioned earlier, increases in muscle sizes, particularly those concerned with flight, are necessary to carry the extra weight. As with other energy sources, when fat is metabolised to yield energy during migration, water is produced as a by-product of the chemical reactions. In fact, each gram of fat yields slightly more than one gram of water. This seeming anomaly is explained by the fact that most of the weight of the water comes from the oxygen that is breathed in and combined with the hydrogen in the fat. This water is of great value since it reduces the need to drink and enables long flights to be undertaken without a break. Considerable heat is generated by the pectoral muscles during flight, so by migrating at altitude birds are able to take advantage of the lower air temperatures found there to keep cool. None the less, considerable quantities of water are still lost from the lungs and air sacs by evaporation.

Some waders have no choice but to undertake migrations by means of long flights. For example, Bristle-thighed Curlews wintering on the Polynesian Islands in the Pacific and breeding in western Alaska undertake a minimum journey of 2700 km (1700 miles) between Alaska and the one staging area that is available – Hawaii. In fact they probably undertake direct flights of 3700 km (2300 miles) (Johnston & McFarlane, 1967).

Migratory flight range can be considerably lengthened by undertaking journeys with a following wind and by selecting the height with the maximum favourable wind speed (Alerstam, 1981). However, winds can change, and although birds seem to be excellent weather forecasters, they can finish up in the wrong places, as when North American waders heading south sometimes arrive in Europe instead of South America. Some idea of the flight ranges of spring migrant waders passing through Britain can be calculated from their weights just prior to departure (Table 12). These estimates are broadly consistent with what we know about their destinations and routes. Although some Turnstones heading for Greenland and Canada are known to stop off in Iceland to replenish their reserves, those in Table 12 have adequate reserves to make such a journey in a single flight. Most Ringed Plovers make a similar journey without stopping in Iceland. Ringed Plovers breeding in Iceland pass through Britain at much lower weights earlier in the spring.

Some birds, e.g. Whimbrels, may arrive on the breeding grounds with reserves of energy, and of protein, in the pectoral muscles (Table 12). There may be a general tendency to overload with fat during the spring migration because of the increasing uncertainty of feeding conditions further north (Gudmundsson, Lindström & Alerstam, 1991). The reserves can then be used if feeding conditions are unfavourable when the birds arrive, as they often are in the Arctic (Green, Greenwood & Lloyd, 1977). Such reserves might also make a contribution to the formation of the eggs. In many species of high-Arctic geese, such as Brent and Barnacle Geese, virtually the whole of the clutch has to be formed from bodily reserves since the birds start nesting early, when the main feeding areas are still covered by snow (see p. 187). To nest later would leave insufficient time for their young to grow big enough to fly by the end of the season (Owen, 1980a). This is why the body condition of geese in late winter, when they are laying down reserves, is one of the factors affecting subsequent breeding success (Boer & Drent, 1989).

One of the most carefully studied shorebirds, from the point of view of its migration strategy, is the Knot (Fig. 59). This is because it tends to

Table 12. *Departure weights and estimated flight ranges of some spring migrant shorebirds passing through the Severn Estuary. The departure weights were calculated as the average of the heaviest 25% of the birds, and the flight ranges were calculated using Davidson's (1984) formula*

| Species and subspecies or sex | Fat free weight (g) | Take-off weight (g) | Amount of fat (g) | Flight speed[a] (km/hour) | Flight range (km) | Presumed breeding areas and distance to nearest by shortest great circle route (km) | |
|---|---|---|---|---|---|---|---|
| Ringed Plover | 46 | 72 | 26 | 51 | 2200 | Iceland, Greenland, Canada | 1600 |
| Dunlin | | | | | | | |
| *Calidris alpina alpina* | 48 | 81 | 33 | 51 | 2600 | Central Siberia | 3500 |
| *C. a. schinzii* | 42 | 61 | 19 | 51 | 1700 | Iceland, Greenland | 1600 |
| *C. a. arctica* | 38 | 58 | 20 | 51 | 1900 | Greenland | 2300 |
| Whimbrel – female[b] | 440 | 520 | 80 | 72 | 2200 | Iceland | 1600 |
| Turnstone | 95 | 153 | 58 | 55 | 3200 | Greenland, Canada | 2300 |

[a] From Noer (1979), Johnson & Morton (1976).
[b] An approximate fat free weight for females (which should constitute most of the upper quartile of the weight distribution) has been used.

**Fig. 59.** Wintering areas (dashes), breeding areas (filled), spring migration routes (arrows) and staging areas (squares and circles) of Palearctic wintering populations of Knots. Siberian breeders (continuous arrows, filled squares) are distinguished from Nearctic breeders (dashed arrows, open circles). Numbers refer to the estimated sizes of the major wintering concentrations (Siberian breeders in roman type and Nearctic in italics).

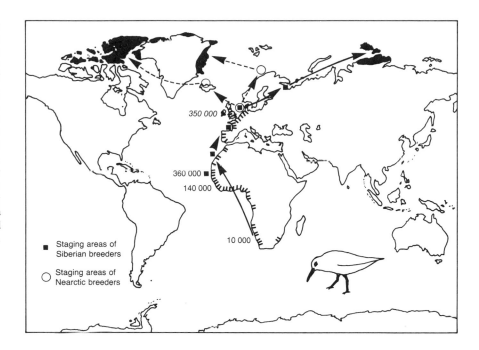

winter and stage in large numbers at a few sites. Knots wintering in southern Africa and breeding in Siberia could make the journey in many small steps, but in fact they do so in a small number of large ones. The first stage, of about 6900 km (4300 miles) is probably to Mauritania, for which they put on about 60 g weight at a rate of about 3 g/day. While the next stage to the Wadden Sea is only half as far at 3300 km (2000 miles), they apparently find it difficult to put on sufficient fat in Mauritania and gain weight at only 0.5 g/day. As a result, they may rely on favourable winds to make the journey, as do Bar-tailed Godwits (Piersma & Jukema, 1990), and if these fail they may fall short and stop off at sites such as the Vendée on the west coast of France. The final leg of 4300 km (2700 miles) is usually made in a single step, although some birds may stop again in the White Sea. It used to be thought that northern Norway was a staging post for these birds, but it now appears that this site is used by European winterers heading for Greenland and Canada (Uttley *et al.*, 1987) (Fig. 59). There are still a number of unresolved questions about Knot migration, such as whether or not there is a further, or alternative, staging area between South Africa and Mauritania, perhaps in Guinea Bissau (Roselaar, 1983; Dick, Piersma & Prokosch, 1987; Smit & Piersma, 1989).

# 5

## The coastal fringe

There is a range of habitats, lying just in or above the intertidal zone, which provides important feeding and nesting sites for coastal birds. These include some of the last remaining areas of coastal wilderness in Britain – saltmarshes, tidal lagoons, sand dunes, machair, grassland and scrub. Situated as they are at the very edge of the land, these habitats are particularly vulnerable to reclamation and development. This is partly because they are seen as places with no commercial value and partly because their intrinsic beauty and importance to wildlife are not as well appreciated as they should be.

In the late 1960s there were estimated to be about 448 km$^2$ of saltmarsh in Britain and the current estimate is 444 km$^2$ (Burd, 1989). The closeness of these two figures does not mean that there has been little change over the last 30 years, but rather that the reclamation of saltmarshes has so far been balanced by the spread of new marsh down the shore, as has occurred on the Wash (Hill, 1988). The largest area of saltmarsh is in East Anglia (about a third of the British total) and in the estuaries around the Irish Sea (about a quarter) (Fig. 60). Brackish water lagoons are no longer very widespread in Britain, the only ones of appreciable size occurring at the Fleet in Dorset (see p. 25), Slapton Ley in Devon and Loe Pool in Cornwall. There are also a number of oyces, the local name for lagoons enclosed by shingle ridges, in Orkney and Shetland. About 7% (500 km) of the total coastline of England and Wales is backed by sand dunes (see Fig. 68, p. 196), which in turn stretch inland to occupy a considerable area. The area of dunes in Great Britain as a whole is about 560 km$^2$, of which over 70% is in Scotland (Doody, 1985). Sand dunes are less frequently reclaimed than saltmarshes, but they are subject to intense recreational pressures. Machair is a habitat limited to westerly facing coasts with an abundant offshore supply of shell sand. It occurs in the Western Isles of Scotland and Ireland with the great majority being in the Outer Hebrides. Coastal grassland (of which machair is a special type) and scrub are both widespread habitats and usually grade imperceptibly into completely terrestrial habitats. We shall only consider areas of grassland and scrub that are influenced by salt spray, which occur in areas reclaimed from the sea, or which are used for breeding by significant numbers of coastal birds.

The main groups of birds that occur in the coastal fringe are those which we have encountered already. Some wildfowl, particularly the geese, are herbivores which specialise in feeding on the plants of the coastal fringe. With one or two exceptions, such as the Wigeon and Teal, most of the ducks are associated with more distinctly marine or freshwater habitats. Gulls, terns and skuas frequently exploit fringe habitats for nesting, as do a number of other coastal birds, at least when such habitats are relatively free from disturbance by humans and other predators.

**Fig. 60.** Distribution of saltmarshes in Great Britain. Some of the marshes referred to in the text are indicated. Redrawn from Burd (1989).

## Saltmarshes and sea-grass meadows in winter

Saltmarshes are regularly exploited as feeding areas by few waders, but they do provide a major roosting habitat for the hundreds of thousands of birds that feed on the mud and sand flats further down the shore. As roosting sites they offer two main advantages – they are close to the feeding areas, thus minimising the energy expenditure required to reach the roost, and they are relatively free from disturbance. Small pools and patches of mud within the marshes also allow some feeding to continue for those individuals which may have had difficulty in finding enough food at low tide.

Two species of waders that feed regularly in saltmarshes are the Redshank and Snipe. About 15% of the Redshank population at Lindisfarne remains in the marshes throughout the tidal cycle (Millard & Evans, 1984). In most estuaries, the whole of the Snipe population remains hidden in saltmarshes (Fig. 61) and the only reliable way of assessing their numbers is to walk through the marshes and flush them. Since they only flush at a short distance this is hard work, and extremely time consuming, so the species is very much under-recorded by the Birds of Estuaries Enquiry. The numbers in the relatively small saltmarshes of the Taff Estuary (total area *c.* 23 ha, dominated by *Spartina*) were assessed in the winter of 1983/84 by walking a series of transects about 80 m apart every few days (Kalejta, 1984). Even at this spacing, it is likely that not all individuals were found, especially in cold weather when the flushing distance was reduced. The counts reached a maximum in midwinter. Larger numbers of Snipe were also present during cold weather when there were heavy frosts inland but the saltmarsh remained unfrozen. These increases were probably due to a coastal movement of birds which normally fed inland. When the saltmarsh itself froze (this did not occur in 1983/84), Snipe left the area completely, perhaps flying to frost-free areas in Ireland or France.

The average Snipe population of the Taff saltmarshes in midwinter was about 25 individuals and if a similar density of birds occurred on saltmarshes throughout Britain, this suggests a population of about 40 000. Given the large numbers that also winter inland, this would go some way towards explaining why the annual shooting bag of Snipe is about 85 000 and yet so few birds are actually counted by bird watchers!

In recent years, *Spartina anglica* has been spread so widely by man, that nearly all British saltmarshes are partly or wholly dominated by this species. It has been blamed for the recent decline in Dunlin numbers (p. 133) and there is no doubt that it has invaded important bird feeding areas at the top of the shore. For example, on the Dovey Estuary in mid Wales, although the cover of *Spartina* has only increased by about 10% in

15 years, there has been a dramatic decline in the numbers of Oystercatchers, Ringed Plovers and Dunlins because the area colonised was previously used by these birds for feeding when the rest of the intertidal mudflats were covered by the tide (Davis & Moss, 1984). In Langstone Harbour, which was one of the first sites to be colonised by *Spartina anglica*, a spontaneous dieback has occurred for unknown reasons, though waterlogging and lack of oxygen in the root layer are suggested causes (Tubbs, 1984). Mud in the dead areas has slumped, been eroded by wave action, or simply been colonised by algal mats or *Zostera*. Invertebrate densities there are generally lower than on open

**Fig. 61.** Group of Snipe in a saltmarsh pool. The generally skulking habits of this species mean that it often goes unnoticed. Chris Rose.

mudflats, but they may nevertheless be exploited by some foraging waders (Haynes, 1984). When herbicides were used to eradicate *Spartina* at Lindisfarne, the densities of feeding waders were actually higher than on open mudflats for the first year or two, but they dropped to lower levels after three or four years. This appears to have been due to initially high densities of *Corophium* and the presence of wet patches which increased prey activity (Evans, 1986).

Considerable numbers of Teal, Mallard, Wigeon and other dabbling ducks often feed at the water's edge in *Spartina* marshes. Seeds of salt-marsh plants, and the snail *Hydrobia*, formed the main diet of Teal, Mallard and Pintail shot on the Medway Estuary in Kent (Olney, 1965). Wigeon from the Medway consumed mainly the algae *Enteromorpha* and *Ulvae*, together with smaller numbers of seeds. Teal have the ability to sort out preferred seeds, such as those of glassworts, sea-spurreys and buttercups, from unwanted ones and inedible detritus. Other dabbling ducks can presumably do the same. Provided that the particles are of different sizes, and are suspended in water, large ones which are not wanted can be avoided by closing the bill sufficiently to keep them out. Small ones which do get in can be trapped by the bristles on the tongue (Fig. 62) and then ejected. The lateral bristles and spines on the tongue are suitable for straining out small items. The backward pointing spines at the base of the tongue help to propel material to the back of the mouth prior to swallowing. The diet of the Cinnamon Teal shown in Fig. 62 consists of 80% plant material, mainly seeds (Palmer, 1976*b*), which is quite similar to that of the Teal and the tongues of the two species are therefore quite similar.

Although cattle and sheep eat *Spartina*, the tough stems and leaves do not seem to be a preferred food of grazing wildfowl. For example, Owen (1973) found that the Wigeon in Bridgwater Bay in the Severn Estuary spent only 23% of their time in *Spartina* marshes despite the large area that these marshes occupied. Of a sample of 60 shot birds, only four contained substantial quantities of *Spartina* leaves or shoots and only one *Spartina* seeds. Instead, Wigeon showed a preference for feeding on inland pastures, or in freshwater habitats, spending 60% of their time in such areas. The birds flew about 15 km (10 miles) to reach suitable sites on the Somerset Levels, where their diet was dominated by the stems and leaves of the plants of damp grasslands. In the intertidal region the Wigeon took mainly algae (*Enteromorpha*), bents (*Agrostis*) and the Common Saltmarsh-grass. Wigeon need to drink regularly whilst grazing, in some cases as frequently as every 10 minutes or so, and this may influence their choice of feeding site. Geese have a similar requirement, but the interval between drinks is about 30 minutes (Loonen & van Eerden, 1989).

Wildfowl are generally much more selective about the grasses and herbs that they eat than are many grazing mammals because they are unable to digest cellulose. Cattle and horses can digest as much as 60% of typical grasses, but they require a very large gut to provide the right environment for the bacteria that carry out the cellulose digestion. As flying animals, few birds can afford to have such large and heavy guts weighing them down. Geese typically eat only those plants, and parts of plants, which have a high protein and carbohydrate content (Owen, 1976; Boudewijn, 1984). Plants with a flaccid growth habit are often preferred because they contain relatively small amounts of the indigestible stiffening tissues that hold most plants upright. Many aquatic grasses, from both brackish and fresh waters, are designed to float with minimal water resistance and so are ideal in this respect.

When feeding on green plants, both geese and ducks pass food through the gut very rapidly because so little of it is digested. The average time taken for grass to pass through the gut of a domestic goose

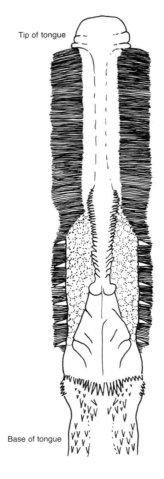

**Fig. 62.** The tongue of the Cinnamon Teal (a North American species) seen from above, based on a drawing by Gardner (1925).

Tip of tongue

Base of tongue

is about two hours (Mattocks, 1971) and similar figures have been suggested for both Brent and Barnacle Geese (Ranwell & Downing, 1959; Ebbinge, Canters & Drent, 1975). In all geese, the oesophagus acts rather like a crop and can store about 10% of the daily food requirement (more when feeding on grain), for later digestion in the safety of the roost (Owen, 1980a).

The preference for digestible, nutritious plants means that wildfowl are often attracted to crops, such as winter wheat, which have been selected by man for their high yield. This is because fewer of the plant's resources are channelled into devices that provide physical and chemical protection against grazing. Similarly, because they often take the underground parts of plants, such as clover stolons, it is not surprising that geese sometimes consume root crops e.g. Pink-footed Geese in Lancashire feeding on carrots (Greenhalgh, 1975). In other cases, mechanical harvesting of crops such as sugar beet results in more waste than harvesting by hand and thus provides a food source which they can exploit without doing any harm, e.g. Bean Geese in southern Sweden (Nilsson, 1989). Both geese and Wigeon now tend to winter in increasing numbers in agricultural areas. While the more widespread availability of suitable autumn sown cereals and root crops has undoubtedly played a part in this (Kear, 1990), it can also be attributed to two changes that have occurred in the intertidal area. One is the spread of *Spartina*, which has already been referred to, and the other is the decline of *Zostera*.

*Zostera* (eel-grass) is a genus of flowering plants which includes three species growing on muddy sand and gravel in the intertidal and subtidal zone. It forms grass-like stands, which are often referred to as sea-grass meadows. They are amongst the most productive shallow-water plants in the world (Barnes & Hughes, 1986). In the 1930s, much of the *Zostera* in European and eastern American estuaries disappeared, probably as a result of infection with a fungal parasite. This in turn caused a decline of about 75% in the Brent Goose population of Europe, which was heavily dependent on *Zostera* at this time (Cramp & Simmons, 1977). Eel-grass began to recover during the 1950s but was knocked back again by the freezing conditions of the 1962/63 winter. *Spartina* now grows in much of the area formerly occupied by *Zostera* and although the latter is once again thriving, it is unlikely ever to be as extensive as it once was. Fortunately, it is recolonising some areas where *Spartina* has regressed, such as Langstone Harbour (Tubbs, 1977). A further complication is an increase in the growth of algae (e.g. *Enteromorpha*) in the intertidal zone, as a direct consequence of nutrient enrichment from treated and untreated sewage (see p. 257). About a third of Langstone Harbour is now occupied by algal mats rather than *Zostera* (Tubbs & Tubbs, 1983;

Haynes, 1984), which at least provide more grazing for wildfowl than *Spartina*. The switch from coastal to inland feeding that has occurred this century has been largely prompted by the decline of *Zostera*. In the early years of the present century, the Wigeon was an almost exclusively estuarine species, but by the 1960s a considerable portion of the population wintered inland (Owen & Williams, 1976). Even in areas where the opportunities for grazing in the intertidal area are increasing again, such as the Hampshire and Sussex Harbours, permanent pastures inland remain an important feeding area for Brent Geese (Round, 1982).

In most cases, there is a seasonal sequence in the exploitation of various foods. For example, Brent Geese in Norfolk first consume *Zostera*, followed by *Enteromorpha*, and only when the supplies of these are exhausted do they switch to grazing the saltmarsh (Fig. 63). Here

**Fig. 63.** The percentage of Brent Geese between Wells and Blakeney in Norfolk feeding on different categories of plants in 1974/75. Tall saltmarsh vegetation was only consumed when all the shorter material had been grazed (the latter began to regrow in late February and early March). Redrawn from Charman & Macey, 1978.

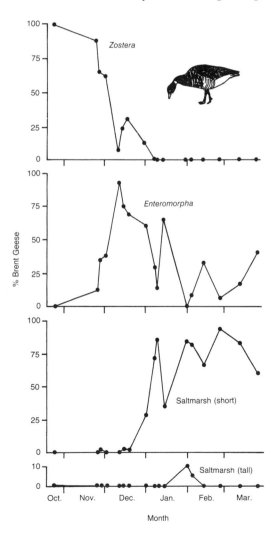

they take Common Saltmarsh-grass, Sea Aster and Sea Purslane (Ranwell & Downing, 1959). As with the Wigeon, *Spartina* is consumed only occasionally. Although *Zostera* is far more digestible than *Enteromorpha*, Charman (1979) found that saltmarsh plants were more digestible than either of them. Drent, Ebbinge & Weijand (1979), on the other hand, argue convincingly that this sequence of food choices (which also occurs in the Netherlands) is due to the relative profitability of the different plant species, based both on foraging effort and digestibility. They found digestibilities of 37% for *Zostera*, 34% for *Enteromorpha* and 31% for Common Saltmarsh-grass.

Despite the fact that inland pastures support three times as many grazing birds per unit area as do saltmarshes (White-Robinson, 1982), both Brent Geese and Wigeon seem to prefer to feed intertidally as long as food resources are available there (Williams & Forbes, 1980). This could be because such areas are relatively undisturbed. One difference between the two species is that Brent Geese can remain on the *Zostera* beds for longer because they are able to feed on the roots and rhizomes when all the leaves have been consumed (Madsen, 1988). They do this by paddling with their feet on the mud to loosen the plant material and then grubbing it up with their bills.

The time when many Brent Geese do move to inland pastures is late winter (St Joseph, 1979), corresponding to the tall saltmarsh vegetation stage in Fig. 63. This tendency has increased as the numbers of Brent Geese have risen and intertidal foods become exhausted more rapidly. There is a movement back to the saltmarshes in the early spring as new young growth becomes available. In the Netherlands, the geese graze the plants in this young sward in proportion to their abundance, with the single exception of Sea Plantain. The young shoots of this species are preferred because they are particularly digestible and nutritious. Once grazed down, they regrow after an interval of about a week, recycling some of the nutrients from the goose droppings in the process and thus the geese can return to the same spot to feed after a suitable interval. The same process occurs with stands of Common Saltmarsh-grass in Denmark (Madsen, 1989). The overall digestibility of the diet in the Netherlands reaches a peak first on pasture, but then declines as the grasses begin to produce flowering shoots (though their maturation is delayed by the intense grazing pressure; Ydenberg & Prins, 1981). The geese then switch to saltmarsh plants as they reach their peak digestibility, which occurs a little later than in pasture. These changes in digestibility are vitally important to the geese as they are trying to achieve the best possible body condition for migration and subsequent breeding. It is probably only Sea Plantain and Sea Arrowgrass that can provide an adequate intake to allow migration and breeding at all. The digestibility of plants on the lower saltmarsh rises to about 36% just before the geese leave.

The other species of goose that depends on saltmarshes for grazing at many of its wintering sites is the Barnacle Goose. In the Solway Firth, the diet consists of the seeds of Saltmarsh Rush and other plants (44% by weight), the green parts of Common Saltmarsh-grass, Red Fescue, Creeping Bent and other species (42%), and the stolons of White Clover (14%) (Owen & Kerbes, 1971). Again, *Spartina* is consumed only occasionally. On Islay, where most of the feeding takes place inland, the most important single species is Perennial Ryegrass (Owen, 1980a).

In 1969, a large section of the Wadden Sea known as the Lauwerszee, was enclosed by a dyke. During the first few years, this area developed an almost pure stand of glasswort (*Salicornia*). The seeds of this plant were extensively exploited in the autumn by Barnacle Geese and Wigeon which stripped them directly from the plants. Teal on the other hand, filtered the seeds from a thin layer of water after they had fallen off. As the polder became less saline over a period of about ten years, the grazing wildfowl gradually switched to feeding on the grasses that replaced the glasswort (van Eerden, 1984).

In other parts of the Netherlands, Barnacle Geese consume glasswort seed early in the autumn, and later switch to inland pastures where they spend about 80% of the daylight hours feeding. They defaecate once every $3\frac{1}{2}$ minutes on average and assimilate about 22% of the ryegrass, which is their main food (Ebbinge, Canters & Drent, 1975). Although most geese are diurnal feeders, some populations of Barnacle Geese are at least partly nocturnal. On Schiermonnikoog (one of the Frisian Islands) during the full moon period, most feeding takes place at night, and very little by day (Ydenburg, Prins & van Dijk, 1984). This may be because the island has no Foxes (which normally provide a threat at night), and the geese are able to avoid human disturbance by feeding at night (Owen, 1980a). However, Ydenburg and colleagues suggest that the geese might conceivably be feeding on invertebrates. Most geese do not normally consume much animal food, though Canada Geese take small quantities of earthworms and Brent Geese a few ragworms.

Geese are certainly capable of selecting plants with higher digestibilities or protein contents. For example, captive young Barnacle Geese show a preference for fescue and for fertilised over unfertilised ryegrass. These plants had a higher protein and water content than other grasses (Owen, Nugent & Davies, 1977). Barnacle Geese may also feed more on those parts of saltmarshes where gulls have bred because the grass has been fertilised by gull droppings (Bazely, Ewins & McCleery, 1991). Visual selection is obviously important both on a large scale (when selecting which area to feed in) and on a small scale (when selecting which individual plant to consume). Young goslings show a marked preference for the colour green and prefer bright green to pale green (Kear, 1964). Geese searching for feeding areas in spring choose darker and more uniformly green areas where plants usually have a higher protein content. They also recognise stubble fields in the autumn, since nearly all species feed to some extent on spilt grain at this time. Touch also plays a role in the selection of individual plants. Exerting just the right amount of grip with the bill tip will break off only the more fragile (and therefore digestible) leaves, leaving the tougher, more fibrous ones behind (Owen, 1980a). Taste has not yet been shown to be an important factor. For example, Barnacle Geese do not choose those saltmarsh plants with the lowest salt content, which might have been expected had they wished to minimise consuming too much of it (Owen, 1979).

Greylag Geese have much heavier, stronger bills than the other species of grazing wildfowl so far considered. Their bills are well suited to digging up the starch-rich roots of saltmarsh plants such as Sea Club-rush and *Spartina*. Usually, they loosen the underground parts of these plants first, by paddling at the surface with their feet. In a few places, Greylag and Pink-footed Geese still graze almost exclusively on saltmarshes, but both are now heavily dependent on agricultural land, especially stubble fields, root crops, pastures and winter cereals.

Disturbance is a major factor influencing where geese will feed. Their relatively indigestible diet has lead to the evolution of a large body size to accommodate a large gut, and their large flight and leg muscles make them a desirable source of food for man. Many of the best goose haunts in the country, such as the White-fronted Goose site on the banks of the Severn Estuary near Slimbridge and the Barnacle Goose site on the Solway saltmarshes, provide a combination of good feeding areas with relative freedom from disturbance. This does not necessarily mean total protection from shooting (though this would be ideal), since both of these sites are traditional goose shoots, but they are areas where shooting is well organised and restricted to a handful of occasions during the winter. The choice of feeding sites on a smaller scale is also strongly influenced by disturbance levels. For example, White-fronted Geese prefer to feed in large fields well away from roads and farm buildings. They also keep away from field edges where hedgerows provide cover for potential predators (Owen, 1972).

In addition to the waders and the wildfowl, saltmarshes support significant wintering populations of gulls, pigeons, Skylarks, pipits, wagtails, Starlings, finches, buntings, sparrows and birds of prey (Fuller, 1982). For example, the Wash saltmarshes, with an area of over 40 km², support populations of passerines, estimated during the 1985/86 and 1986/87 winters at roughly 30 000 Skylarks, 7000–17 000 Twites, 2000

Rock Pipits, 1000 Reed Buntings and 500 Lapland Buntings (Davies, 1988). The Lapland Buntings, which have relatively short wing lengths, corresponding with those of the Scandinavian population, occurred mainly at the top of the saltmarshes in the vicinity of the sea walls. The Twites were mainly concentrated in the glasswort and Sea Aster zones well down the marsh. The large difference in the population size of this species in the two winters (i.e. 7000 compared with 17 000) may have been due to a difference in the size of the glasswort seed crop, with birds moving to Essex or the continent when the crop was poor. The Twites have been shown, from many ringing recoveries in the area, to originate from the relatively isolated breeding population in the Pennines.

Skylarks are also the commonest wintering species in the small Taff saltmarshes, but, because of the greater dominance of *Spartina*, they occur at about half the density of that found on the Wash. The next most common species in the Taff are the Chaffinch, Starling and Meadow Pipit. The diet of both Skylarks and Chaffinches (as revealed by their droppings) consist mainly of the seeds of Annual Sea-blite in early winter. In late winter, Skylarks take mostly Sea Aster seeds and Chaffinches those of Spear-leaved Orache. Skylarks forage for seeds on the ground and thus have to wait for them to fall off the plant or to be dislodged by other species such as Goldfinches (Kalejta, 1984).

**Breeding of coastal geese**

Three species of geese breed in the British Isles – Greylag, Canada and Egyptian – but only the first of these is indigenous. Even in this species, the situation is complicated by the release of feral birds by wildfowlers, and only in the Scottish Highlands and Western Isles is the population completely wild. In the latter area, coastal habitats are used for nesting, especially offshore islands and the machair, but the inland range is extensive as well. In winter there is also a considerable influx into Britain of Greylag Geese that breed in Iceland. Canada Geese and Egyptian Geese breed mainly in freshwater habitats.

The remaining species of geese that utilise intertidal areas for feeding are arctic breeders (Fig. 64). All have well-defined winter haunts and their movements to and from the breeding grounds follow well-established and traditional routes, usually with just one or two major staging areas and often a special moult migration. This helps to make the size of the various populations relatively easy to assess (Table 13). The numbers of nearly all species have increased as hunting mortality has decreased in recent years (Owen & Black, 1991). The Bean Goose (not shown in Fig. 64) has a broadly similar breeding range to the European White-fronted Goose except that it extends further south and west (into Finland, Norway and Sweden, for example). It was once a common wintering species in Britain, but now occurs at only a few localities on the east coast. Being relatively tolerant of cold conditions, the main part of its range is in the Netherlands, southern Sweden and Germany.

The Pink-footed Geese which winter in Europe breed in Iceland and Greenland. Saltmarshes are used to a lesser extent than previously, but are still an important feeding habitat in spring. The adults that breed in Iceland undergo a remarkable northward moult migration to Greenland in July and August. Presumably, Greenland offers richer and less disturbed feeding areas at this time of year. The breeding population of Svalbard winters mainly in Denmark and the Netherlands.

Two populations of White-fronted Geese winter in Britain, one from the former USSR and one from western Greenland. The main British haunt of the European White-front is the New Grounds at Slimbridge. Numbers have declined there as conditions have improved in the Netherlands (mainly as a result of agricultural compensation schemes), but large numbers still come in hard winters. The Greenland White-front winters mainly on Islay in Scotland and in Ireland.

**Fig. 64.** Breeding ranges (hatched and filled) and migration routes (arrows) of coastal wintering geese belonging to the genus *Anser* (*a*) and *Branta* (*b*). EWF = European White-fronted Goose, GWF = Greenland White-fronted Goose, NAWF = North American White-fronted Goose, B = Brant, DBB = Dark-bellied Brent Goose, LBB = Light-bellied Brent Goose. Based on information in Owen (1980*a*).

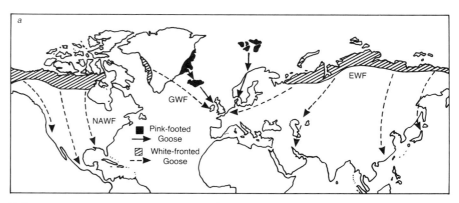

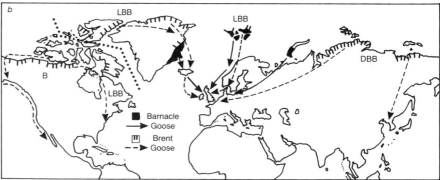

Table 13. *Current population sizes of those races of European wintering geese which depend at least partly on saltmarshes and coastal grasslands. Counts greater than 100 000 rounded to nearest 10 000, 1000–100 000 rounded to nearest 1000, 100–1000 rounded to nearest 10. From Owen, Atkinson-Willes & Salmon (1986), Salmon, Prys-Jones & Kirby (1989), Batten et al. (1990) and Kirby, Waters & Prys-Jones (1991)*

| Species | Race or population | British wintering population | Population in rest of Europe |
|---|---|---|---|
| Bean Goose (*Anser fabalis*) | Western Bean Goose | 450 | 80 000 |
| Pink-footed Goose (*Anser brachyrhynchus*) | Greenland/Iceland | 176 000 | – |
| | Svalbard | – | 30 000 |
| White-fronted Goose (*Anser albifrons*) | Greenland White-fronted Goose | 14 000 | 13 000 |
| | European White-fronted Goose | 6000 | 300 000 |
| Greylag Goose (*Anser anser*) | Iceland | 110 000 | – |
| Barnacle Goose (*Branta leucopsis*) | Greenland | 35 000 | – |
| | Svalbard | 12 000 | – |
| | Siberia | – | 50 000+ |
| Brent Goose (*Branta bernicla*) | Dark-bellied Brent Goose (Siberia) | 90 000 | 76 000 |
| | Light-bellied Brent Goose (Canada/Greenland) | 16 000 | – |
| | Light-bellied Brent Goose (Svalbard) | 4000 | 1000 |

There are three major breeding areas of the Barnacle Goose – east Greenland, Svalbard and Novaya Zemlya. Only the first two populations winter in Britain, the former in western Scotland (especially on Islay) and Ireland, and the latter in the Solway Firth. Prior to their return journey to the Solway, Svalbard Barnacle Geese move from the valleys where they breed and moult, to the scree slopes beneath Little Auk and Brünnich's Guillemot colonies. Here they build up their fat reserves by feeding on the autumn flush of vegetation which occurs as a result of the downwash of guano from the auk nests above (Prop, van Eerden & Drent, 1984). Spring, and occasionally autumn, staging areas occur in Iceland for the Greenland population, and along the Norwegian coast for the Svalbard one.

The Brent Goose breeds further north than any other species of goose, including the Snow Goose with which it shares some of its breeding range. Three subspecies or races are widely recognised (Fig. 64*b*), but there is in fact a continuous gradation of characters from the very dark Black Brant, with a broad white neck collar (which usually meets at the front) and pale flank patches, to the very pale Light-bellied Brent Goose, with separate small neck patches and browner plumage (Boyd & Maltby, 1979). The dotted line in Fig. 64*b* indicates the dividing line between the breeding ranges of Light-bellied Brent Geese wintering in North America and Europe.

As mentioned on p. 131, there is clear evidence that the breeding success of Dark-bellied Brent Geese, as measured by the proportion of young birds arriving at the wintering areas, is higher during years when lemming abundance in Siberia is high and predation on goslings is consequently lower (Fig. 65). This was first noticed by Roselaar (1979), and the link between the fluctuations of wader and Brent Goose populations from the same breeding areas was soon noted by others, e.g.

**Fig. 65.** Percentage of first-year birds in Dark-bellied Brent Goose populations wintering in Britain. The letters indicate lemming abundance on the Taimyr in the preceding breeding season (H = high, M = moderate, L = low). Based on data in Summers & Underhill (1987), updated from Salmon, Prys-Jones & Kirby (1988, 1989) and Kirby, Waters & Prys-Jones (1991).

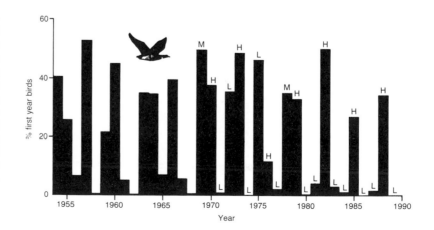

Robertson (1981). This link is thought to exist because wader eggs and chicks also suffer less predation in good lemming years. The idea was developed by Summers (1986) and this has led to an upsurge of interest and controversy. Dhondt (1987) pointed out that if such a link existed, it would not be simple, since the geese might also be expected to do well if lemming numbers were intermediate and predator numbers low, as is sometimes the case in years following very low lemming numbers. Boyd (1987) pointed out that Brent Geese breed poorly in seasons when the snow melt is late, and Owen (1987) suggested that the relationship between breeding success and lemming abundance could be spurious if both animals were responding to the same favourable climatic conditions.

Summers & Underhill (1987) responded by showing that the percentage of juveniles in the wintering populations of Brent Geese, Sanderlings, Curlew Sandpipers and Turnstones breeding on the Taimyr coast were significantly correlated with one another, and with lemming abundance (classed as either high or low). There were also some significant, though weak, correlations with temperatures in the Taimyr, with slightly better breeding occurring in warmer summers preceded by cooler winters. As predicted by Dhondt, the weakest correlation between breeding success and lemming numbers occurred two years after the previous lemming peak, when predator abundance was most variable. Curiously, there is a significant correlation between lemming abundance in the Taimyr and the spring departure weights of Brent Geese in the Wadden Sea, and it may therefore be difficult to disentangle the effects of these two factors on subsequent breeding success (Ebbinge, 1989). The fact that they are correlated at all does suggest the existence of a common climatic factor, such as warm springs producing good plant growth. Boer & Drent (1989) analysed breeding performance data from 21 populations of nine species of geese and concluded that climatic factors, rather than lemming abundance, were most important in producing synchrony in goose breeding success. A further complicating factor is that strong tail winds during migration enable the geese to reach the Taimyr in better condition, thus allowing them to breed more successfully, other conditions permitting (Ebbinge, 1989). Predation pressure appears to account for 60% of the variation in breeding productivity and favourable tail winds a further 13% (Underhill & Summers, 1990).

Geese are quite different from waders, in that they are much more dependent on body reserves for breeding. In Canada Geese, when the females arrive on the breeding grounds, they mobilise some of the protein from their main flight muscles and use it to form eggs (Hanson, 1962). They no longer require such large muscles since they fly much less once they reach the breeding grounds, and they also become lighter as they live off their fat reserves and consequently can fly efficiently with smaller muscles. This is one of the reasons why it is so important for geese to achieve the best possible body condition before they reach the breeding grounds. To what extent they are able to do this depends partly on the quality of food available to them, but also on other factors. For example, White-fronted Geese spend about 3% of the daylight hours with their head up looking out for danger (Owen, 1972). By feeding in a flock, each individual can reduce the amount of time it has to spend in vigilance rather than feeding, since the task can be shared with other flock members.

Geese that have bred successfully in the summer, forage in family groups during the first part of the winter (Fig. 66). These groups often move to the edge of large flocks so that the youngsters can benefit from a greater choice of food ungrazed by the rest of the flock. Whilst feeding, they rely on the vigilance of their parents. When the family groups break up in late winter, the males then invest a great deal of time in guarding their mates. They do this to protect them from the attentions of unpaired males and to allow them to concentrate on feeding to achieve the best possible body condition. Thus, in the spring, the males effectively defend a mobile territory around their mates. This allows female Barnacle Geese, for example, to feed for 90% of the 18 hours of daylight available at the spring staging areas in Norway (Owen, 1987). The males benefit from mate guarding because females in good condition lay more

**Fig. 66.** Family of Brent Geese feeding. The male is more alert than both his mate and their offspring. The juvenile can be recognised by the lack of white neck stripes and the presence of pale wing bars formed by the tips of the coverts. Chris Rose.

eggs and so the males' reproductive output is increased. One of the parts of the body in which geese store fat is the lower abdomen, and birds in good condition can be recognised by the bulging profile of the area between the tail and the legs (Owen, 1980b). When they arrive at the breeding areas, those geese in the best body condition not only lay larger clutches, but also leave the nest less often in order to feed, thus decreasing the vulnerability of their eggs to predators. This may also be the case if they choose nest sites with good feeding areas nearby (Owen, 1987).

**Breeding birds of saltmarshes**

It seems curious that areas covered by the tides, even if only the highest ones, should be used for nesting by birds. However, saltmarshes do have a distinct community of breeding birds (Fuller, 1982). There is a set of particularly high spring tides (equinoctial tides) in April followed by a similar but less extreme set in May, but once these have finished, the upper saltmarsh is not normally inundated again until autumn. There is thus sufficient time for nesting to take place safely, provided it starts rather late.

The breeding community is large and includes several waders (Oystercatcher, Ringed Plover, Lapwing, Dunlin, Snipe, Black-tailed Godwit), gulls (all the British species except the Kittiwake), wildfowl (Shelduck, Teal, Mallard, Garganey, Shoveler, Eider and Red-breasted Merganser), passerines (Skylark, Meadow Pipit, Rock Pipit, Yellow Wagtail, Wheatear, Sedge Warbler, Linnet, Reed Bunting) and others (Grey Partridge, Water Rail, Moorhen, Coot, Cuckoo, Short-eared Owl). In a survey of the Wash saltmarshes in 1973, the peak densities of breeding passerines, in pairs per km$^2$, were Skylark 105, Reed Bunting 80 and Meadow Pipit 76 (Cadbury et al., 1975). The colonial nesters, such as the gulls and terns, sometimes occur in very large numbers in the small number of suitable areas. For example, many of the Black-headed Gull colonies in southern England are coastal ones. This gull builds its nest on a mound of vegetation which helps to avoid inundation by the tide.

Dense stands of *Spartina* are not particularly attractive as a nesting habitat and consequently the breeding densities of most species on saltmarshes are generally lower than the peak densities on the Wash (c. 200 pairs per km$^2$ for all species combined) compared with other habitats (Fuller, 1982). One species for which saltmarshes are of major importance is the Redshank (Fig. 67), which occurs at densities of c. 100 pairs per km$^2$ at the best sites. An RSPB survey in 1985, covering about 8% of the total area of saltmarsh in Britain, led to an estimate for the total population of about 17500 pairs (Cadbury, Green & Allport 1987). About half of these Redshanks were situated on the East Anglian salt-marshes, and another quarter on those in north-west England. This compares with only about 3000 on the machair of the Outer Hebrides (see later in this chapter). Thus 60% of the British breeding population occurs on saltmarshes. The British population represents about 20% of the European total.

At least a quarter of the Redshank pairs in the RSPB survey had young, indicating moderate breeding success despite the problems of nesting in the intertidal zone. The highest breeding success tended to occur at sites with high densities of birds. Redshanks prefer saltmarshes where grazing intensity is reasonably light, so that the vegetation is several centimetres high, but not too rank. Lapwings, on the other hand, prefer heavily grazed saltmarshes with a very short sward.

Many of these waders start nesting and egg-laying before the high spring tides in May, and thus many nests are inundated. If the eggs are washed out of the nest, they perish and the birds usually re-lay in a new nest. If the eggs remain in the nest, and the nest is not badly damaged,

**Fig. 67.** Redshank wing-stretching after a spell of incubation. As in many British breeding Redshank, summer plumage is only partially developed. Harold E. Grenfell.

then incubation may be continued. Such eggs often hatch successfully provided they are relatively undeveloped when flooded (well-developed eggs are sensitive to chilling). In a detailed study on the Wash saltmarshes, Green (1987) found a peak Redshank nest density of 80 pairs per km$^2$. The May spring tides flooded most of the study plots and destroyed about 45% of the nests known to have been present. However, half of the flooded nests did survive and, provided the birds resumed incubation, their hatching success was no different from that of nests which remained dry. Grass growing around the rim of the nest cup prevented many eggs from floating away, and this may be one reason why Redshanks often nest in clumps of grass which completely cover the incubating parent. Overall, only 14% of nests during the whole season failed as a result of flooding, and an average of 1.8 chicks per pair were produced, which is as good as at most inland sites. Redshank chicks, like those of all waders, swim strongly so that flooding of the marsh after hatching is tolerated well. The major cause of nest loss in Green's study was predation by gulls, which affected about 59% of nests; a further 35% were trampled by cattle. These rates of loss were so high that the true nesting density was closer to 130 pairs per km$^2$ than the maximum figure of 80 nests per km$^2$ actually present at any one time. Other species in the study with a breeding density of more than 10 pairs per km$^2$ were Shelduck, Mallard and Linnet.

On the Ribble Marshes in Lancashire, Redshanks are very site faithful, about 40% of individuals returning to breed in the same area each year. Birds were more likely to return if they had bred successfully rather than failed, and this is presumably a mechanism by which they avoid areas prone to heavy flooding, predation or trampling by cattle. Females tended to move greater distances between nesting attempts than males, which is quite often the case for birds in general (Greenwood, 1980). Of the chicks known to have survived beyond 20 days of age, 23% also returned to breed in the same general area. Half of these young Redshanks bred in the year after they hatched, which is unusual for waders, since many species do not breed until they are two years old or more (Thompson & Hale, 1989).

Other species do not tolerate flooding as well as Redshanks. For example, tern and gull colonies often fail badly if this occurs. Black-headed Gulls have been known to swim vertically above their flooded nests in an attempt to continue incubation, though their clutches are usually washed away (Tinbergen, 1974)! Reed Bunting nests with young also usually fail if flooded (Green, 1987).

**Lagoons**

Coastal lagoons are bodies of salt or brackish water, separated from the sea by a bar of sand or shingle. There are only 41 such lagoons left in the British Isles (excluding the major offshore archipelagos) (Barnes, 1989). Many have been lost in the past and some are currently under threat. There have been few systematic studies of the birds of the mainland lagoons, but the oyces of Orkney (excluded from the above total) have been surveyed. They are 0.02–0.04 km$^2$ in size and usually have a short channel connecting them to the sea. Salinity is very variable, depending on the frequency with which they are replenished with sea water. They have muddy bottoms, very variable vegetation and their communities of waders and wildfowl do not differ appreciably from those of the nearby freshwater lowland lochs which they closely resemble (Berry, 1985).

On shores where material continually builds up on the seaward side, lagoons can become so distant from the sea that their waters become fresh. This has happened at the Loch of Strathbeg in eastern Aberdeenshire, which is now over a kilometre from the sea. The loch and surrounding farmland is owned by the RSPB and forms one of the most important autumn staging areas for Icelandic Pink-footed Geese. Considerable numbers (c. 19 000) winter there, along with Icelandic Greylag Geese and Whooper Swans. It is one of the few sites in Britain where an escaped Chilean Flamingo has managed to survive for a prolonged period (nine years). Unfortunately, this bird is reported to have been shot when it migrated to Iceland with some Whooper Swans (Chapman, 1986).

The Fleet in Dorset retains a very strong maritime influence since sea water enters at the south-eastern end and produces a tide of over a metre. South-easterly winds can pile up water in the north-west to a depth of a further metre. Overall, the water depth is greater at the south-east end (c. 5 m) and gradually shallows to less than half a metre at the Abbotsbury end (at low tide). The salinity declines from about 60% of normal seawater to completely fresh water at the north-west end. The Abbotsbury swannery is the best known ornithological feature of the Fleet and has been in existence for nearly a thousand years. In 1591, 500 Mute Swans were present – very similar to the numbers there today, though they have fluctuated between about 400 and 1600 during this century. The swannery was established to provide food for the monks at the nearby monastery. The herd is semi-domesticated and nests in a dense colony covering about a hectare, with the nests as little as 2 m apart. Most of the cygnets are fed and raised artificially in pens and are released onto the Fleet in September, rather than being eaten. The number of nests in recent years has varied between 19 and 100 (Prendergast & Boys, 1983).

The 4.8 km² of the Fleet support extensive growths of *Zostera* and Beaked Tasselweed, which comprise the main food of the swans in winter, and also provide grazing for about 5000 Wigeon and 3000 Coot. There are good numbers of several other species of wildfowl, including Brent Geese, Pintail, Pochard, Tufted Duck and Goldeneye. The numbers of Red-breasted Mergansers are also high, sometimes exceeding 400. The Coot used to be shot regularly, as a method of conserving the food resources for the swans, with bags as large as 850 in a day. On the whole, the swans do not appear to be in particularly good condition, since they are lighter, breed later, and have smaller clutch sizes than swans elsewhere in the country.

Other wildfowl also breed in the Fleet, including Mallard and Shelduck regularly, and Garganey and Shoveler occasionally. All these species, including the swans, may be adversely affected by the equinoctial spring tides in April. If the birds breed late (as in 1979) and have not started laying when these tides occur, there is little disruption, but if they nest early (as in 1980) many birds lose their eggs and have to re-lay (Perrins & Ogilvie, 1981). Grebes, herons and waders are also common in the Fleet. On Chesil Beach about 100 pairs of Common Terns, 100 pairs of Little Terns, 30 pairs of Ringed Plovers and 20 pairs of Redshanks breed, along with Oystercatchers and Lapwings.

Small coastal lagoons can be extremely attractive to birds, especially when they are undisturbed. For example, Brownsea Island in Poole Harbour has a brackish lagoon as a legacy of attempts to reclaim intertidal land during the last century. It has a total area of about 0.15 km². Considerable numbers of waders from the whole of Poole Harbour congregate in the lagoon at high tide – both to rest and in some cases to continue feeding. Counts of 1000 Oystercatchers, Redshanks and Curlews, 600 Black-tailed Godwits and 200 Spotted Redshanks are typical during the autumn. Both Spotted Redshanks and Avocets now winter in the area and use the lagoon for feeding. Common and Sandwich Terns have established breeding colonies amongst the Black-headed Gulls at the lagoon (Prendergast & Boys, 1983).

The availability of feeding sites which waders can exploit at high tide (when intertidal feeding areas are covered) is an important factor influencing the number of birds that an estuary can support (Davidson & Evans, 1986). Coastal lagoons provide such sites. Poole Harbour is exceptional in this respect, since in addition to the Brownsea Island lagoon, it also has a number of other brackish and freshwater lakes nearby, together with low lying grasslands which often flood in winter, e.g. Wareham Watermeadows. The latter site supports large flocks of Dunlins, Curlews, Redshanks, Black-tailed Godwits and Ruffs at high tide in winter (Green & Cade, 1989). In addition, the area has a relatively long and convoluted coastline with extensive saltmarshes, intersected by small creeks in which waders can feed at high tide. For a site with only 12 km² of mudflats, Poole Harbour appears to support an exceptionally high density of waders, but this is no doubt due to the 9 km² of saltmarsh, 105 km of shoreline and the 11 km² of lagoons and damp grasslands which are also available. Although this site is unusual, there is a general tendency for small estuaries to have a relatively greater area of peripheral feeding sites than large ones and this may explain the higher densities of birds that they support (see p. 137).

Food supplies in shallow saline lagoons can be unpredictable because of the fluctuations in salinity and temperature which can occur, especially in the summer months. For example, the salinity in the Doveys, one of the six main lagoons on Havergate Island where Britain's largest breeding colony of Avocets is situated, can become high due to evaporation. So high in fact that the numbers of ragworms, *Corophium* and the larvae of *Chironomus salinarius* (a midge) start to decline. The Opossum Shrimp, which forms the main food of Avocets at Minsmere, does not occur at all at Havergate because of the high summer salinity (Mason, 1986). This situation can reduce chick rearing success at Havergate. As soon as the problem was identified, the salinity was reduced in summer by simply clearing out some silted up ditches and opening sluices to allow water from the River Alde to enter during high spring tides. This produced an improvement in food supplies and a marked increase in chick survival (Cadbury *et al.*, 1989).

Lagoons were used during Roman times to obtain salt by the evaporation of sea water. This was especially common in eastern England, but these lagoons have long since disappeared. Salt pans or salinas are still widespread in other parts of the world such as Spain, Portugal and Africa. The most saline of these support no invertebrates, but the less saline ones have good populations of species such as brine-shrimps (*Artemisia*), and are used by birds for feeding. Many in the Algarve, Portugal, for example, are used throughout the tidal cycle by Kentish Plovers, Dunlins and Black-tailed Godwits, whilst other birds use them as supplementary feeding areas at high tide (Rufino *et al.*, 1984). Some species, such as the Kentish Plover and Kittlitz's Sand Plover, frequently nest in the vicinity of salt pans. The Camargue colony of some 6000–13 000 pairs of Greater Flamingos is dependent on salt pans for breeding sites and for feeding (Johnson, 1975). This is also true of flamingo colonies elsewhere in the world. Unlike other shorebirds, both flamingos and shelducks do occur quite regularly in the more salty salinas because they can feed directly on the algae which are often present even when there are no invertebrates (Britton & Johnson, 1987).

## Sand dunes

The more mature and stable stages of sand dune succession further inland tend to support greater densities of breeding and wintering birds than do the early, more mobile stages. For example, Fuller (1982) recorded only eight pairs of breeding waders per km$^2$ in unstable dunes in the Outer Hebrides, compared with 43 pairs per km$^2$ on stable dune grassland. The Ringed Plover is one of the most characteristic waders of sand dunes, but this habitat is probably not ideal for them. This was demonstrated when an area of stony ground at Lindisfarne, where Ringed Plovers had bred successfully at high density for a number of years, was invaded by sand. The density of breeding birds dropped and their breeding success appeared to decline (Pienkowski & Pienkowski, 1989). These authors argued that the Ringed Plovers increased their territory sizes in order to reduce the predation rate on their eggs, which are less well camouflaged on sand than on stones, hence causing a reduction in density.

The most abundant breeding passerines of sand dunes are Skylarks and Meadow Pipits, with Reed Buntings and Stonechats also numerous in some areas. Given the abundance of these species, it is not surprising that sand dunes constitute an ideal breeding habitat for Merlins, for whom Meadow Pipits often make up the bulk of the diet. However, most sand dune systems are now subject to such intense recreational pressure that many traditional Merlin sites have been abandoned. The lack of vegetation cover, and the absence of a well-developed litter layer, means that passerines are also less abundant in mobile dunes than in fixed dunes. One of the classic studies of a bird community was carried out on an area of sand dunes at South Haven near the mouth of Poole Harbour (Fig. 68) by Lack & Venables (1937). They found that the Meadow Pipit was dominant in all habitats. A few pairs of Lapwings and Ringed Plovers nested in areas of burnt heathland on the dunes, and Shelducks and Wheatears were associated with the presence of Rabbits, in whose burrows they nested. The area remains much the same today, with the lagoons (including Little Sea) forming one of the subsidiary feeding areas for the waders of Poole Harbour and often attracting exotic species such as Spoonbills and Little Egrets.

While the densities of breeding passerines in sand dunes are generally low, some other species frequently nest in large numbers in dune systems. This is particularly true of the few remaining systems that are relatively inaccessible and therefore relatively free from disturbance and predation (Fig. 69). The largest Black-headed Gull colony in Britain in the late 1950s was situated in the sand dunes at Ravenglass on the coast of the Lake District near Sellafield. This once consisted of 13 000–16 000 pairs (Gribble, 1962). The largest colony is now Needs Oar Point in Hampshire (an alluvial island in the mouth of the River Beaulieu), with 20 000 pairs. The latter site also has 350 breeding pairs of Common Terns, 100 of Sandwich Terns, 15 of Little Terns and occasionally a pair of Mediterranean Gulls. Of the 185 Black-headed Gull colonies located in England and Wales in 1958, 22% were in coastal dunes and marshes. Some Lesser Black-backed Gull colonies in sand dunes are also very large. Walney Island in Cumbria, for example, supported a peak population of 19 000 or so pairs in 1978 (as well as 32 000 pairs of Herring Gulls), though this has now declined to about 15 000 pairs of Lesser Black-backed Gulls (and 8000 pairs of Herring Gulls) (Dean, 1991). Common Gulls sometimes breed in sand dunes, as in the Wadden Sea (Goethe, 1981), and most species of terns will also do so – Sandwich Terns and Common Terns especially.

**Fig. 68.** Distribution of sand dunes in Great Britain. Some of those referred to in the text are indicated. Redrawn from Doody (1985).

Some of the early experiments designed to try and understand why gulls and other birds should nest in such dense colonies were carried out in the Black-headed Gull colony at Ravenglass (Tinbergen, 1974). Rows of hens' eggs were laid out at right angles across the colony boundary. Those outside the colony were taken first by predators, and those in the middle of the colony last, if at all. Gull nests outside the main colony had a very low breeding success and this has been shown to be the case in a great many other colonial species such as Black-tailed Godwits. Social nesting clearly offers advantages by reducing predation through communal defence. However, in the Ravenglass colony, there were many occasions when Foxes caused widespread mortality of adults and

**Fig. 69.** Sand dunes at Barley Cove on the very south-west tip of Ireland. Unless they are very isolated, small sand dune systems like this do not usually support many breeding birds.

their eggs at night, 230 adults being killed during the worst night. Most of these incidents occurred on very dark nights when gulls were unable to detect the approach of the Fox. On lighter nights, the birds left in a mass panic flight. Early and late nests also tended to suffer higher predation and this suggests that breeding synchrony may have an advantage through its swamping effect on predators. By producing a mass of eggs and young at the same time, a small number of predators can only consume a tiny proportion of them, reducing the risk for well-synchronised birds. Nevertheless, because the gulls and their offspring were vulnerable to predation and to losses caused by tidal inundation, Tinbergen considered that sand dunes and tidal saltings were suboptimal habitats and the colony has now disappeared.

Tinbergen (1974) also studied breeding colonies of Herring and Lesser Black-backed Gulls amongst sand dunes in the Netherlands. In this case egg cannibalism was quite a frequent phenomenon, caused by a few gulls which specialised in this type of feeding. Although it seems contradictory, Oystercatchers, Pheasants, Grey Partridges and Nightjars all nested successfully in the gull colony, presumably deriving some benefit from the anti-predator activity of the gulls or just by sharing the risk of predation with a larger number of individuals.

Sand dunes are also widely used as breeding sites by Eiders and Shelducks. One of the largest Eider colonies in Britain is situated in the 7 km$^2$ of dune ridges at the Sands of Forvie in Aberdeenshire. Some 2000 pairs nest in this area, moving into the dunes in mid-April. Prior to this the birds feed on the mussel beds in the nearby Ythan Estuary. The males defend an area around their females that enables them to feed three times as fast as males and thus accumulate the necessary body reserves for nesting (Milne, 1974). The nest sites vary from open areas in vegetated dunes, to some with a complete canopy of vegetation over the nest. Clumps of rushes in ditches are particularly favoured. Females tend to return to the same general area every year, but not necessarily the same nest site. Egg laying is generally later, and the eggs smaller, in cold springs such as those of 1964 and 1969. The average clutch size is four. During the whole of the 26 day incubation period the females do not feed, but merely drink, losing some 40% of their initial body weight during this time. Hatching success was quite low in Milne's study, varying from 42% to 74% in different years, depending on the level of predation by Carrion Crows and Herring Gulls.

Chick mortality was also very high, exceeding 90% during the first seven weeks, in seven out of the ten years of Milne's study. In the other three years it was 83%, 68% and 59%. Most of the mortality did not occur during the apparently hazardous journey of the young from the dunes to the shore where they assemble in crêches. Instead it was caused by gull predation during bad weather when the crèches tended to split up (Mendenhall & Milne, 1985), or by high levels of internal parasites causing diseases such as coccidiosis.

In a nearby area, Gorman (1974) found that Eiders showed a preference for nesting in long heather and rushes, and that these were subject to much lower levels of predation, presumably because of their greater cover. Following breeding, Eiders concentrate into a small number of areas with good food supplies for the annual moult (see p. 99). Some 14 000 to 20 000 birds moult in eastern Scotland, mainly in the Firths of Forth and Tay, and off the Aberdeenshire coast (Campbell & Milne, 1983). There are also large moulting concentrations in Orkney and Shetland (Hope Jones & Kinnear, 1979).

Given their high egg losses, it is not surprising that Eiders often nest where they obtain some protection from predators. For example, in Scandinavia, they may nest in gull and tern colonies, or near human habitations. This helps explain why they take so readily to nesting on Eider farms (p. 62). A successful colony even managed to establish itself within a few metres of tethered huskies in Greenland (Meltofte, 1978) – safe as long as the dogs were not moved! In gull colonies, they may run the risk of a higher rate of predation if they lay earlier than the gulls (Götmark, 1989).

One of the best studied Shelduck breeding colonies is also situated in sand dunes, in this case near Aberlady Bay in the Firth of Forth. Here the birds nest in Rabbit burrows and in artificial nest boxes provided for them. In March and April, the adults prospect in the dunes for suitable nest sites. The mean clutch size in this area is nine and the incubation period 32 days. Between a third and a half of all nests contain eggs laid by more than one female. Many of these nests are eventually deserted (60%), even if they manage to avoid being taken by predators. This high rate of nest parasitism and desertion may be a consequence of the partly artificial nature of this colony combined with its high density (Pienkowski & Evans, 1982a). However, there is also a significant tendency for Shelduck on the Ythan to hatch fewer young at higher population densities (Fig. 70), so overcrowding is clearly an important factor under more natural conditions too.

Although Aberlady Bay appears to be a favoured feeding area for Shelducks, both in the winter and in the breeding season (Jenkins, Murray & Hall, 1975), this is disputed by Evans & Pienkowski (1982). The birds certainly have a reduced breeding success compared with more dispersed populations which feed and nest in apparently less favoured sites further along the coast. In the more dispersed pairs, duckling survival was six times higher than in Aberlady Bay, and it is doubtful if the latter population is sufficiently productive to be self supporting. The main cause of high chick mortality in Aberlady Bay is predation by Herring Gulls during periods when the adults are involved in territorial disputes with other Shelducks on the mudflats. The gulls attacked chicks much less frequently in the absence of such disputes (Pienkowski & Evans, 1982b). This seems to be another consequence of the high density of breeding birds at Aberlady Bay. Most of the duckling mortality in Shelducks occurs during the first ten days of life. As with Eiders, it is highest during low temperatures, high winds and high rainfall. On the Ythan Estuary, there is also a higher mortality when there are more chicks in the nursery areas (Makepeace & Patterson, 1980). The chicks are most vulnerable if they become separated from their parents, which often happens if they are forced to dive and swim underwater. Wildfowl may benefit from reduced nest predation by nesting in gull colonies, but they also run a greater risk of losing chicks as they leave the nest sites. Thus Lesser Scaups and Gadwalls nesting in gull and tern colonies on islands in a Canadian lake generally had a good hatching success, but if they nested amongst California Gulls, their fledging success was very low because these gulls attacked their young (Vermeer, 1968).

The Sands of Forvie support a breeding colony of Black-headed Gulls (c. 200 pairs in 1974), mixed in with about 500 pairs of Sandwich Terns, 300 pairs of Common Terns and 80 pairs of Arctic Terns. There appears to be plenty of alternative space available for the terns to nest well away from the gulls. This suggests that the terns chose to nest close to a

**Fig. 70.** Relationship between Shelduck breeding abundance and breeding success on the Sands of Forvie. Fewer pairs hatch young when more pairs breed, perhaps because a greater amount of time is spent prospecting and competing for the available nest sites. Redrawn from Patterson & Makepeace (1979).

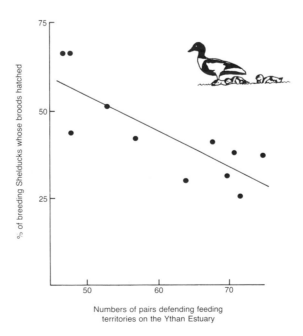

potential predator of their eggs and young because of the net benefit that it provides by way of protection against more serious predators such as Foxes, though this remains to be proven. The advantage must be considerable as the terns are also subject to food robbery by the gulls. This starts early in the breeding season when the terns are courtship feeding. Although attacks by Black-headed Gulls on Sandwich Terns are successful on 20–30% of occasions, the overall level of food robbery is quite low with only 1% of the terns that return to the colony with food losing it to gulls early in the season, rising to about 6% when they have large chicks. Terns carrying larger fish are more likely to be attacked than those with small fish (Fuchs, 1977).

Terns may also rob one another. Roseate Terns in particular are adept at grabbing fish from other terns' bills. On Coquet Island, off the Northumberland coast, Roseate Terns attack all other species of terns, with a success rate of 2–18% on Common Terns (Dunn, 1973). Since only about 20 pairs of Roseate Terns currently nest on the island, their impact on nearly 3000 pairs of the other species is relatively small.

Dune slacks frequently support a wide range of breeding aquatic birds, such as the following, which are listed in decreasing order of abundance, based on Fuller (1982): Reed Bunting, Mallard, Moorhen, Snipe, Sedge Warbler, Redshank, Grasshopper Warbler, Reed Warbler, Teal, Shoveler and Water Rail. Like saltmarshes, sand dunes may also support game birds. For example, the Sands of Forvie has a small breeding population of Red Grouse (Moss, Watson & Parr, 1975). Wintering birds of sand dune systems have not been well studied, perhaps because of their relatively impoverished nature. Two species that occur equally

**Fig. 71.** Distribution of wintering Snow Buntings and Lapland Buntings in the British Isles in 1981–84. Redrawn from the BTO's Winter Atlas results (Lack, 1986), enclosing and filling those areas in which more than five birds per day were seen in a 10 km grid square.

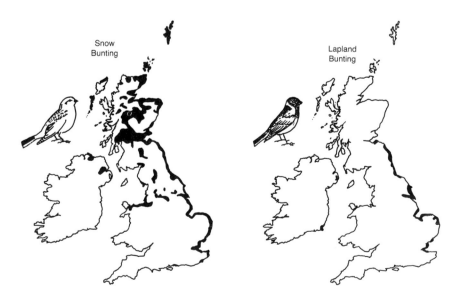

often on saltmarshes, sand dunes and in coastal grasslands are Snow Buntings and Lapland Buntings (Fig. 71). About 10 000–15 000 of the former winter in Britain, whilst there are less than 500 of the latter in most winters (Lack, 1986).

## Machair

The sand that makes up machair is calcareous, and consequently it can give rise to much more productive soils than siliceous sand dunes. Machair also differs from ordinary sand dunes in having a relatively flat, well-vegetated surface (Fig. 72). A Wader Study Group/Nature Conservancy Council survey of the breeding wader populations of the western seaboard of the Outer Hebrides, which includes the majority of the machair in the British Isles, was carried out in 1983. The peatlands, which cover the majority of the land in the Hebrides, support only very low numbers of birds. Blackland, which is the area of peat immediately adjacent to the machair, supports moderate numbers, though densities can be quite high close to lochs and in marshes near the machair. Moving westwards from the blackland, there is usually a zone of wet machair which can support exceptionally high densities of waders. For

**Fig. 72.** Typical western seaboard of the Outer Hebrides, with white shell-sand beach backed by a narrow ridge of dunes leading to the flat, fertile plain of the machair. Chris Rose.

example, Etheridge (1982) recorded breeding Dunlin densities of 300 pairs per km$^2$ in this type of habitat. However, densities half as great as this are more typical for the total populations of Redshanks, Dunlins and Lapwings combined (Fuller *et al.*, 1986).

The majority of wet machair habitats remain uncultivated because of the difficulty of getting machinery onto them early enough in the growing season. Dry machair has been extensively cultivated, however, and this has had an important influence on wader density and diversity. There are a variety of cultivation systems in operation. Typically, small blocks of land are subjected to a three or four year rotation in which the land lies fallow for two or three years after a crop of cereals or potatoes has been harvested. Cultivated dry machair generally supports higher densities of waders than uncultivated machair, and it has been suggested that this is a consequence of its greater habitat diversity. In fact, it seems more likely that it is due to differences in soil fertility or moisture content (Fuller, *et al.*, 1986). For example, the highest average density of breeding Ringed Plovers is on dry, cultivated machair (60 pairs per km$^2$), but the peak density of over 100 pairs per km$^2$ is actually recorded on damp, cultivated machair, which is an uncommon habitat type.

Saltmarsh supports quite high breeding densities of birds, particularly Dunlins and Ringed Plovers, but sand dunes only have large numbers (particularly of Lapwings) in the dune slacks. Overall, Ringed Plovers and Dunlins are heavily dependent on machair and dunes, whereas Oystercatchers, Lapwings, Redshanks and Snipes are more or less equally distributed across both habitats and across the machair to blackland or machair to moorland transition zones.

Overall population estimates for the western seaboard of the Outer Hebrides (see the first column of Table 14, which consists almost entirely of Hebridean birds) show that they support about a quarter of British breeding Ringed Plovers and Dunlins, and about 7% of Oystercatchers and Redshanks. Other species that nest in this area are Eiders, Shelducks and Wheatears amongst the dunes, and Skylarks, Corn Buntings, Little Terns and Common Gulls on the dry machair. Wet habitats support several species of nesting ducks, Moorhens, Coots, Little Grebes, Mute Swans, Arctic Terns and Black-headed Gulls. Red-necked Phalaropes are now restricted to just a few sites. The machair remains a key site for the dwindling Corncrake population of the British Isles. Indeed, it supports 13% of the European breeding population, probably because of the late cutting of hay and the proximity of fen vegetation which provides cover (Stowe & Hudson, 1988).

In winter, the machair bird community is dominated by Rock Doves, Skylarks, Corn Buntings, Twites, House Sparrows and Starlings. There are also significant populations of Whooper Swans and Greylag Geese.

Table 14. *Estimated breeding populations of waders (in pairs) in different habitats in Britain and Europe. Where ranges are given in the main sources, the midpoints of these have been used. The estimates are very approximate. Upland moorland and grassland are the most significant habitats not included in the table. + = a few pairs*

| Species | Machair[a] | Saltmarsh[b] | Lowland wet grassland[c] | Total for British Isles[d] | Total for Europe[d] |
|---|---|---|---|---|---|
| Oystercatcher[e] | 2700 | 4600 | 500 | 41 000 | 208 000 |
| Avocet[f] | | 340 | | 340 | 19 300 |
| Ringed Plover[g] | 2400 | 350 | | 8 600 | 50 000 |
| Lapwing | 5000 | 2600 | 7000 | 215 000 | 869 000 |
| Dunlin | 3500 | 150 | | 9400 | 66 000 |
| Ruff | | | 7 | 7 | 247 000 |
| Snipe | 1600 | + | 3700 | 40 000 | 534 500 |
| Black-tailed Godwit | | + | 70 | 70 | 112 500 |
| Whimbrel | | 2 | | 215 | 45 000 |
| Curlew | | 400 | 500 | 35 500 | 125 000 |
| Redshank | 2600 | 17 500 | 3000 | 32 100 | 168 000 |
| Red-necked Phalarope | 2 | | 13 | 30 | 75 000 |

[a]Fuller *et al.* (1986), Nairn & Sheppard (1985).
[b]Allport, O'Brien & Cadbury (1986) (excluding Ireland), Green & Cadbury (1987).
[c]Smith (1983) (England and Wales only).
[d]Including British Isles: Piersma (1986).
[e]Oystercatchers also have a substantial breeding population on sand and shingle beaches.
[f]Most pairs of Avocets breed in saltmarsh habitat in coastal lagoons, but a few pairs also nest on other saltmarshes.
[g]A further 4100 pairs of Ringed Plovers are estimated to nest on beaches and 800 in sand dunes (Prater, 1989).

There are extensive areas of fresh flood water on the machair plain which provide feeding for Ringed Plovers, Sanderlings, Dunlins, Redshanks and Turnstones. These are particularly important during the frequent westerly gales which sweep the intertidal area with waves, even at low tide, and make feeding on the shore very difficult.

The high productivity of the machair for breeding waders has been attributed to a number of factors including the high calcium content of the shell sand. Farming activities, by adding organic matter in the form of kelp, have enhanced the productivity of soil invertebrates even further, and wader densities are highest in the cultivated areas. However, highly productive grasslands cover much of Britain and yet support few breeding birds because of the lack of cover to hide nests. Clearly, another important factor on the machair is that the less intensive grazing pressure allows sufficient cover to develop to provide suitable nest sites. Crofting farmland between the coast and the moorlands of the Shetlands is both drier and only locally influenced by shell sand, and so supports lower breeding densities of waders than the Outer Hebrides, but none the less higher densities than agricultural grasslands on the mainland. Less intense grazing pressure is again also important. About 800 pairs of Oystercatchers and Lapwings, and 200 pairs of Redshanks and Snipes nest in this habitat in Shetland (Campbell, 1989).

## Coastal grassland

Sand dunes and machair represent particular types of coastal grassland, but there is another category growing on non-sandy, neutral soils, which is even more widespread. This includes low lying grasslands on silt or clay deposits in areas that have been reclaimed from the sea, and cliff top grasslands where strong winds and salt spray are significant environmental factors. In both these cases, grazing by livestock, or occasionally Rabbits, is the most important factor maintaining the habitat and preventing it from undergoing succession to scrub or wood-land.

Low lying grasslands where the water table is close to the surface and where water accumulates at times of high rainfall are also known as fens, and the vegetation in such areas consists of species that tolerate flooding. Two such areas, both of them reclaimed from estuaries, have already been referred to – the Somerset Levels and the Lancashire Mosses. In the few areas of the Somerset Levels that still flood regularly in winter, there are good numbers of wildfowl as well as inland winter-ing populations of Dunlins and Redshanks. These same areas support good breeding populations of waders in summer.

The largest remaining area of fenland in Britain is in East Anglia. This once occupied over 3000 km², but has now been reduced to less than a tenth of this (p. 212). Areas protected from shooting, and specifically managed to optimise conditions for birds, support large populations of wintering and breeding wildfowl and waders. For example, nearly 50 000 are recorded on part of the Ouse Washes. Other areas of the British Isles in which there are still reasonably extensive wet coastal grasslands are Essex, north Kent, the Lough Erne and Lough Beg basins in Northern Ireland and the Shannon Callows in Eire.

Key factors that influence the numbers of waders which breed are the timing of spring flooding, the timing of the introduction of grazing stock, the maintenance of a high water table and the provision of shallow pools. The needs of different species are not necessarily identical. For example, summer flooding is bad for most species, including Black-tailed Godwits, since nests may be destroyed and earthworms drowned (Green, Cadbury & Williams, 1987), but it is good for Snipes since they are dependent on soft, penetrable soils, and they are able to breed as such floods recede (Green, 1988).

Wet grasslands support a substantial proportion of Europe's breeding population of waders, especially Lapwings, Black-tailed Godwits, Curlews and Redshanks (Piersma, 1986). However, the proportion is decreasing as agriculture intensifies, and in the British Isles, coastal habitats support all our breeding Avocets and Black-tailed Godwits (Davidson, 1991), though most of the latter occur on grasslands that were reclaimed from the sea some time ago and are now several miles inland. In addition, 90% of Ringed Plovers, 40% of Dunlins, 13% of Snipes, 7% of Lapwings and 3% of Curlews occur in coastal areas (Table 14).

The mating systems of waders exhibit a remarkable diversity, most notably amongst those species which nest on arctic tundra, but there is also considerable variability within the grassland breeders. These variations probably represent adaptations to extreme climatic variability in the north and the generally ephemeral nature of grasslands in the south. They allow effort to be concentrated on rearing a small number of young in unfavourable years, but permit maximum output to be achieved when conditions are good. Monogamy is generally a suitable strategy where conditions are relatively stable from year to year, but various forms of polygamy, and the associated production of multiple clutches, allows individuals to take advantage of good years and of newly appearing patches of suitable habitat. Coastal grasslands along the tops of cliffs provide a very different kind of habitat from low lying ones. The growth rates of the grasses are usually low because of exposure to wind and salt spray, and consequently they do not provide particularly good grazing. However, this habitat does appear to be important to our dwindling populations of Choughs. This species is now restricted to parts of the coast of Wales, the Isle of Man, a few sites in western Scotland and along all but the eastern coast of Ireland. The last colonies on the north Cornish coast disappeared in about 1950. Choughs do much of their foraging by digging in bare ground and short vegetation for leatherjackets, beetle larvae and Yellow Meadow Ants, for which clifftop grasslands and heaths are an ideal habitat. They also feed on machair, dune grasslands and heavily grazed pastures, though unlike other corvids, avoid arable fields. Low vegetation is critical, as was demonstrated on both Bardsey Island and the Calf of Man, where the cessation of grazing caused a population decline which has been reversed by the reintroduction of stock (Bullock, Drewett & Mickleburgh, 1983). Many of the areas from which Choughs have disappeared no longer have any clifftop grassland because arable cultivation has been pushed right to the edge. Populations continue to decline in areas such as south-west Wales

Table 15. *Size and location of some of the largest current breeding seabird colonies in the British Isles. From Lloyd, Tasker & Partridge (1991)*

| Species | Site | Number of pairs | Habitat |
|---------|------|-----------------|---------|
| Fulmar | St Kilda | 62 800 | Cliff |
| Manx Shearwater | Rhum | 150 000 | Grassland |
| Storm Petrel | Inishtearaght | 15 000 | Boulders and coastal grassland |
| Gannet | St Kilda | 52 000 | Cliff and coastal grassland |
| Shag | Foula | 2400 | Boulder fields |
| Great Skua | Foula | 2500 | Moorland |
| Black-headed Gull | Lough Neagh | 33 000 | Freshwater marshland |
| Common Gull | Correen Hills | 19 000 | Moorland |
| Lesser Black-backed Gull | Skomer | 13 200 | Coastal grassland |
| Herring Gull | Copeland Islands | 7000 | Coastal grassland |
| Kittiwake | Bempton | 83 000 | Cliff |
| Sandwich Tern | Scolt Head | 3100 | Saltmarsh and sand dune |
| Guillemot | Handa | 98 700 | Cliff |
| Razorbill | Handa | 16 400 | Cliff |
| Puffin | St Kilda | 155 000 | Coastal grassland |

where the loss of grasslands continues. Although grassland is a critical habitat for the remaining 1000 pairs of Choughs in the British Isles, the shoreline is also an important feeding habitat in winter (Roberts, 1983).

Several species of seabirds that normally breed on cliffs and steep slopes may spread over the top of the cliff and breed on coastal grasslands provided that these are relatively undisturbed. Thus offshore islands are favoured, and some of our largest seabird colonies are located in this type of habitat. For example, the 30 000 Gannets on Grassholm now cover a substantial portion of what was once a grass-covered island. Large colonies of most species of gulls, terns, skuas, Cormorants and Fulmars occur in similar areas. Such grasslands are the regular breeding habitat of Manx Shearwaters and Puffins. Puffins have the inner toe on each foot modified for burrow excavation by the possession of a large claw which lies flat on its side during walking. Occasionally, burrow-nesting seabird colonies may be situated well inland, as with Britain's largest colony of Manx Shearwaters, which is 650 m above sea level and 3 km from the sea on the Isle of Rhum. In fact, most of the largest breeding colonies of all species of seabirds in Britain are sited on grasslands (Table 15) or sand dunes.

Arctic Skuas nest on clifftop slopes, as well as on grassland and moorland further inland. Most British colonies are situated on offshore islands, the strongholds being Shetland with 60% of the population and

Orkney with 30%. The breeding population, which only numbers about 3400 pairs, relies almost entirely on food robbery to obtain its food, the favoured victims being Arctic Terns and Kittiwakes. This is very different from Arctic Skuas in the Arctic where lemmings are the staple diet. The recent decline in the numbers of terns and Kittiwakes in Scotland (p. 108) has had a significant impact on the breeding success of Shetland Arctic Skuas (Furness, 1987a, 1989). To a certain extent the Great Skua overlaps in breeding habitat with the Arctic Skua, though it has a larger population of about 7900 pairs and a more restricted distribution, with about two-thirds of the population occurring on Foula. When both species increased in numbers in the past, there was a tendency for Arctic Skuas to be displaced by Great Skuas from the best breeding areas to less favourable habitats such as clifftop grasslands, old cultivated fields and peat hags (Fig. 73). Arctic Skuas are also more tolerant of human disturbance and have concentrated along roads and near buildings (the main road runs along the eastern side of the island). This may have caused a reduction in breeding success, but the population is not in decline as a result of these changes (Furness, 1987a). Great Skuas are large, powerful predators, relying less on food robbery, but they have been directly affected by the decline in sandeels which has also affected a number of other species of Scottish seabirds (p. 264).

The presence of large numbers of birds during the breeding season often exerts an influence on the vegetation of coastal grasslands. The most extreme cases are gannetries that revert to bare rock, covered with nesting material and guano, the latter being commercially exploited in some parts of the world where it does not wash away in winter rains (p. 14). Some species that nest less densely, such as Herring and Lesser Black-backed Gulls, can also bring about considerable vegetational

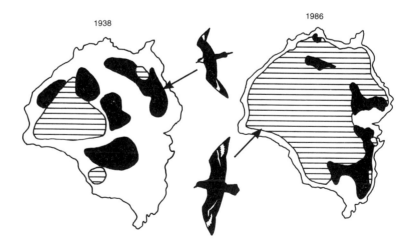

**Fig. 73.** Changes in the breeding distribution of Arctic Skuas (filled) and Great Skuas (hatched) on Foula, Shetland, between 1938 and 1986. Redrawn from Furness (1987a).

changes. They indulge in grass pulling during territorial disputes and also trample the ground near the nest, adding faecal material and regurgitated pellets to the soil. The net result is that gull colonies contain an excess of ruderal species (fast growing opportunists which colonise bare ground), such as Spear-leaved Orache, Yorkshire-fog, Common Sorrel, Curled Dock, Common Chickweed and Sea Mayweed. Guano enrichment encourages the development of patches of Common Nettle and Small Nettle. The resultant plant community is thus due to the interaction between the changes brought about by the gulls during the summer and those caused by grazing and salt spray during the winter (Sobey & Kenworthy, 1979).

### Coastal scrub

Where the cessation of grazing by stock or Rabbits allows coastal grass-land to undergo natural succession, scrub is the first type of habitat to develop. This may in turn be replaced by woodland, but there are several factors that tend to prolong the scrub stage near the coast. One is wind, which can be sufficiently severe to stunt trees into a scrub-like growth form. This is often seen on exposed southerly and westerly cliff tops. In dune habitats, soil instability combined with a limited supply of water and nutrients may initially limit vegetation to the scrub stage. Characteristic species are Sea-buckthorn, Gorse and Blackthorn. In dune slacks, the water-logged soil may limit vegetation to dense thickets of species such as Creeping Willow, which may be anything from a few centimetres to a few metres high.

A study of the breeding bird community of the sand and shingle scrub at Dungeness revealed high densities of Whitethroats, Linnets, Dunnocks, Redpolls and Sedge Warblers (Morgan, 1978). The overall density of the 30 breeding species in this survey was 1300 pairs per km². The scrub consisted largely of Sea-buckthorn with some Elder and Wild Privet. Densities almost as high as this have been recorded from Sea-buckthorn scrub in north Ireland, where the commonest species were Willow Warbler, Chaffinch, Blackbird and Robin (Nairn & Whatmough, 1978). The main reason for the considerable difference in these two communities was that at Dungeness the scrub was open and 3 m or so tall, whereas in Ireland it was up to 5 m tall with many Sycamores.

Two species that show a marked coastal component to their breeding and wintering distribution in Britain are the Stonechat and Cirl Bunting. Neither have any obvious requirement for coastal habitats, although the open scrub which occurs there is an ideal habitat for both species. The Cirl Bunting's stronghold is the coastal areas of South Devon, where the winters are milder than in the rest of the country, partly because of the ameliorating influence of the warmer nearby sea (Sitters, 1985). This seems to explain why a number of passerines also have a coastal distribution. Black Redstarts are one species which frequent the coast itself, especially cliffs. Song Thrushes, Meadow Pipits and Skylarks are very widely distributed inland during winter, but a significant proportion of the population is coastal and feeds in coastal habitats. For many other species, coastal scrub and woodland provide the right kind of habitat combined with suitably mild conditions. Stonechats defend winter territories in coastal areas from October onwards. For both Chiffchaffs and Firecrests, the critical factor is probably the low incidence of winter frosts. The small number of British wintering birds of the latter two species probably build up during a series of mild winters and drop after a cold one.

Coastal scrub also provides good cover and a source of food for migrants that follow the coastline. The wealth of berries usually produced by Sea-buckthorn in autumn and early winter attracts feeding thrushes and *Sylvia* warblers, and the impenetrable thickets provide a secure roost. In fact, most of the ringing studies carried out at British bird observatories are based on birds utilising coastal scrub at such sites as Dungeness, Gibralter Point, Portland Bill and Spurn.

Large parts of the coastline in tropical regions are dominated by mangroves, and these are important to some local birds as well as to migrants from arctic and temperate areas. They provide nesting sites for boobies, frigatebirds, noddies and Fairy Terns (e.g. Feare, 1974), but the greatest importance of mangroves to coastal birds is probably as a feeding area. The mud between the mangrove roots and in open spaces between the trees supports large populations of invertebrates, particularly fiddler crabs (Chapman, 1977). These provide food for such species as Green Herons, Little Egrets, Roseate Spoonbills, as well as considerable numbers of wintering sandpipers. For example, the mangrove dominated coasts of Guinea-Bissau, Sierra Leone and Guinea in West Africa support respectively one million, 160 000 and 400 000 Palearctic waders, with Ringed Plovers, Grey Plovers, Little Stints, Curlew Sandpipers, Redshanks, Common Sandpipers, Black-tailed and Bar-tailed Godwits being commonest (Zwarts, 1984; Tye & Tye, 1987; Altenburg & van der Kamp, 1988). These birds feed and roost amongst the mangroves for much of the time, and even at low tide, when they usually feed in the band of open mudflats beyond the mangroves, are difficult to observe.

# 6

# Threats to coastal birds

The threats to our coast, and therefore to coastal birds, are many and varied and to deal with them all is quite a daunting task. They range from the incidental effects of environmental pollution, which may themselves be as small and localised as a pool of spilled oil or as all-pervasive as global warming, to the wholesale losses of coastal habitats which are a direct result of their exploitation for industrial and recreational purposes. Significant lengths of our coast are of great natural beauty, and in such cases the landscape needs conserving. Some areas are of particular value for their wildlife, and these too are worthy of protection. At the same time, there is a legitimate desire of more and more people to take part in a range of recreational pursuits which may conflict with landscape and wildlife conservation and with one another. Unfettered and uncoordinated free market development cannot be expected to produce a satisfactory resolution of these conflicts. Some planning is clearly called for. At the moment, there is no policy providing even guidance on how these differing interests can be resolved.

## Habitat loss

A good deal of habitat loss in coastal areas took place at a time when its effects were poorly documented. Thus, while it is often quite easy to quantify the areas that were lost, it is more difficult to determine the effects on birds and other wildlife. The largest losses during the last two millenia have been associated with the reclamation of low lying coastal marshes in estuarine areas and their replacement with intensively worked farmland of little value to conservation. For example, until the seventeenth century such marshes occupied over 3000 km² of East Anglia. Given the prevalence of malaria in the region and the potential for substantial profits to be made from reclaimed land, it is not surprising that there was pressure in some quarters (though not from the fen people) to drain the marshes. Between 1650 and 1825, about 71% was reclaimed for agriculture by the construction of the Hundred Foot River and various other drainage works (Purseglove, 1989). Subsequent drying and shrinkage of the peat meant that the ground level dropped, the drains had to be deepened, and windmills employed to remove the water. The introduction of steam powered pumps during the industrial revolution lead to another phase of accelerated drainage during which a further 27% of the area was lost (Fig. 74). When diesel pumps were introduced earlier this century there was a further decline, so that by 1930 only 100 km² or so were left. Since the Second World War, further reductions have occurred and now only about 1% of the original area of fens remains. This last phase of drainage was only a small one in East Anglia (its effects are scarcely visible in Fig. 74), but in England and Wales as a whole 20 000 km² of land was drained, largely with taxpayers' money through both direct and indirect subsidies. Ironically, further taxpayers' money is now being paid to farmers in some areas, such as the Somerset Levels, to forego further drainage and adopt traditional farming techniques of value to wildlife. However, much of this money is wasted because the Internal Drainage Boards (which are dominated by farming interests) continue to take measures (such as reducing water levels in the main drains) that result in an overall lowering of the water table.

During Medieval times, the East Anglian fens and the Somerset Levels were utilised for summer grazing, and as a source of fish (Salmon, Eels, Roach, Perch, Bream, Pike, Flounder and Tench), waterfowl (swans, geese, herons and seasonal migrants), rushes, reeds and peat (Coles & Coles, 1986, 1989). The Victorian phase of drainage resulted in the disappearance of several wetland breeders from the British Isles, notably the Avocet, Bittern, Black-tailed Godwit, Black Tern and Ruff, as well as a considerable diminution in the populations of several other species (Williams & Bowers, 1987).

The major phases of sea wall construction around our coasts are believed to have taken place during the Roman and Medieval periods. In

**Fig. 74.** Estimated rate of decline in the area of fens in East Anglia and the introduction of various pumping methods. Redrawn from Thomas, Allen & Grose (1981).

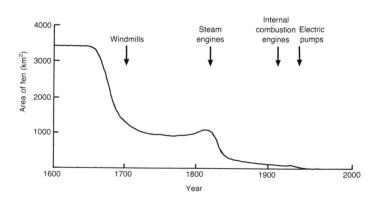

most cases, these were an essential preliminary to the reclamation of the fens and coastal marshes discussed above. One of the main reasons for constructing the sea walls was to prevent losses of human life and livestock due to unpredictable floods resulting from the coincidence of high winds and high tides. Even after such walls were in place, flooding still occurred occasionally. In 1607, the sea walls around the Severn Estuary were breached and 26 parishes on the Welsh side were inundated, resulting in the loss of many people and cattle (North, 1955). This Medieval wall and the present day one coincide for much of their length, but a Roman sea wall once existed at least a kilometre further out to sea. The bottoms of the drainage ditches can still be seen on the present mudflats. The Roman wall may have been breached suddenly, or may have fallen gradually into decay (Allen, 1987). We do not yet understand why the Romans were able to reclaim so much more land than proved possible in the Middle Ages, though this may have been due to changes in the wave, current and sediment regime within the estuary, rather than changes in sea level.

More recent losses of intertidal habitat associated with the expansion of ports have been considerable, as an examination of the two examples in Fig. 75 illustrate. As usual, there is very little information available on the effect these reclamations had on birds. In the case of the Taff, the only clue is a statement by a committee of the Cardiff Naturalists' Society in 1925 that 'the mudflats and salt marshes around the Port of Cardiff, owing to the extension of the city, no longer attract the great

**Fig. 75.** Losses of intertidal habitats in two British estuaries. Teesmouth has now been reduced to only 6% of its original area and the Taff to 44%. Redrawn from Evans (1979) and Ferns (1987).

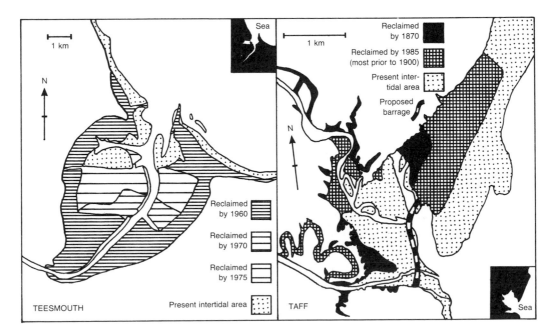

numbers of wildfowl and waders which formerly visited them in winter'. Major reclamations for industrial purposes have taken place in virtually all estuaries adjacent to towns and cities, including the Thames, Firth of Forth, Clyde, Mersey, Dee, Severn and Southampton Water (Prater, 1981). Significant losses still occur as a result of port expansion, as at Felixstowe Docks, even though these sites are nominally protected against such developments. Present proposals for a barrage across the mouth of the Taff (Fig. 75) would result in the loss of most of the remaining mudflats and nearly all of the birds. Losses of intertidal habitat have also been sustained as a consequence of developments associated with the extraction, transportation and processing of oil. The latter are often in less industrialised estuaries such as the Cromarty Firth, Dornoch Firth, Sullom Voe, Milford Haven and Poole Harbour.

Many power stations are situated close to the coast because of the ready availability of water for cooling, and the easy access to cheap, imported fuels, notably coal. On the whole, the discharge of cooling waters is not a major problem, since the total heat input is small compared to that caused by sandbanks heating up at low tide on a sunny day. Specialised plant and animal communities sometimes develop at hot spots. Indeed, discharge points can become favoured feeding places for gulls and terns as at Dungeness, and at West Thurrock on the Thames, because the upwelling warm water brings food items to the surface. However, this heat represents the loss of a valuable resource which could be put to better use than merely warming up the sea. The burning of coal results in the production of large quantities of fuel ash. This can be used to construct building blocks and other materials, but a great deal is also dumped locally. Fuel ash settlement lagoons sometimes provide good wader roost sites, but they are of no value as feeding sites. Intertidal areas, freshwater marshes and other natural habitats are prime targets for ash dumping because of their low commercial value. Other forms of industrial waste, and even domestic refuse, may also be used to reclaim the foreshore as has happened near Edinburgh in the Firth of Forth (Furness, 1973a).

While major industrial reclamations usually receive public attention (Fig. 76), and can be objected to by conservationists, there is also a much more insidious process of 'nibbling' away as a result of small developments, such as roads, marinas, slipways, housing, rubble disposal and so on. Small losses of this sort represent the major cause of the reduction in intertidal habitats at sites such as Langstone Harbour (Prater, 1981).

Reclamation of intertidal land for agricultural purposes also continues to be a major threat to estuaries in rural areas. The expense of reclamation means that intensive cultivation must be introduced as quickly as possible and thus these areas are well drained from the outset. Around the Wash, some 320 km² of land have been reclaimed from the sea since the Middle Ages (Fig. 77). Despite agricultural surpluses, it has been argued that continued reclamation of the Wash would be economically viable if the land were used to produce crops that are not being over-produced, such as field vegetables and bulbs (North, 1987). This fails to take account of the fact that more of such crops would be produced on existing land if there were quotas or a lower guaranteed price for cereals.

Unlike the situation on the Severn, the present sea walls on the Wash are some 10 km or so to the seaward side of the Roman walls. Since new saltmarsh has usually formed in front of each sea wall after it has been

**Fig. 76.** Almost total destruction of a coastal habitat. This part of Port Talbot in South Wales once supported 2000 wintering White-fronted Geese and many other coastal waders and wildfowl.

constructed, this has led people to suggest that an equally large intertidal area accumulates after each new phase of reclamation and so there
is no net loss. This would of course be true if the intertidal area was
reclaimed at about the same rate as fresh sediment accumulated.
However, the rate of coastal erosion, which supplies a good deal of the
sediment, has been reduced due to human activities (e.g. the construction of sea defences) and thus deposition has not kept up with reclamation on the Wash. As a result, the low water mark has not shifted
appreciably and the shore profile has steepened. The intertidal area has
thus been reduced and there has been a complete loss of the intermediate sand flats in some places (Evans & Collins, 1987). The first birds to
have been affected are those dependent on the lugworms in these areas,
namely Bar-tailed Godwits, Oystercatchers and Knots (Goss-Custard
*et al.*, 1987).

The largest reclamations of intertidal land in Europe have occurred in
countries bordering the Wadden Sea, a site of critical importance to
British wildfowl and waders. Prior to the construction of the first dykes
about 1000 years ago, people built their houses on raised mounds to
escape occasional floods. The tidal range is so small in this area (1–2 m)
that flooding usually occurs as a result of storm surges which may pile
several metres of water up on top of normal tide levels. As on the Wash,

**Fig. 77.** Reclamation of the
intertidal area of the Wash
during the last millenium or
so (present mudflats not
shown). Redrawn from
Doody & Barnett (1987).

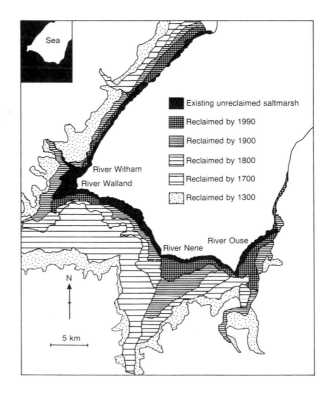

Table 16. *Relative changes in the numbers of the commonest species of waders and wildfowl following partial enclosure of mudflats by a new dyke near the Danish/German border in the Wadden Sea (from Laursen, Gram & Frikke, 1984)*

|  | Denmark | Germany |
|---|---|---|
| Area reclaimed (ha) | 1100 | 550 |
| Loss of high level mudflats (%) | 92 | 1 |
| Bird population changes (%) |  |  |
| Brent Goose | −81 | −37 |
| Shelduck | −55 | −14 |
| Mallard | −60 | +211 |
| Wigeon | −30 | +93 |
| Oystercatcher | −12 | +118 |
| Bar-tailed Godwit | −77 | 0 |
| Dunlin | −82 | −5 |

there has been a steady seaward progression of successive dykes. At first only saltmarsh was reclaimed, but the introduction of steam engines meant that it was possible to pump rainwater out of even lower lying land and mudflats began to be reclaimed. In the German section of the Wadden Sea, which has about 230 km of coastline, some 1070 km$^2$ of land was embanked between 1570 and 1870, and a further 102 km$^2$ between 1870 and 1970. The coastline has thus been moved about 15 km seaward during this period. Present new accumulations of sediment are only sufficient to permit saltmarsh to accumulate in a few areas. In addition, the barrier islands which form the outer limit of the Wadden Sea are slowly migrating landwards, and this is also reducing the intertidal area (Drijver, 1983).

During the last 40 years, 350 km$^2$ of intertidal land in the Wadden Sea has been embanked, but not all of this has been immediately reclaimed. For example, the damming of the Lauwerszee has been used to create a large freshwater lake. One of the most recent reclamations straddles the German/Danish border where some 17 km$^2$ of saltmarsh and mudflat were enclosed in the early 1980s. This area formerly supported some 19000 waders and 9000 wildfowl. Since reclamation, there have been considerable declines in most species (Table 16), especially in the Danish sector where a large area of high level mudflat has been lost. Some of the birds moved into the German sector (where mostly saltmarsh was reclaimed), or to other parts of the Wadden Sea. None the less, there has been an overall decline in the numbers of Brent Geese, Shelducks and Dunlins. Brackish and freshwater lagoons have been created to the landward side of the new dykes as a combined measure for flood protection and conservation, but they do not provide full compensation for the lost feeding areas, tending to benefit species that were previously present in small numbers, such as Teal, Pintail and Shoveler (Laursen, Gram & Frikke, 1984).

In the Delta area of the Netherlands a somewhat different approach has been adopted in recent flood protection works. Part of the Delta was dammed several years ago but the latest barrier (across the Ooster-schelde) consists of a series of gates, rather like the Thames barrage. The gates are normally open, but can be closed to keep out the highest tides and storm surges, so as to protect the land from flooding. The whole Delta scheme is designed to prevent another disastrous flood like that of 1953 when almost 2000 people and many livestock were drowned. Several large freshwater lakes, plus one brackish and one saltwater one, have now been created in the Delta out of what were previously arms of the sea. The damming of the two westernmost arms in 1971 destroyed 135 km$^2$ of intertidal land and as a consequence large numbers of waders and wildfowl were lost, despite an increase in bird diversity resulting from the availability of the new permanent lakes (Meire *et al.*, 1989).

The final Oosterschelde storm surge barrier was expected to reduce the intertidal area only a little because of a decline in tidal range from 3.4 m to 3.0 m (Leewis, Baptist & Meininger, 1984). In fact, when the barrier was finished in 1986, currents in some navigation channels where found to be dangerously strong. In order to reduce them, the barrier was kept closed for longer than planned and this had the effect of reducing the intertidal area by 35%. The Oystercatcher population had already increased in the Oosterschelde because of the displacement of birds from further east in the Delta, and the same was true for Curlews and Dunlins. The area had thus reached its apparent carrying capacity (Lambeck, Sandee & Wolff, 1989). During severe weather in February 1986, the Oystercatchers in the Oosterschelde weighed about 40 g less than in the previous winter (before closure), whereas birds in the unaf-fected Westerschelde maintained their weight. During this cold spell over 5000 Oystercatchers died in the Oosterschelde whereas none did so in the Westerschelde (Smit, Lambeck & Wolff, 1987). This example demonstrates vividly how habitat loss and reduction of tidal range can concentrate birds into the remaining feeding areas to such an extent that their mortality rate increases as soon as conditions become difficult.

The Delta experience also sheds light on one of the most important fundamental questions underlying the effects of losses of intertidal habitat, i.e. are estuaries already full of birds or can more of them be packed in? Experimental studies of Oystercatchers subjected to artificial tides only a third as long as normal, show that they do have the capacity to maintain their overall rate of food intake by simply speeding up

(Swennen, Leopold & Bruijn, 1989). It is tempting to generalise from such a result to suggest that more Oystercatchers could fit into smaller areas, and the fact that the British population is currently increasing seems to support this view. However, existing food supplies may not be sufficient to support the faster feeding rates required for survival in colder weather, especially if they have been depleted by an excess of birds earlier in the season.

In some species, winter food supplies do not appear to be limiting and, in those which are no longer shot in such large numbers, populations are generally increasing, e.g. most species of geese. There is some evidence, however, that nesting habitat is beginning to be limiting. For example, the rate of increase of Svalbard Barnacle Geese is slowing because the quality of food supplies around the remaining nest sites is poor and because the vegetation of brood rearing areas is beginning to deteriorate (Prestrud, Black & Owen, 1989).

In the British Isles, tidal power barrages have been proposed for a number of estuaries, notably the Severn and Mersey. The currently favoured scheme for the former area (Fig. 78*a*) would cost in the region of £10 billion and would generate 16.7 TWh per year of electricity on the ebbing tide, i.e. about 7% of present consumption. Sluices and turbines would allow the flood tide to enter and there would be some pumping at

**Fig. 78.** The Severn tidal power barrage (*a*) and some other tidal reclamation schemes in Great Britain (*b*). S = sluices, T = turbines (numbers indicated in both cases), L = small vessel locks, Lock = major shipping lock. Redrawn from Ferns (1989).

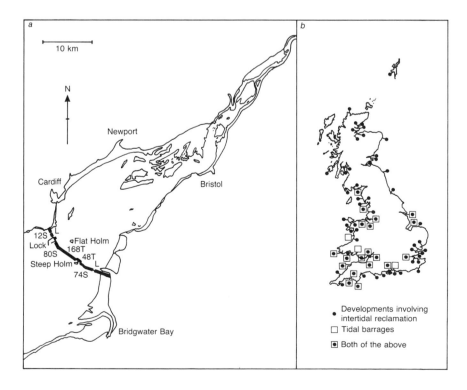

high tide to increase the water level behind the barrage. There is much to be said for a scheme that could produce electricity without generating any atmospheric pollution, but there are difficulties in dealing with the large surges of power which a barrage would produce. Also it does not reduce the need for other power stations much, since alternatives have to be available when the barrage is not generating. The environmental effects of tidal barrages are difficult to predict. In the Severn, the large amount of sediment suspended in the water column results in invertebrates being generally smaller than in other estuaries. This is one reason why it is presently one of the most important British estuaries for our smallest coastal wader, the Dunlin, which has an average peak wintering population there of about 45 000. A barrage would roughly halve the size of the intertidal area, but might allow it to become more productive (since the currents would be reduced), so that higher densities of birds on the remaining mudflats could possibly offset some of the losses.

At present, no fewer than 41 estuaries are considered suitable for tidal power generation by the Department of Energy (1989), of which 14 have been selected for more detailed study. These include the top five estuaries in the country for shorebirds – the Wash, Morecambe Bay, Thames, Dee and Humber (Fig. 78b). Perhaps the most likely tidal power barrage to be constructed first is the Mersey, a site that supports more Teal and Pintail than any other estuary in the country (over 9000 of each species). Tidal power barrages retain some intertidal habitat, albeit a reduced amount, and do at least offer a non-polluting source of energy. The same cannot be said for the plethora of amenity barrages which are currently fashionable. Some of these have the explicit function of covering 'unsightly' mudflats in order to encourage developers of waterside housing and offices whom, it is claimed, would otherwise be put off. This blinkered view is slowly being replaced by the realisation that a changing tidal scene, complete with mudflats and birds, is just as attractive to those who live and work there, as well as being of greater ecological value.

In some parts of the world, such as the Mediterranean, tourism constitutes a major threat to intertidal habitats. As well as replacing coastal fringe habitats, the construction of hotels and other tourist facilities causes intense disturbance in the intertidal zone. The answer to these problems has to be a more enlightened zoning of the coast and the declaration of certain inviolable conservation areas. However, it is not easy to convince countries with hard pressed economies, and a need for foreign currency, of the merits of such proposals. Only when the wildlife that such areas protect become tourist attractions in themselves do the benefits become apparent.

Against all these losses of coastal habitats have to be set some changes brought about by man which have been favourable to coastal birds. One of the most striking has been the accidental provision of cliff-like nesting sites on various types of buildings (see also p. 104). Such new habitat has been extensively exploited by many gulls, birds of prey and Rock Doves. The Rock Dove was domesticated for food around 4500 BC, probably by the provision of special buildings (columbaria) for breeding which simulated a series of rocky niches (Simms, 1979). Kittiwakes were one of the first gulls to exploit buildings as nest sites. The colony on a warehouse on the banks of the Tyne, studied by John Coulson and his students, has been in existence since 1949. Larger gulls rarely nested on buildings prior to the 1940s, but BTO surveys in 1969 and 1976 revealed that Herring Gulls had increased at 17% per annum during this period, with 3000 rooftop pairs by the latter date. Herring Gulls were widely distributed throughout Great Britain, whilst Lesser Black-backed Gulls were only abundant on rooftops around the Bristol Channel (Fig. 79) and Great Black-backed Gulls were quite rare.

One of the possible reasons that buildings became so attractive to gulls was their close proximity to refuse tips which were an important

**Fig. 79.** Distribution of Herring Gulls (*a*) and Lesser Black-backed Gulls (*b*) nesting on buildings in 1976. There are rather few colonies far from the coast, except in the south west. Redrawn from Monaghan & Coulson (1977).

Number of pairs

- ● > 100
- ● 26–100
- ● 6–25
- • 1–5
- · Prospecting

source of food during the major gull population increase that occurred in Europe and North America during the 1960s and 1970s (e.g. Drury, 1965; Mudge & Ferns, 1982b). This population increase probably occurred because the ready availability of food resulted in an increased survival rate of chicks and adults. Closeness to the feeding areas allows a faster rate of provisioning of the older chicks than is possible at the traditional colonies on cliffs and offshore islands, at least in sites such as the Bristol Channel where the majority of the population is dependent on refuse tips. Rooftops thus offer the potential for rapid chick growth and better survival because of more regular adult attendance. On the other hand, gutters and drains may become blocked with nesting material. Workmen on the roof, and even people in the street below, may be harrassed by the adult gulls. These factors in turn result in action being taken to prevent nesting. Thus, in Cardiff, for example, although chicks on rooftops grow faster, the overall number of chicks produced per pair is lower than at natural colonies.

Rooftop nesting is quite widespread elsewhere in Europe, with for instance no less than 2500 pairs of Yellow-legged Herring Gulls in Bulgaria (D. Nankinov, cited in Patterson, 1989). Oystercatchers also nest on buildings occasionally (e.g. Munro, 1984), their young presumably having more chance of survival than most other waders in these circumstances because they are fed by their parents.

## Habitat manipulation

The distinction between habitat loss and habitat manipulation is rather a fine one. Reclamation of the foreshore for agriculture, considered above, really represents the replacement of one type of coastal habitat by another. The generally low value of such agricultural land for coastal birds is the reason it has been treated as habitat loss.

Large sections of our coastline are protected by means of groynes, breakwaters and concrete facings. The objective of breakwaters is usually to provide shelter, but groynes are specifically designed to prevent the erosion and removal of sandy beaches near holiday resorts or to collect beach material in order to protect an artificial or natural coastline from the worst of wave action. Groynes work by intercepting longshore drift (see p. 21), which means that they deprive areas further along the shore of their source of beach material and so create problems of erosion somewhere else. Thus the construction of a large breakwater to provide shelter for the entrance to Newhaven harbour in Sussex has deprived beaches at Seaford of their source of shingle to such an extent that sea protection work has become necessary. Such changes in the location of accumulated beach deposits do not usually have significant implications for birds, though the protection of the upper shore with concrete facings may result in the loss of some breeding sites.

Sand and gravel extraction also result in changes in the sediment regime of our coasts. Coupled with the above engineered changes, and long-term natural fluctuations in the supply of beach materials (Smyth & Jennings, 1988), it is not surprising that large changes in coastal erosion can and do occur. Unfortunately it is very difficult to ascertain which of the above factors may have contributed in any particular case. For example, significant loss of sand is occurring at Sker, part of the Kenfig Burrows National Nature Reserve in South Wales, but it is very difficult to know if this has anything to do with the offshore dredging of sand in the Bristol Channel or if it is just the result of an increased frequency of storms in recent years. On the other hand, the destruction by wave erosion of a village in Start Bay, Devon, has been directly attributed to the dredging of an offshore shoal (Carter, 1989). Some material is dumped at sea, including industrial and domestic wastes and sewage sludges, but the amount of such material is small compared to the 20 million tonnes or so lost by dredging.

The potential conflict between coastal protection and sand and gravel extraction led the Hydraulics Research Station (now Hydraulics Research Ltd.) to recommend that dredging should not take place in water that is less than 18 m deep at low tide (Ranwell & Boar, 1986). This recommendation could be enforced easily since most such areas are under the control of the Crown Estate Commissioners, but it is widely disregarded. Much aggregate dredging takes place even in the intertidal area itself (at high tide), with at best a very superficial assessment of the environmental effects.

Sand is also taken directly from the beach in some places, e.g. in Shetland for building purposes, in the Outer Hebrides for the liming of blackland and peatland pastures, and at Southport for glass making. The consequences for coastal erosion in these cases can be serious. It can lead directly to increased wave erosion of the shore (as in eastern Swansea Bay), or it may reduce the supply of sand to dune systems, resulting in a reduced dune height and consequently reduced flood protection. Removal of sand from dunes themselves often remobilises a stable dune system and places nearby land and property at risk from inundation both by the sea and sand. Most of these types of sand extraction are fortunately now on the decrease because of the importance of sandy beaches and sand dunes for recreation and conservation. However, the widespread availability of machinery such as JCBs means that small numbers of ignorant people can do large amounts of damage in a short period of time.

Agricultural intensification of land near the coast, or that reclaimed from the sea many centuries ago, still exerts a significant adverse influence on coastal bird populations. Low lying wet grasslands in England and Wales have nearly all disappeared and with them their breeding wader populations. For example, between 1982 and 1989, breeding Lapwings declined by 37% and Snipes by 9%. Oystercatchers, which prefer drier nesting sites, increased during the same period (O'Brien, 1990). As the density of stock increases, so the number of nests and young destroyed by trampling increases. Sheep have even been recorded eating Arctic Tern chicks, and Red Deer the heads of Manx Shearwater chicks, probably to combat calcium deficiencies (Furness, 1988). The machair in Scotland is one of the most important British habitats threatened by agricultural change. The Integrated Development Programme (IDP) for the Western Isles, which ran from 1982 to 1987, aimed amongst other things to improve the productivity of some 20% of the machair and 50% of the 'inbye' land by spending public money on drainage, fencing, herbicides and reseeding. Naturally enough, the crofters welcomed the proposals to increase their standard of living and were very resentful when conservationists attempted to interfere (Robertson, 1982). Had ecological advice been sought at an earlier stage, and had suitable 'environmental strings' been attached to the spending of such large sums of taxpayers' money from the outset, then the story might have been a different one. Unfortunately, the Department of Agriculture and Fisheries for Scotland (DAFS) does not take much notice of advice from conservationists in England or anywhere else. The concern which arose out of the IDP, closely followed by the furore over the afforestation of the flow country in Caithness (also involving large sums of taxpayers' money) have been two factors that caused resentment amongst the Scottish establishment and led to the punitive dismemberment of the Nature Conservancy Council.

The effects of the IDP on bird populations are as yet difficult to assess. It has certainly accelerated the overall process of agricultural improvement which was already occurring, and which will presumably continue in the future. Breeding wader populations are apparently decreasing (Jackson, 1988; Wells, 1991), especially in the areas of dry machair. If the machair is not to be completely destroyed in its present form, a new development plan is needed which will encourage traditional and non-intensive agricultural methods, while at the same time diversifying the economy and improving incomes without environmental damage.

Habitat changes can, of course, be extremely beneficial to birds. A spectacular example is the island of dredged spoil in Chesapeake Bay on the Atlantic coast of the USA. This was created merely as a convenient way of disposing of material dredged from the approaches to Baltimore Harbour. The mud around the edges of the island soon become colonised by intertidal invertebrates and the area was utilised by many thousands of waders during migration, mainly Semipalmated Sandpipers, as well as over 20 000 Scaups.

Positive management undertaken specifically to benefit bird populations can be enormously successful in increasing both the numbers and diversity of bird species. There is no better example of this than the work of the RSPB in Britain. At breeding seabird colonies, protection from erosion, tidal inundation, predation and disturbance can all be provided by appropriate management techniques. Little can be done to improve the food supplies of those birds which feed at sea, other than by applying political pressure to try and ensure the health of the marine environment by the prevention of pollution and over-fishing. However, for coastal birds that feed intertidally or inland, there is great scope for improving their feeding conditions. For example, at Elmley on the Isle of Sheppey, the RSPB has increased the amount of both winter and spring flooding, to the benefit of both wintering and breeding wildfowl (Fig. 80). The creation of brackish water lagoons, either by excavation, or by the damming and subsequent flooding of low lying land, has proved successful in increasing both the numbers and diversity of waders and wildfowl. Breeding populations, for example, have increased by 42% (Everett, 1987). Since shallow water areas provide most of the feeding opportunities, the length of the water's edge, rather than the total water area, is the most important factor affecting the numbers of many species (Hill, 1989).

**Fig. 80.** Positive management for birds by the RSPB at Elmley in Kent, mainly involving the controlled retention of water on the land. 1 = storage reservoirs; 2 and 3 = areas kept mostly wet in winter. s = sea wall; c = wall to retain floodwater. Redrawn from Everett (1987).

## Hunting

Deaths of birds caused by habitat loss and habitat modification are seldom direct and visible, even though they are still the major cause of

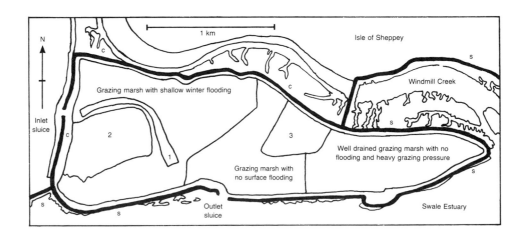

the population changes that have occurred in Europe during the historical period. Most of this mortality occurs when birds are excluded from the reduced areas of breeding or wintering habitat, and either fail to produce sufficient offspring to maintain population size, or die when conditions become difficult (as in the case of Oystercatchers in the Delta area of the Netherlands described earlier).

Hunting and accidental killing are more dramatic and noticeable forms of mortality and tend therefore to attract much more public attention than habitat loss. Prior to the invention of the gun, most hunting of adult birds was carried out by means of crossbows and a variety of traps, nets, springs, snares and bird lime. It is very difficult to derive any realistic estimates of the extent and impact of such hunting activities since so few records are in existence. Firearms were invented in the fourteenth century, but did not become effective as weapons for shooting birds until the invention of the flintlock in the seventeenth century. They were not very quick and convenient to use, however, until breach loading and the cartridge system were perfected in the mid-nineteenth century. We can gain some idea of the impact of firearms in the last century both from the diaries of the collector naturalists, and from the records at major seabird breeding colonies.

A short extract from the diaries of R.N. Dennis, the rector of East Blatchington in Sussex, gives some idea of the casual attitude to the killing of birds even amongst intelligent and compassionate people. Part of his entry for 20 July 1849 reads 'I flushed a small flock of (I believe) six Turnstones killed one of them. I then fired into a small flock of sandbirds and picked up a Turnstone three Purres [Dunlins], then shot at a pair of birds busy feeding among the seaweed at high water and killed them together with a Purre which was feeding beyond them' (Dennis, 1925). The overall impact of such shooting would have depended upon the numbers of people with firearms and the frequency with which they used them, but it is evident that such shooting was quite widespread, involved large numbers of virtually all species of birds, and usually destroyed any rare visitors to an area. Dennis describes both Chough and Osprey as having been shot in his part of Sussex.

In the case of coastal species, both adults and young are particularly vulnerable during the breeding season. Smith (1869) refers to whole boatloads of Puffins, Guillemots and Razorbills from Caldey Island being left to rot on the beach at Tenby, after being shot by sporting tourists. The flightless Great Auk was easy prey, being herded into compounds or sometimes directly up the gangplank and on board ships at the huge colony on Funk Island (Lucas, 1890). Although Great Auks had always been taken for food by mariners, it was probably the feather trade that was responsible for the major decline. The last few pairs were killed by specimen hunters with clubs in Iceland in 1844. The great decrease in the world population of Gannets was primarily due to human activities. Most of the decrease occurred in the nineteenth century and subsequent protection has seen a steady increase this century. For example, Bird Rocks in the Gulf of St Lawrence, Canada, probably supported about 100 000 pairs in the first half of the eighteenth century. This declined to less than a thousand at the end of the century, largely because of the use of adult flesh as bait by fishermen. During this period, the world population of Gannets was reduced by over 60% in 60 years (Fisher & Lockley, 1954). The population on Bird Rocks has now recovered to about 5000 pairs, but the area of the island has been greatly reduced due to erosion by the sea.

Some forms of seabird exploitation have continued over long periods in a regulated fashion without affecting population size, e.g. Gannets, in the past, on Ailsa Craig in the Firth of Clyde and the Bass Rock in the Firth of Forth. On St Kilda and the Faeroes, fowling has also been carried out in a sustainable fashion. The St Kildans, who left the island in 1930, usually ate a Fulmar, Gannet or Puffin every day, mostly young birds taken from the nest. The total harvest of Fulmars possibly amounted to half of the breeding output of the islands. St Kildans also took a great many eggs and sold feathers for the millinery trade. Young Fulmars were banned as a source of food following an outbreak of psittacosis, a disease communicable to human beings.

The trade in bird feathers during the eighteenth century was one of the major factors responsible for the wholesale destruction of seabird breeding colonies. It is hard for us to appreciate how widespread the use of bird plumage was in women's fashion, not simply limited to headgear, but also forming the trimmings of dresses, coats and muffs. A wide range of species was used, from ostriches, storks and egrets to Robins and Wrens. Yarrell (1885) describes how some 9000 Kittiwakes were killed each season on Lundy for their plumage – 'In many cases the wings were torn off the wounded birds before they were dead, the mangled victims being tossed back into the water'. The introduction of the various seabird protection acts, which were designed to end such slaughter, during the breeding season at least, merely switched the pressure to birds overseas. At its peak, in the first ten years of this century, the importation of feathers into Britain, mainly from France, South Africa and the Netherlands, averaged 650 tonnes a year and was worth £2 million (Samstag, 1988). The London market declined in the 1920s and there are clear signs of a population increase in most of Europe's seabirds as a consequence.

The seabird protection acts were effective at limiting the 'sport' shooting at breeding colonies and most species have thus been able to increase in numbers during the present century. For example, the Black-headed Gull came back from the brink of extinction in Britain, and the Kittiwake population has been able to increase by about 50% in every decade since the 1920s (Cramp, Bourne & Saunders, 1974). Such increases cannot be expected to go on indefinitely. Sooner or later, food supplies or breeding sites will run out and populations will stabilise. For example, there are signs that the rate of population increase in Gannets (2% per year) is beginning to ease off (Wanless, 1987). Populations may even decline if there are natural cycles in food supplies, or if fish stocks are overexploited by man (see later). Some people have argued, however, that populations such as that of the Great Skua (Furness & Monaghan, 1987), have risen to far higher levels than those which existed in the past, mainly because the fishing industry has increased the availability of food.

Egg collection for human consumption has occasionally been responsible for slowing population recovery. For example, during the Second World War, there was a distinct reduction in population growth at some gull colonies because of egg collection, though those in more remote areas may actually have received less attention during the war years. Eggs and young birds still represent an important source of food for a few local communities. As many as half a million Short-tailed Shearwaters, or mutton birds, are still taken each year from breeding colonies in Tasmania, and special licences allow for traditional hunting by Eskimos at seabird colonies in countries such as Canada and Greenland. The impact of these activities is often small, provided that traditional methods are adhered to and repeating shotguns, rifles and speedboats are not employed. However, regulations banning the use of the latter are difficult to enforce and are widely ignored (Evans, 1984).

The most intensive exploitation of seabird eggs has taken place in the tropics at colonies of Sooty Terns. During the peak of such activities in the 1940s, between one and two million eggs were harvested from the Seychelles annually. Overexploitation occurred with, for example, the Amirante Islands yielding 5 million eggs in 1930, but only a million in 1954 (Ridley & Percy, 1958). These huge crops were used in the production of barrelled yolk for export, which is now banned because of its wastefulness. None the less, Sooty Terns no longer breed on the islands of Rodriguez and Agalaga and many species continue to decline because of habitat change and unregulated egg taking by fishermen (Feare, 1984). Egg collection remains the major threat to seabirds in the Caribbean, even in areas where it is officially prohibited (van Halewyn & Norton, 1984).

The hunting of adult terns on the African coast, mainly by children using baited hooks, is believed to have been a factor in the decline of breeding Roseate Terns in northern Europe. This species is in fact Britain's rarest regularly breeding seabird, and the population has fallen by 80% over the last 15 years (Fig. 81). Attempts to conserve the species have involved an educational project in Ghanaian schools, funded by the RSPB and ICBP, and supported by the Ghanaian government, as well as wardening of the breeding sites in Europe (Everett *et al.*, 1987). Habitat manipulation is also being carried out at some breeding colonies. The planting of Tree Mallow provides cover which Roseate Terns prefer to nest in, and the provision of nest boxes in exposed parts of colonies provides shelter for the chicks.

Egg collection is a particularly serious threat for rare species. Sandwich Terns were probably eradicated from the vicinity of Sandwich, and Kentish Plovers from Kent (and from many other areas) because their eggs were taken for food when they were abundant and by collectors when they became rare.

During the present century, legislation for the protection of birds from indiscriminate shooting, killing and egg collecting has improved matters in a number of European countries. Indiscriminate shooting of waders and wildfowl continued in Britain even after the 1954 Protection of Birds Act, but was gradually replaced by more responsible wildfowling simply because there were so few sites where the casual wildfowler, not part of an organized and self-policed club, could indulge in his sport. This, combined with the decline in the use of punt guns (large shot guns mounted in boats, capable of killing and crippling dozens of birds with a single shot), is probably one of the main reasons why the populations of many waders have increased during recent years (Tubbs, 1977).

**Fig. 81.** Pair of Roseate Terns in flight. This species is in danger of extinction as a breeding bird in Britain, but intense efforts are being made to conserve it both in Europe and in its African winter quarters. Chris Rose.

Table 17. *Estimated annual numbers of wildfowl killed by shooting in Great Britain. Based on figures in Owen, Atkinson-Willes & Salmon (1986), taken from British Association for Shooting and Conservation statistics*

| Species | Numbers |
|---------|---------|
| Wigeon | 60 000–120 000 |
| Teal | 180 000 |
| Mallard | 840 000–900 000 |
| Total | 1 080 000–1 200 000 |

Restrictions in the list of legitimate quarry species have continued, and these have often served to demonstrate how the population was kept at an artificially low level by sport shooting. When the Brent Goose, Greylag Goose and Shelduck became protected in various European countries, their population sizes increased dramatically. The same has happened in North America where shooting bags can be more easily controlled. For example, the North American Atlantic Brant population fluctuated between 100 000 and 250 000 birds in the 1960s and this allowed annual shooting bags of 15 000–33 000 to be sustained. After a bag of 71 000 in 1971 the population dropped to 73 000. Nowadays, the annual bag is set on the basis of each year's breeding success and shooting can be suspended completely until the population recovers (Owen, 1980*a*).

Wildfowling organisations themselves have become much more concerned about monitoring and regulating the impact of their sport. The British Association for Shooting and Conservation (formerly the Wildfowlers Association of Great Britain and Ireland) established a National Shooting Survey in 1979 and this can be used to estimate annual kills. BASC membership stood at about 63 000 in 1984, and the estimated numbers of ducks shot each year (Harradine, 1985) is about a million (Table 17). This includes an estimate of the number shot by non-BASC members, but there are difficulties with these figures since a survey in 1982 showed that 28% of the 850 000 or so shotgun certificate holders in Britain shot ducks and geese, mostly the former. This means that BASC members include less than half of the wildfowlers in the country. Despite this, the annual kill is unlikely to have been underestimated, and yet, in the cases of Mallard and Teal, more appear to be shot than are counted during the National Wildfowl Counts!

While such figures suggest that the shooting bag is quite high relative to population sizes, it is clearly not having any adverse influence. The populations of all target duck species, except Pochard, are currently at a higher level than they were in 1970. However, regular monitoring is required, since the situation could change rapidly if a series of poor breeding seasons were to occur. Fortunately, wildfowling organisations have become much more conservation minded in recent years. For example, they have designated most of the mudflats of the Wash for which they lease the shooting rights as non-shooting, sanctuary areas. There is thus every reason to expect that they would respond well to shooting quotas or bag limits were it necessary to introduce them. The only current blot on the copy book of British wildfowlers is their continued defence of night shooting, which is now banned in most civilised countries because of the high risk of killing protected species.

Information is rather patchy concerning the impact of hunting throughout the rest of Europe, but an annual bag of about 10 million ducks has been estimated, leaving a surviving January population of about 15 million. Countries such as the states of the former USSR, France and probably Italy (though both Italy and Greece refuse to provide statistics) take a disproportionately large share of this total and the level of hunting pressure which they exert would almost certainly not be sustainable were it applied across Europe as a whole. A shortening of the hunting season and a ban on night shooting have been suggested as possible mechanisms by which these hunting pressures could be reduced. Some reduction seems advisable since the wildfowl population of Europe as a whole, despite large differences between species, has declined over the last 20 years or so (Tamisier, 1985).

Except in Denmark, most of the sea-ducks are not quarry species in Europe. In Britain, three such species are on Schedule 1, Part 1, of the Wildlife and Countryside Act and are therefore given special protection (Table 18). There are several other shorebirds and seabirds on this Schedule. Many coastal birds are game species, in the sense that they can be killed during the hunting season, and three of the largest gulls are on the list of pest species. Of the legitimate quarry species, wild geese, Moorhens, Gadwalls and Goldeneyes are given a small level of protection in that they can only be hunted for personal use and may not be sold. The protection afforded to most geese during recent years has led to large population increases in most species which have in turn led to occasional agricultural damage and calls for population culls (Owen, 1990).

During severe weather conditions, legitimate quarry species (like any other birds) may become very approachable. This is because birds at risk of death through starvation continue to feed, rather than flying away as they would normally do when approached by a predator. In extreme

Table 18. *Coastal birds included in special sections of the 1981 Wildlife and Countryside Act. Other coastal species are afforded ordinary protection under the act, i.e. they cannot be taken (shot or caught, etc.) without a special licence*

*Species protected by special penalties at all times*

| | |
|---|---|
| Red-throated Diver[a] | Little Ringed Plover[a] |
| Black-throated Diver[a] | Kentish Plover[a] |
| Great Northern Diver[a] | Temminck's Stint[a] |
| Slavonian Grebe[a] | Purple Sandpiper[a] |
| Black-necked Grebe[a] | Ruff[a] |
| Leach's Petrel[a] | Black-tailed Godwit[a] |
| Purple Heron | Whimbrel[a] |
| Spoonbill[a] | Greenshank[a] |
| Bewick's Swan | Green Sandpiper[a] |
| Whooper Swan | Wood Sandpiper[a] |
| Scaup | Red-necked Phalarope[a] |
| Long-tailed Duck[a] | Mediterranean Gull |
| Common Scoter[a] | Little Gull |
| Velvet Scoter[a] | Roseate Tern[a] |
| White-tailed Eagle[a] | Little Tern[a] |
| Merlin[a] | Shore Lark[a] |
| Peregrine[a] | Chough[a] |
| Black-winged Stilt[a] | Lapland Bunting[a] |
| Avocet[a] | Snow Bunting[a] |

*Birds protected by special penalties in the close season*
(21 February–31 August below high tide mark, and 1 February–31 August inland)

| | |
|---|---|
| Pintail[b] | Goldeneye |

*Birds that may be killed in the hunting season specified above, but not at other times*

| | |
|---|---|
| Pink-footed Goose | Shoveler[b] |
| White-fronted Goose[c] | Pochard[c] |
| Wigeon[b] | Tufted Duck[b] |
| Teal[b] | Golden Plover[b] |
| Mallard[b] | Snipe[b,d] |

*Birds that may be killed by authorised people at all times*

| | |
|---|---|
| Lesser Black-backed Gull | Herring Gull |
| Great Black-backed Gull | |

[a]Species that must be registered and close ringed if in captivity.
[b]Species that may be sold in the hunting season.
[c]Protected in Scotland.
[d]Close season 1 February–11 August.

conditions, they may metabolise part of their wing muscles in order to remain alive, thus becoming incapable of flight. Not only is it unsporting to shoot them under such circumstances, but very large numbers would be shot if it were permitted. Thus a set of formal meteorological criteria were agreed in January 1979 which would automatically result in the banning of shooting in the British Isles. For the ban to start, over half of 13 meteorological stations across the country have to record the ground as being frozen at 9.00 a.m. on 14 successive days. These conditions would have been met in the winters of 1961/62 and 1962/63 had the system been in operation, and there were bans in 1978/79, 1981/82 and 1984/85. The bans are publicised through BASC and the news media. Under these circumstances, cannon-netting and mist-netting of birds for ringing are also restricted, and bird watchers are asked to show restraint by not disturbing vulnerable birds (Batten & Swift, 1982).

This system has worked very successfully, but it is of limited value unless similar regulations are adopted by the other countries in Europe to which some of these waders and wildfowl migrate during cold weather. At present, no other European country operates such an automatic shooting ban and the result is large kills of wildfowl in hard weather.

Shooting also exerts a powerful influence on the behaviour of birds. Wigeon, for example, may become entirely nocturnal feeders in those parts of the Ribble and Ouse Washes where shooting takes place, concentrating on unshot reserves during the day. In parts of Denmark, Brent Geese and Wigeon may be unable to exploit all of the available *Zostera* because of disturbance from wildfowling (Fig. 82), with shooters in mobile punts being the most disruptive. Since the late 1960s and 1970s, there has been a marked change in the behaviour of both target and non-target species in the Solent, in that all have become much tamer and more approachable. Brent Geese in particular will now feed within 20 m of traffic or humans, which was certainly not the case when they were a quarry species. These changes coincided with a decline in the numbers of punt gunners in the area (Tubbs, 1977). At the present time, migrant waders and wildfowl that breed and winter in areas where they are shot at infrequently are noticeably more approachable than our own wintering flocks, suggesting that shooting influences the behaviour of even those birds which are not supposed to be quarry species.

The enforcement of shooting regulations is poor in some European countries. For example, 34% of 272 live Bewick's Swans X-rayed at Slimbridge by the Wildfowl and Wetlands Trust had lead shot in their bodies, excluding the gizzard, with the average number of shot per bird being three. This is very similar to the levels found in quarry species such as Canada Geese, Pink-footed Geese and Mallard, suggesting that they are all subjected to the same amount of shooting. This is despite the fact that Bewick's Swans are protected in the breeding areas in the former USSR, in the wintering areas and in all countries along known migration routes in-between (Evans, Wood & Kear, 1973). This must largely be due to law-breaking, rather than accidental misidentification, since there are no large white quarry species with which they might be confused. Some of the birds in the above study gained lead pellets during the course of the winter and so must have been shot near Slimbridge itself. In the case of Arctic Skuas breeding on Fair Isle, numbers are directly limited by illegal shooting by crofters (Furness, 1987a). Shooting is probably playing a significant part in the extinction of the Slender-billed Curlew, since it frequently associates with Curlews (a legitimate quarry species in many countries), but is rather less wary. The world population of this species is probably less than 100, and only 4–6 birds were found recently in one of the main wintering haunts in Morocco (van den Berg, 1988).

### Accidental killing

The killing of birds as an incidental result of other activities can be a significant cause of mortality. Seabirds have probably drowned in fish-

**Fig. 82.** Distribution of eelgrass (*a*) and feeding flocks of Brent Geese and Wigeon (*b–d*), in relation to wildfowling in part of the Wadden Sea. Wildfowl abundance in 200 × 200 m squares is expressed as a percentage of the abundance in the densest square. Redrawn from Madsen (1989).

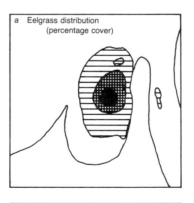

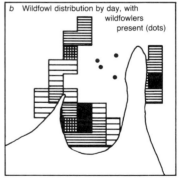

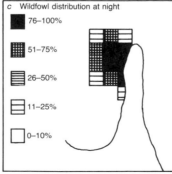

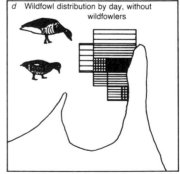

ing nets for as long as man has used them. Hemp, sisal and wound or braided man-made fibres are relatively conspicuous materials and are also quite expensive to purchase and maintain. Consequently, bird mortality in such nets is generally low, except where they are set close to breeding colonies. Relatively little netting of these types is lost or thrown away at sea. The introduction of monofilament nylon netting with welded joints in the 1950s altered the situation dramatically. It is cheap, almost invisible under water, and fishermen are able to deploy enormous lengths of it. It is not suitable for the construction of trawl nets but is used mainly to provide a curtain of netting held near the surface by floats or near the bottom by weights (both known as drift nets). It is highly efficient at catching virtually anything of the right size that swims into it, including tuna, salmon, Mackerel, Herring, squid, turtles, dolphins and seabirds. Considerable quantities are left in the sea because fishermen are unable to relocate it or simply cannot be bothered to collect it. This is often because they have deployed so much that they cannot keep track of it all. Viewed from space at night, the Pacific Ocean often has lines of lights stretching for distances of some 48 000 km (30 000 miles). These are the lights which the fishing fleets of Japan and other nations use to attract squid to the vicinity of monofilament drift nets. Altogether, they set a staggering 2 850 000 km (1 750 000 miles) of net each year (DeGange & Day, 1991).

It is difficult to estimate how many birds are killed in drift nets since fishermen are not keen to provide information that might lead to controls on their activities. Very limited data suggest that perhaps 300 000 birds a year (mostly Black-footed and Laysan Albatrosses) are killed in squid nets, along with 100 000 dolphins, seals and turtles (Smith, 1989). Japanese, Korean and Taiwanese squid fisheries kill an estimated 875 000–1 660 000 birds a year (DeGange & Day, 1991). Japanese salmon fisheries in the North Pacific kill an estimated 96 000–750 000 seabirds, mostly Short-tailed Shearwaters and Tufted Puffins (Jones & DeGange, 1988). A systematic study of mortality in fishing nets off the Newfoundland coast found 100 000 bird and mammal casualties caused by 24 fishermen during a four-year period. This was probably the greatest single mortality factor for a number of the local species (Piatt & Nettleship, 1987). In the South Atlantic, the effects of similar kinds of mortality are beginning to be detected at breeding colonies. For example, declines in the numbers of Wandering Albatrosses are associated with an increase in the number of ringed birds caught by tuna fishermen, in this case on baited long lines (Croxall & Prince, 1990).

A United Nations voluntary ban on drift nets in the South Pacific was implemented in 1991, and while Japan agreed to conform to it by 1992, Taiwan and South Korea (the other major drift netting nations) did not.

Table 19. *Changes in the numbers (percentage in brackets) of recoveries of British ringed auks (from Mead, 1989)*

| Cause of death | Guillemot | | Razorbill | |
| --- | --- | --- | --- | --- |
| | Ringed in 1970 or earlier | Ringed July 1987–June 1989 | Ringed in 1970 or earlier | Ringed July 1987–June 1989 |
| Found dead | 226 (45) | 283 (43) | 307 (50) | 83 (58) |
| Oiled | 142 (28) | 100 (15) | 109 (18) | 22 (15) |
| Shot/hunted | 112 (22) | 35 (5) | 155 (25) | 2 (1) |
| In fishing gear | 24 (5) | 244 (37) | 40 (7) | 37 (26) |

This has resulted in some Japanese fishing companies registering their vessels in North Korea and in the relocation of drift net fishing to the South Atlantic with consequent mortality of Rockhopper Penguins (Ryan & Cooper, 1991).

The mortality in Europe is also considerable. In February 1988, two fishing boats off the Cornish coast were observed hauling in monofilament nets containing 453 dead seabirds, mostly Razorbills. The boats left the area, abandoning another net with 83 auks still in it, 15 of which were still alive. Their splashes at the surface attracted other birds which also became entangled (presumably they thought the activity indicated an abundance of food). Other fishermen were seen filling plastic bags with dead auks, weighting them and dumping them at sea. Storms may have brought these birds further inshore than normal and made them unusually vulnerable to entanglement, but such a level of mortality must have had a serious effect on local breeding colonies, if they were local birds (Ewins, Birkhead & Partridge, 1988). In one abandoned (and illegal) Salmon drift net off the west coast of Scotland, 360 dead seabirds and many dead fish and mammals were counted (Anonymous, 1989). Many British breeding auks, especially Razorbills, die in fishing nets off the coast of Portugal and Spain, e.g. 700 dead Razorbills recorded in one year near Cadiz and 445 in a small village south of Lisbon (Castro, 1984; Teixeira, 1985). These birds are often used as an indicator of the presence of fish and the nets set around them. There has been a significant change in the pattern of mortality of British ringed auks, with a decline in the numbers of birds oiled and shot, and an increase in the proportion drowned in netting and other fishing gear (Table 19). Declines of 45% in the breeding population of Guillemots in Shetland and of 90% in northern Norway have been at least partly attributed to deaths in fishing nets (Heubeck, Harvey & Okill, 1991; Strann, Vader &

Barrett, 1991). In 1990, Britain was instrumental in stopping an outright ban on the use of drift nets in the Mediterranean, on the grounds that the ban might be extended to the Atlantic and the North Sea, even though Britain does not use drift nets in those areas now. It was argued that we might wish to do so in the future!

Small fragments of netting also provide a hazard for seabirds and other marine animals. For example, 3% of Gannets observed flying past Helgoland near the Wadden Sea coast were entangled in fragments of netting or line (Schrey & Vauk, 1987). Such fragments are frequently used as nesting material by Gannets and other birds and this can result in chicks becoming entangled. Colonies often have to be visited towards the end of the fledging season in order to cut free birds that have become anchored to their nests by nylon.

## Introductions

The colonisation and spread of alien species in new areas can provide a threat to bird populations and may induce change in whole ecosystems. Such events occur quite naturally, but at relatively infrequent intervals. However, the great increase in the mobility of humans, especially during the last 100 years, has accelerated the rate of introduction of new species enormously. Introductions often fail because the colonising species are poorly adapted to the new conditions, including different sources of foods, microclimates, predators and parasites. It may require some time before introduced organisms can adapt, and repeated introductions may have to occur before they are successful. The way in which *Spartina alterniflora* from North America adapted to British conditions, in the form of the new species *Spartina anglica*, has already been described (p. 31). Once adapted, colonists may be very vigorous and successful because the specialist predators and parasites that occur in their natural range are absent. The new ones to which they do have to adapt are usually generalists, able to attack a range of species but not particularly successful as control agents.

Another plant species recently introduced to Europe is the seaweed *Sargassum muticum*. This was predicted by Druehl (1973) as an inevitable consequence of the importation of Japanese oysters into French oyster-beds which occurred in the 1960s. It was first discovered in Britain in 1973 on the Isle of Wight and in Portsmouth Harbour. It is a rapidly growing weed which can detach itself and form free floating masses which foul propellers, fishing gear and cooling water intake screens. Attempts were made to eradicate it from the Solent (Gray & Jones, 1977), but it is now common along both sides the English Channel and has spread to Norfolk, the Netherlands and Norway. Its major impact on bird populations is likely to be through the smothering of *Zostera* beds thus reducing wildfowl feeding resources as has already happened in North America.

Much more serious, from the point of view of coastal birds, is the introduction of rats, cats and other predators to islands where breeding colonies are situated. Whilst some of the larger species can coexist with both Norway and Black Rats, smaller burrow-nesting ones cannot, and lose both eggs and young to these predators. The colony of Bonin Petrels on Midway Island in the Pacific is likely to become extinct as a consequence of rat infestation (Vermeer & Rankin, 1984). Dark-rumped Petrels on the Galapagos Islands declined by over 30% between 1977 and 1982 as a result of the loss of 90% of eggs and chicks to introduced Black Rats. In the latter case, adults are also killed by farm dogs, cats and pigs (Moors & Atkinson, 1984). Cats have been deliberately introduced to a number of islands (often to control accidentally introduced House Mice around settlements), as have Mink, Stoats, Ferrets and mongooses. Dogs are introduced as working animals; Foxes and Mink for fur and as pets; and pigs, Polynesian rats and rabbits as a source of food. There are many examples of cases in which such introductions have caused declines or extinctions of seabirds populations. The Guadelupe Storm-petrel is believed to have been made extinct by introduced cats on Guadelupe (Moors & Atkinson, 1984). The islands of St Helena and the Ascensions once supported tens of millions of pairs of breeding seabirds judging by the archaeological evidence. They are the only land in some 15 million km$^2$ of ocean. A combination of introduced predators and direct human disturbance and exploitation has now reduced the population to very small numbers, with the exception of Sooty Terns (Olson, 1977).

We in Britain must take our share of responsibility for the predator introductions that have occurred in many parts of the former British Empire, not to mention the overexploitation of seabirds for food and feathers. It seems we have not learnt from these mistakes, for introductions continue to take place even in the British Isles. For example, Hedgehogs were introduced to South Uist in 1974 and to Steep Holm in 1975 (Legg, 1978). In the former case, the Hedgehogs' proclivity for consuming the eggs of ground nesting birds has produced extra losses at a time when the wader populations of the machair are already declining. On Shetland, Hedgehogs are responsible for the loss of about 10% of Arctic Tern eggs at colonies where they are present (Uttley, Monaghan & Blackwood, 1989). Mink also pose a threat in many parts of Britain. Wherever they are ranched, escapes inevitably occur, and such escaped animals can damage breeding seabird colonies by taking both adults and young. Despite this, there have been repeated attempts to obtain planning permission for Mink ranches on Orkney, Shetland and other Scottish islands. Escaped Mink were responsible for exterminating Britain's largest colony of Common Terns in 1989 as well as a number of other tern colonies on small islands in north-west Scotland (Craik, 1990). Other species, such as Common and Black-headed Gulls, Black Guillemots, Eiders and Red-breasted Mergansers are also being affected (Craik, 1991).

On the positive side, attempts to clear islands of introduced resident predators can be successful if the island is relatively small. Norway Rats have been removed from Cardigan island of the coast of Wales, allowing Puffins and Manx Shearwaters to return (see p. 3). This does not work in the case of small islands in lochs or sea lochs where the problem is visiting predators from the mainland. ICBP have listed some of the world's islands supporting important seabird populations where introduced predators ought to be eliminated or controlled (Croxall, Evans & Schreiber, 1984).

### Pollution

#### Oil

For coastal birds, oil contamination is undoubtedly the most regular and obvious evidence of the adverse effects of pollution. There can be few people who walk along our beaches who have not encountered the pathetic sight of a dead or dying oiled seabird. Regular small spills of oil, coupled with deliberate discharges of tank washings and other wastes, mean that there is always sufficient oil around to contaminate some birds, even in the absence of major incidents. Indeed, small discharges may be responsible for a greater overall mortality than the more noticeable major incidents. Public concern at this overt evidence of bird suffering led the RSPB and the Seabird Group in 1966 to expand an existing survey of bird corpses along the coast. In this Beached Bird Survey, volunteers walked stretches of coast every month from September to March counting the numbers of dead and dying seabirds and their degree of contamination with oil. Extra counts took place during periods of high mortality. The total distance covered in the survey, which was suspended in 1986 (except for a single winter count throughout the whole of Europe), was about 2000 km for each count.

The numbers of corpses found during the 1970s (Table 20) indicate that gulls and auks represent the major part of the mortality, followed by waders and wildfowl. However, the percentage of corpses oiled was greatest in the auks, divers, Gannets and seaducks. February was the month with the highest mortality for nearly all species except gulls.

Table 20. *The 25 most numerous species recorded during the Beached Bird Survey in 1971–79 along 80 000 km of the UK coast, excluding those killed in major mortality incidents. From Stowe (1982b)*

| Species | Total found dead | % oiled |
|---|---|---|
| Herring Gull | 9850 | 9 |
| Guillemot | 8415 | 58 |
| Black-headed Gull | 3749 | 9 |
| Razorbill | 3055 | 61 |
| Common Gull | 2233 | 11 |
| Great Black-backed Gull | 2053 | 10 |
| Oystercatcher | 1815 | 15 |
| Kittiwake | 1785 | 18 |
| Fulmar | 1461 | 13 |
| Cormorant | 1305 | 17 |
| Gannet | 1289 | 29 |
| Shag | 1121 | 15 |
| Eider | 735 | 27 |
| Redshank | 698 | 4 |
| Common Scoter | 688 | 51 |
| Curlew | 528 | 2 |
| Lesser Black-backed Gull | 526 | 9 |
| Shelduck | 508 | 6 |
| Pink-footed Goose | 402 | 1 |
| Dunlin | 309 | 4 |
| Mallard | 303 | 5 |
| Red-throated Diver | 283 | 67 |
| Puffin | 279 | 47 |
| Wigeon | 252 | 4 |
| Greylag Goose | 242 | 2 |

Most of the gull corpses were found in September and were probably juveniles that were unable to feed themselves adequately. The proportions of birds that were oiled were highest in the south and east, and in the Irish Sea, where the major shipping traffic and the most contaminated beaches were found (Fig. 83).

The Beached Bird Survey did not reveal all the oiled birds on our beaches, but it is the only systematic source of information on bird mortality from oiling. Recovery rates of ringed corpses deliberately jettisoned at sea in order to measure how many turn up on our beaches, suggest that each corpse on the beach represents 2–10 dead birds at sea (Stowe, 1982b). Even allowing for this, the overall background mortality of coastal birds caused by oil pollution does not appear to have been sufficiently large to have had a significant impact on population trends. This does not mean, however, that nothing needs to be done to reduce small oil discharges, for they are detrimental to marine life generally and may have a serious local impact on populations. A strong plea has also been made for the resumption of the full Beached Bird Survey because it is the most reliable indicator of changes in the winter mortality of sea-birds (Bourne, 1990) and the RSPB has agreed to revive it. An extension of the European survey is also being considered.

Major incidents in the UK, and their effects, are well catalogued (Table 21), but these represent only a fraction of the occasions on which oil is released into the sea. For example, there were 673 pollution incidents in Milford Haven between 1967 and 1983, involving a total of 3000 tonnes of oil (Little *et al.*, 1987). There was a marked increase in the number of oil spills in Britain in 1989, and this may be evidence of a recent fall in standards.

The *Torrey Canyon* incident in March 1967, involving the release of 124 000 tonnes of crude oil off Land's End, was one of the best monitored incidents. Only a few thousand auks were thought to have been oiled, but the use of powerful and unsuitable detergents and emul-sifiers did serious damage to marine life (Smith, 1970). When the *Amoco Cadiz* grounded off the Brittany coast in March 1978, 227 000 tonnes of light crude oil were released, and an estimated 4500 birds (mostly Puf-

**Fig. 83.** Proportions of oiled corpses (filled segments) recorded by the Beached Bird Survey. The figure in the centre of Great Britain is the overall percentage of that species found oiled. Redrawn from Stowe (1982*b*).

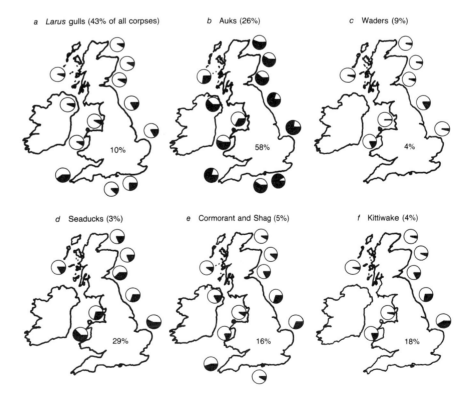

*a*  *Larus* gulls (43% of all corpses)  10%

*b*  Auks (26%)  58%

*c*  Waders (9%)  4%

*d*  Seaducks (3%)  29%

*e*  Cormorant and Shag (5%)  16%

*f*  Kittiwake (4%)  18%

Table 21. *Oiling incidents affecting 50 or more birds in the United Kingdom, July 1966–June 1983. From Stowe & Underwood (1984)*

| Year | Total number of incidents | Minimum number of birds known to have been affected | Number of major incidents affecting > 1000 birds |
|---|---|---|---|
| 1967 | 3 | 11 623 | 3 |
| 1968 | 2 | 1668 | 1 |
| 1969 | 3 | 4838 | 1 |
| 1970 | 9 | 18 892 | 3 |
| 1971 | 8 | 3113 | 1 |
| 1972 | 13 | 2035 | 0 |
| 1973 | 10 | 1825 | 0 |
| 1974 | 9 | 1931 | 0 |
| 1975 | 6 | 798 | 0 |
| 1976 | 8 | 3292 | 1 |
| 1977 | 6 | 3348 | 1 |
| 1978 | 12 | 8112 | 1 |
| 1979 | 16 | 11 847 | 3 |
| 1980 | 10 | 6286 | 2 |
| 1981 | 8 | 3204 | 1 |
| 1982 | 15 | 4929 | 0 |
| 1983 | 6 | 1472 | 1 |
| Total | 144 | 89 213 | 19 |

fins, Razorbills and Guillemots, but including over 100 divers) were killed (Hope Jones *et al.*, 1978). Little was done to establish and enforce a greater degree of safety in the tanker routes to Milford Haven following the grounding of the *Torrey Canyon*, and the result was two further incidents off the Dyfed coast – the *Christos Bitas* (3700 tonnes of crude oil in October 1978) and the *Bridgeness* (200 tonnes of fuel oil in June 1985). The *Bridgeness* incident occurred near the major seabird breeding colonies on Grassholm, Skomer and Skokholm and it was fortunate that the amount of oil released was so small. None the less it did result in the deaths of about 2000 auks.

Oil kills birds in a number of different ways – by the inhalation of toxic fumes, by the swallowing of oil during preening (causing acute poisoning or chronic liver or kidney damage), by hypothermia following the physical collapse of the insulating air pockets in the plumage, and by starvation because the birds are unable to feed. Unfortunately, even tiny amounts of oil, which may cause only a slight staining of the plumage, can be sufficient to kill, by decreasing the insulating and waterproofing properties of the feathers. When food is abundant, birds may be able to cope with this, but at times of stress due to poor feeding conditions it may be sufficient to tip the balance (Levy, 1980). Even a single dose of weathered crude oil fed to nestling Black Guillemots caused a 20% reduction in weight gain (Peakall *et al.*, 1980). The difficulties of cleaning oiled birds (p. 72) are such that few of them survive. However, their chances would improve if better facilities were available to those who take on the thankless task of trying to clean and rehabilitate them. The oil companies and the oil using industries seem reluctant to take responsibility for the effects of their activities. For example, the RSPCA wildlife rehabilitation centre at West Hatch in Somerset receives no regular funding from the oil companies despite the fact that about a third of their normal workload consists of coping with oiled birds.

One of the most recent major oiling incidents was that caused by the grounding of the *Exxon Valdez* (as a consequence of bad seamanship) in Prince William Sound in the Gulf of Alaska in March 1989. Some 35 000 tonnes of crude oil were released and although considerable efforts were made to clean up the spillage, the delay in attempts at containment in the early stages (as a result of faulty and unavailable equipment) meant that these efforts had little effect. The use of hot water to clean beaches seems now to have been more damaging than leaving the oil untouched. Oiled beaches are now difficult to distinguish from unaffected ones, whereas those treated with hot water have not yet recovered. It was estimated that the spill killed nearly 400 000 birds, along with 5000 Sea Otters, and the local Herring fishery had to be suspended. Oil reached the tip of the Alaska Peninsula 900 km away, fortunately missing the millions of waders and wildfowl in the Copper River Delta and the main population of Sea Otters in the east of the sound. The slow rates of bacterial degradation of oil in cold Alaskan conditions means that it will probably take longer to decay than in warmer southern waters, yet the oil company involved threatened to pull out of the area and discontinue with the clean up operations (Pain, 1989) though, fortunately, it did not do so. The amount of crude oil deliberately released into the Persian Gulf in 1991 during the recent war may have been as much as ten times as great as that in the Alaskan incident. In this case, only 20 000 birds were oiled (Dennis, 1991), though most of the grebes and cormorants in the affected areas were completely wiped out, including a high proportion of the world population of Socotra Cormorants.

The development of the oil industry in the North Sea has not so far been responsible for any incidents resulting in the deaths of very large numbers of birds. This is despite the fact that there have been some quite large leakages and that the baseline level of beach pollution, and the number of birds recorded by the Beached Bird Survey in this area, have both risen. However, since the populations of many species in the area have also risen, such evidence is not conclusive. One of the worst incidents involved the *Esso Bernicia* which released 1200 tonnes of heavy fuel oil in Sullom Voe, Shetland, in December 1978. This killed over 100 Great Northern Divers, 250 Long-tailed Ducks and 500 each of Eiders, Shags and Black Guillemots.

Several oil production platforms are situated quite close to the shore, such as those in the Beatrice field in the Moray Firth. It is feared that incidents in this area could have a devastating effect on the large seabird breeding colonies on the cliffs situated not far away – 150 000 individual Guillemots, 70 000 pairs of Kittiwakes, 30 000 pairs of Fulmars and 16 000 individual Razorbills (Mudge, 1986). Equally vulnerable are the moulting and wintering seaducks – 15 000 Long-tailed Ducks, 10 000 Common Scoters and 5000 Velvet Scoters (Campbell, Barrett & Barrett, 1986). Between 1970 and 1984, there were 80 confirmed incidents of oil pollution in the Moray Firth, during which 6000 birds were found dead but these were mainly caused by shore based establishments unrelated to the oil extraction industry (MacLennan, 1986). Were there to be a spillage on the scale of the *Exxon Valdez*, it is a cause for some concern that the Department of Transport's Marine Pollution Control Unit would only have the capacity to deal with about half of the oil released.

### Heavy metals

The pollution of rivers and the sea with heavy metals such as mercury, lead and cadmium is not something that might be expected to have an impact on coastal birds. With the exception of mercury, these metals do not normally accumulate in larger quantities in animals higher up the food chain in the way that happens with many organochlorine compounds.

One of the earliest records of heavy metal contamination in birds came in the 1960s when the deaths of White-tailed Eagles in several countries around the Baltic were attributed to mercury poisoning. The metal could have been ingested via birds and mammals which had eaten seeds dressed with mercury based antifungal agents, but it seems more likely that it was obtained from the eagle's main source of food, namely fish. Cod, Eels, flatfish and Pike from the Baltic all contain elevated levels of mercury (Falandysz, Jakuczun & Mizera, 1988). This is probably a consequence of its occurrence as an antifungal agent in the paper industry's pulping vats. As with the mercury in seed dressings, this form of the metal is readily taken up by living tissues (unlike the raw metal). Even though the use of mercury in paper manufacture has declined greatly because of its effects on fish populations, it may take years for levels in the environment to fall. This is probably beginning to happen in the

Baltic, since the mercury content of Guillemot and Black Guillemot feathers from the area is declining, though it is still high compared with those from museum specimens collected a century or so ago (Applequist, Drabaek & Asbirk, 1985). In Swedish and Finnish Ospreys, feather mercury was low from 1840 to 1940, increased two- or three-fold between 1940 and 1965 and since then has declined a little (Poole, 1989). In both Ospreys and White-tailed Eagles, the highest levels of mercury recorded are considered sufficient to either kill adults or to reduce the hatching success of their eggs.

Detailed work on the mercury content of skua feathers indicates that birds accumulate the metal in their bodies between moults and then deposit it in the first flight feathers to be grown during the moult, especially in the feather tips (which are formed first). Skuas may also combine mercury with selenium to form less toxic bimetallic compounds. A considerable puzzle surrounds mercury levels in skuas. Great Skuas from Foula in Shetland have quite high levels, but those in birds from Gough Island in the South Atlantic (some 2300 km southwest of the South African coast) are even higher. Gough Island was originally selected as a control site for comparison with Shetland, since no mercury contamination was expected so far from any industrialised areas. Mercury levels in 14 other species of seabirds from Gough are also high (Muirhead & Furness, 1988). The explanation for this is unknown, but the source of mercury (and cadmium) in these birds (and in tuna and other wide-ranging fish from the world's oceans) is assumed to be natural. Those species on Gough that feed on squid have the highest cadmium levels, but there are no correlations between mercury levels and diet. Since mercury levels in tinned tuna are sufficiently high that its daily consumption by humans is not recommended, it should be a matter of some priority to all of us that we establish whether the source is natural or man-made.

Fulmars and Kittiwakes from St Kilda also have high tissue cadmium levels and in this case show evidence of damage to the normal structure of their kidney tissues, although this does not appear to prevent them from living normal lives. The presence of damage suggests that exposure to cadmium is a relatively recent phenomenon, since the body has detoxification mechanisms designed to prevent both mercury and cadmium from causing cellular damage. These mechanisms involve the formation of organic complexes with the metals, known as a metallothioneins, which are then locked away in isolated cellular compartments (Cain, 1984). Birds regularly exposed to high cadmium levels would be expected to have evolved metallothionein production rates high enough to have prevented damage.

Heavy metals have also been recorded in waders, though the mercury and cadmium levels never approach those of seabirds. Cadmium and zinc levels tend to be higher in estuaries such as the Severn which receive significant pollution from smelting works. Lead levels, on the other hand, are high in areas in the vicinity of sources of organic lead pollution, such as the Mersey (Evans *et al.*, 1987). Lead has received a great deal of attention as an environmental pollutant, due to the possibility that blood levels in children from schools and homes situated close to busy road junctions may be sufficiently high to cause mental impairment. The source of this lead is petrol, in which it is used as an anti-knock agent. The manufacture of this lead (in the form of organic alkyl compounds) has itself caused pollution. A factory making these products is situated on the banks of the Manchester Ship Canal which is connected to the Mersey. In autumn 1979, 2400 shorebirds were found dead or dying in the middle reaches of the Mersey Estuary. A similar incident, involving mostly gulls (< 1000 birds in all) occurred in autumn 1980 and there were a few more casualties in autumn 1981. The most abundant victim in 1979 was the Dunlin (*c.* 1350), followed by the Black-headed Gull (*c.* 400). Subsequent tissue analyses showed that the birds had been poisoned by alkyl lead. Apparently healthy birds from the Mersey also showed signs of incipient lead poisoning, i.e. greenish bile and enlarged gall bladders (Bull *et al.*, 1983). The 80 kg per day of trialkyl lead, which was discharged into the Manchester Ship Canal is sufficient to account for the amounts found in the estuary and its plants and animals, but the reason for the acute incident in 1979 is still uncertain.

Poisoning as a result of the deposition of lead shot from shotgun cartridges and anglers' weights has not been the major problem in the coastal environment that it has in freshwater areas, though there have been local problems. In 1981 over half the English Mute Swan population died from lead poisoning, but the ban on the use of lead in angling weights has brought a dramatic recovery.

The major sources of heavy metals in the marine environment are sewage outfalls and sewage sludges. For example, 76% of the 41 tonnes of cadmium discharged to estuarine and coastal waters in the UK each year are derived from sewage. Most of the cadmium in this sewage comes from industrial sources (Hutton & Symon, 1987). Several heavy metals adsorb onto the organic fraction of estuarine and marine sediments, especially humic substances (similar to humus in the soil). This makes it all the more likely that they will enter the food chain since invertebrates eat such materials (Gardiner, 1974). Some metals, including zinc, copper and cadmium, may also be absorbed directly from the water by polychaetes and other invertebrates (Mason, Jenkins & Sullivan, 1988). The former two metals are essential nutrients required in small quantities for normal metabolism, but which cause poisoning at high concentrations.

It is not difficult to reduce the amount of heavy metals in effluents to a safe level. In 1980, contamination of American lobsters in the Gulf of St Lawrence, Canada, with cadmium was so great that it constituted a threat to human health and the lobster fishery was closed. The local lead smelting works introduced an effluent treatment process and reduced its annual cadmium discharge from 25 tonnes to less than one tonne. Cadmium levels in the lobsters declined by 50% and the fishery has been able to reopen. Provided that dredging of the nearby harbour does not remobilise buried cadmium in sediments, prospects for the future look good (Uthe *et al.*, 1986). Disturbance of buried sediments can release a number of heavy metal contaminants. This happened in Budle Bay on the Northumberland coast, where increased bait digging for lugworms lead to a rise in cadmium and lead levels in local invertebrates (Howell, 1985).

A heavy metal that finds its way into coastal waters in a quite different way is tin. This is used in alkylated and arylated forms as an anti-fouling ingredient in paints. These organic forms of tin are used because they are sufficiently toxic to marine life to prevent organisms from attaching themselves to boat hulls. When deformities amongst commercial oysters and the death of marine life in marinas, and other enclosed areas with a high density of boats, was attributed to tin from anti-fouling paints, the paint industry's response was to launch a major campaign opposing regulations against the use of their products (Alzieu & Heral, 1984; Stebbing, 1985). Fortunately, the Department of the Environment was able to demonstrate that harmful effects occur at levels of tributyl tin of less than one part per billion (p.p.b.) in sea water, and that enclosed waters in some parts of the UK reached 14 p.p.b. Although regulations have now been introduced, there are exceptions to the ban on organotins, for example, they can still be used on aluminium hulled vessels where the copper based alternative anti-fouling paints cause corrosion, and in fish farm cages and mooring lines where copper might prove toxic to fish. Moreover, copper based paints are allowed to have 0.4 g tin per 100 ml of paint as a 'booster'.

### Organochlorines

Organochlorines have turned out to be a major contaminant of both terrestrial and marine ecosystems. DDT was the first such compound to come into widespread use when it was introduced as an agricultural insecticide just after the Second World War. In the environment, DDT rapidly degrades to DDE and when consumed this accumulates in the fatty tissues of animals. Carnivores thus consume more than herbivores and there is a biomagnification effect. The direct toxicity of DDE to birds is not particularly high, but it does cause egg-shell thinning, and thus more frequent egg breakage and higher embryo mortality. These effects were first appreciated because of a widespread decline in the Peregrine population resulting from a reduction in their breeding success (Ratcliffe, 1970). Many inland and coastal populations of Grey Herons in Britain, and darters, pelicans, herons, egrets, bitterns, ibises and storks in North America, all exhibit egg-shell thinning as a consequence of DDE residues (Ohlendorf, Klaas & Kaiser, 1978; Cooke, Bell & Haas, 1982). Small waders wintering along the coast of Texas had elevated DDE levels (sufficient to impair reproduction in some individuals), as a result of run-off from agricultural land, eight years after DDT was withdrawn from use as a pesticide (White, Mitchell & Kaiser, 1983). Other organochlorine pesticides, such as the cyclodienes (including aldrin and dieldrin), which came into widespread use in the late 1950s, had a greater direct toxicity to adult birds. These pesticides were used mainly on arable crops and were largely responsible for the decline (almost to extinction in eastern Britain) of the Sparrowhawk population (Newton & Haas, 1984). Coastal birds are only occasionally killed by the direct ingestion of pesticides, as opposed to being contaminated through the food chain. Gulls and waders were killed in the Netherlands by ingesting baits (illegally) soaked in parathion (Bruijns, 1963), and several hundred Pink-footed and Greylag Geese have died as a result of consuming pesticide dressed seed wheat in Britain (Stanley & Bunyan, 1979).

A more recent problem is posed by the use of pesticides to control sea lice on Salmon in marine fish farms. The most widely used pesticide for this purpose is Nuvan (active ingredient dichlorvos), even though it has not officially been cleared for use and is supposed to be purchased only by veterinary prescription. Dichlorvos is extremely toxic to marine invertebrates and little is known about its rate of breakdown and routes of transmission through marine food chains (Ross, 1989). The use of wrasse, which are known to act as cleaner fish and can remove both adult and larval sea lice, would obviate the need for pesticides if they could be cultured successfully with the Salmon.

Peregrine populations have now recovered as DDE and HEOD (the breakdown product of cyclodienes) levels have dropped. However, coastal populations recovered later and generally still have a poorer breeding success than inland ones. This has been attributed to the higher levels of PCBs (polychlorinated biphenyls), mercury, and perhaps other pollutants, which these birds accumulate as a consequence of eating contaminated seabirds (Newton, Bogan & Haas, 1989).

PCBs are industrial chemicals used in some adhesives, paints and transformers, because of their oil-like nature, chemical inertness and electrical insulating qualities. Unlike the pesticides considered above, they consist of a complex cocktail of closely related compounds and it is very difficult to determine which individual constituents are the ones responsible for producing PCB poisoning in animals. PCBs were first identified in fish and birds in the Baltic, though unidentified peaks which we now know were PCBs had been present on the gas chromatographs used to analyse organochlorine compounds for years previously.

In the autumn of 1969, a series of incidents occurred which became known as the Irish Sea bird wreck. It started in July near Fort William in Scotland when fishermen noticed that Guillemots were unusually tame and approachable. In the autumn, considerable numbers of unoiled dead birds began to be washed ashore. A total of about 15 000 birds, mostly Guillemots but also a few Razorbills and other species, were recorded. Some of the birds were emaciated, and since the wreck followed a period of autumn storms, one obvious explanation was that the birds had used up all their fat reserves whilst they were prevented from feeding during the storms, and had simply starved to death (Bourne, 1989*a*). Examination of their body tissues for organochlorines and metals, on the other hand, suggested a more complicated explanation. The total body burdens of pollutants were similar to those of a control sample of healthy birds obtained from the North Sea, but the levels in some organs, particularly the livers, were very high, especially in the case of PCBs (10–200 p.p.m.). It appears that when the birds metabolised their body fat during the storms, the organochlorines were released and lodged in their livers or kidneys, and so may have poisoned them. Some individuals did have the liver lesions and swollen, pale kidneys symptomatic of PCB intoxication. The fact that some birds came ashore prior to the storms also supported the poisoning explanation (Holdgate, 1971).

Puffins artificially dosed with PCBs using slow release, subcutaneous implants, developed high body levels but showed no reduction in survival and breeding success. This was despite the fact that, overall, levels reached those at which symptoms have been observed in other species (though their liver levels were lower than birds from the Irish Sea bird wreck). Until we can identify and quantify which PCBs are responsible for poisoning, it will be difficult to interpret all these facts properly (Duinker, Schultz & Petrick, 1988). The dosed Puffins did show higher levels of microsomal monoxygenase than has previously been recorded in any seabird. This is an enzyme involved in the breakdown of organochlorine compounds and consequently their body levels of PCBs did decline a couple of years after exposure (Harris & Osborn, 1981).

Wrecks in general seem to be associated with food shortages rather than pollutants (Fig. 84). For example, 20 000 birds, mostly Razorbills and Guillemots, came ashore along the North Sea coast of Britain early in 1983, following a period of NE gales and a shortage of Sprats (Bourne, 1983a). Many auks and gulls were wrecked along the coast of France, Iberia and Morocco in January 1984, including 50 000 Kittiwakes in France alone. There were apparently no signs of oiling and no unduly high residue levels (Bourne, 1984).

Despite these incidents, the populations of most species of seabirds in much of Britain and Europe have been increasing (e.g. Stowe, 1982a; Lloyd, Tasker & Partridge, 1991), at least until relatively recently (Harris & Wanless, 1988). Such evidence has been used to suggest that the regulations applying to pollutants are adequately enforced (e.g. Cormack, 1984, in the case of oil pollution). The organochlorines are a particularly insidious type of pollutant since they have become wide-

**Fig. 84.** Female Common Scoter. This individual was photographed close inshore in the Bristol Channel following a storm. Scoters do not usually allow a close approach and the bedraggled plumage of this individual suggests that it was not in good condition. Harold E. Grenfell.

spread throughout the world, even in areas where they are not manufactured or used, and they are present in the tissues of all of us. We cannot be happy about this until we are confident that they pose no risk to health, either on their own or through synergistic interactions with other compounds. Coastal birds can continue to provide us with clues to the health of the marine environment providing adequate funds are available to monitor population changes and to regularly analyse corpses for contamination. In particular, they can alert us to new risks, some of which may have a direct bearing on human welfare (Natural Environment Research Council, 1983).

Newly recognised pollutants, such as dioxins and furans, may pose threats to human health through the consumption of contaminated fish or shellfish. These compounds arise during the incineration of plastics and solvents, and also occur as impurities and by-products during various industrial processes. They have probably been widespread in the environment for years, but it is only recently that the analytical techniques capable of detecting them have become available, because they occur (and are toxic) at extremely low concentrations. Over the last ten years, most organochlorine compounds have shown a decline in the environment, but there has been an upsurge in the concentrations of dioxins and furans, probably as a result of their derivation from timber preservatives used by the logging and pulping industries. As with many such compounds, the first clue that these chemicals might represent a risk came from the detection of elevated levels in birds (Holden, 1981). Great Blue Herons in British Columbia, Canada failed to fledge any young in a colony with elevated dioxin levels. Dioxins are highly potent embryotoxins even at concentrations as low as a few parts per billion (Whitehead, 1989). In 1988 and 1989, fisheries for crab, prawn and shrimp, and the harvesting of cultured oysters, were banned in the vicinity of ten pulp mills in British Columbia because the animals contained unacceptably high levels of dioxin. It is thought that the dioxins may have originated during the interaction of chlorine bleaching agents with naturally occurring phenolic compounds in wood pulp (Waldichuk, 1990). PBDEs (polybrominated diphenyl ethers), widely used as flame retardants, are another group of potentially dangerous compounds which are becoming more widespread in the marine environment.

### Flotsam, jetsam and other rubbish

Pollution of the marine environment with materials lost and jettisoned by shipping, or discharged into the sea by rivers and sewers, may represent a hazard to coastal birds. Lost and discarded monofilament netting has already been mentioned, but fragments of plastic, polystyrene and rubber can also cause problems. Such materials are widespread in the sea and frequently accumulate along the fronts which are important sources of food to seabirds (see Fig. 30, p.79). Thus, at the front between the Humboldt Current and the coastal waters just off Chile, a concentration of garbage and of plankton feeding birds has been noted (Bourne & Clark, 1984). A line of debris is often a good indication of the presence of a front in the sea. It is not surprising, therefore, that a range of materials is found in the guts of birds, presumably having been ingested in mistake for food. About 10% of the Puffins in the North Sea have pieces of elastic thread derived from the clothing industry in their stomachs. Usually these fragments are in the form of short pieces, but they can become knotted into balls which obstruct the gizzard. Presumably such pieces of elastic look very like small fish or worms in the water, though how they come to be in the North Sea is not known (Parslow & Jefferies, 1972; Harris, 1984).

In a systematic study of debris in birds' guts, only 8% of Fulmars found dead on the Dutch coast and 20% of Fulmars obtained live on Bear and Jan Mayen Islands in the Arctic did not contain any plastic. The average number of fragments in the stomachs of the Dutch birds was 12, and over a third of them had more than ten. Other items these birds had consumed included paper, aluminium foil, fishing hooks and small tar balls (van Franeker, 1985). Such fragments were also present in Fulmars, Manx Shearwaters and Leach's Petrels, but not in Storm Petrels, at Scottish breeding colonies. In this case, all the fragments floated and the smallest sizes were present in the smallest species. There was some evidence that body weight was lower in those birds which had ingested the most plastic (Furness, 1985). Plastic has also been found in the stomachs of seabirds from North America, the South Atlantic and Antarctica (Hays & Cormons, 1974; Bourne & Imber, 1982; Furness, 1983; van Franeker & Bell, 1988). Whether these fragments cause birds any harm is still not clear. Chickens fed ten polyethylene pellets weighing 30 mg each (a typical amount found in Fulmars) showed reduced rates of feeding and growth because of the smaller capacity of the stomach (Ryan, 1988). There is also a possibility that pollutants adsorbed onto the plastic during manufacture might be assimilated by birds. In Great Shearwaters from the South Atlantic there was a significant correlation between the mass of ingested plastic and body PCB burden, suggesting that the former could have contributed to the latter (Ryan, Connell & Gardner, 1988). Although there are alternative explanations for this correlation, PCBs are used in the manufacture of plastics and do occur adsorbed to polystyrene spherules in the sea (Carpenter *et al.*, 1972).

Those species which are successful shoreline scavengers, notably the gulls and crows, have adapted to feeding on a wide variety of human wastes. That from the fishing industry is a special case which will be dealt with later (p. 259). Since the 1950s, a much greater proportion of domestic refuse has consisted of waste food and most of it is disposed of in large landfill sites near our towns and cities. Gulls and crows have been quick to exploit this sudden increase in the availability of food (Fig. 85). The diet of Herring Gulls on refuse tips consists mostly of meat and fat (58% by dry weight) with smaller amounts of carbohydrate-rich material such as bread and potatoes (37%) (Mudge, 1978). One consequence of this new food source was a marked increase in the populations of Herring Gulls, Lesser Black-backed Gulls and Black-headed Gulls between the 1950s and the mid-1970s in Europe (Harris, 1970a; Sharrock, 1976) and a number of species in North America (Vermeer, 1963; Kadlec & Drury, 1968). This in turn has led to the need to control gull populations in cases where they pose a threat to other seabirds, as on the Isle of May (Gordon, 1976), or where they may contaminate water catchments, e.g. on the Lancashire fells (Wanless & Langslow, 1983).

Since the mid-1970s there have been population declines in most of these species which may have been caused by different factors in different areas. In the north, where a good deal of fish is taken, it is part of the general decline in seabird populations which is currently taking

**Fig. 85.** Gulls feeding at a refuse tip. The widespread availability of food at landfill sites played a significant part in allowing the increase in gull populations which occurred between the 1950s and the 1970s. Chris Rose.

place (Harris, 1989). In some areas, it may be due to a decrease in the amount of refuse available at landfill sites through better tipping practice, the employment of recycling plants, incinerators and pulverisers, and even the deliberate exclusion of birds by the use of nets. However, a major and inadequately understood phenomenon is the mortality of adults during the summer months from botulism. For example, between 1983 and 1985 on Flat Holm in the Bristol Channel, over 500 adult Herring Gulls died from botulism, during a period when the breeding population fell by 185 pairs (56%).

Botulism is a form of poisoning caused by toxin from the bacterium *Clostridium botulinum*, which is widespread in the environment and multiplies in the anaerobic conditions found in rotting organic matter, especially during the summer months. Gulls are mainly affected by the strain (type C) that is found in broiler chickens. Botulism seems to be more widespread in gull populations that are heavily dependent on refuse as a source of food. There would appear to be three possible ways in which they might ingest the toxin. One of the earliest suggestions was that small pools at refuse tips might become contaminated. A more direct route is via poultry litter containing dead chickens used as a top dressing on fields (Hogg, White & Smith, 1990). A third possible source is cooked or uncooked chicken containing *C. botulinum* spores which are activated by the anaerobic conditions found in putrefying refuse. Chicken is one of the commonest food items eaten by gulls scavenging at tips, forming some 30% of the diet (Mudge, 1978).

As gull populations have declined, so the mortality rate from botulism appears to be declining. This is presumably because the birds that are left are those which avoid ingesting infected food, perhaps because of a distaste for putrid meat, or because they concentrate on other food sources. It is also possible that they have developed a resistance to the toxins. It is to be hoped that populations remain at the present more modest levels since gulls can adversely affect other breeding seabirds.

### Sewage and other organic discharges

Sewage is a subject that most of us prefer not to talk about, but the way in which we dispose of it does have implications for birds. It is the major component of the large quantities of organic material that now enter the marine and coastal environments of Europe. The main sources are raw sewage, dumped sewage sludges, other sludges (derived from anaerobic digesters) and agricultural wastes. In addition, phosphates and nitrates represent a direct source of nutrients entering the sea via our rivers.

Raw sewage in Britain is frequently discharged to sea via relatively short outfalls. The effects that this may have on coastal birds varies from beneficial, through the provision of direct sources of food, to adverse, as a consequence of habitat change. Gulls benefit substantially by directly consuming a variety of wastes in the sewage, including fragments of meat, bread, vegetables and seeds. These find their way into the sewerage system via kitchen sinks and food processing factories. Since the size of the fragments is small, it is the small gulls that most commonly exploit them, notably Black-headed Gulls. Spent barley discharged by distilleries and waste potatoes from food processing factories can provide a source of food for wildfowl such as Goldeneyes, Scaups and Mute Swans. For example, up to 25 000 Scaups occurred in the Firth of Forth off Edinburgh in the 1960s and 1970s, where they fed on waste grain from distilleries and on the enhanced invertebrate populations resulting from organic enrichment by sewage (Campbell, 1978) (Fig. 86). This concentration of birds had disappeared by the 1980s because of improved sewage treatment, leaving the whole of Britain and Ireland with a Scaup population estimated to be only 7000 (Salmon, 1988).

**Fig. 86.** Female Scaup. This species benefited from waste barley discharged into the Firth of Forth by distilleries in the 1960s and 1970s, and virtually disappeared when such wastes became unavailable. Harold E. Grenfell.

Generally speaking, the outfalls with the greatest volume of discharge support the greatest number of birds, provided there is no substantial toxic component, as has been demonstrated for wildfowl in the Tay Estuary (Pounder, 1974) and for gulls in the Bristol Channel.

Sludges are the residues of material left after the treatment of sewage or other waste in digesters. One major problem in disposing of it is its contamination with heavy metals and organochlorines. These become concentrated in sludges because they are not broken down by bacterial action during the digestion process. Anaerobic and aerobic digesters are very efficient at reducing a large volume of organic matter, such as human and animal excrement, into a small volume of sludge. Metals in sludges are derived from everyday sources such as copper pipes, lead-rich road dust and run-off from galvanised metal sheeting and roofing materials. Organochlorines are derived from agricultural run-off and various manufacturing processes. Sludges can be spread on the land as a fertiliser, but the presence of the contaminants limits the quantities that can be applied at any one location. About 60% of sludges are used on the land in Britain and there is a code of practice governing how this should be done (Department of the Environment, 1981). Metals persist in treated soils for very long periods (McGrath, 1987).

The remaining sludges, including about 10 million tonnes of sewage sludges, are currently dumped at sea, mostly in relatively shallow coastal waters. One effect of this dumping is claimed to be an increase in the proportion of fish that develop ulcers, tumours and various other diseases (Dethlefsen, 1984). Other studies have suggested that diseased fish are no commoner near dumping grounds than they are elsewhere (McVicar, Bruno & Fraser, 1988), though this may be because abrasion against fishing nets is a predisposing factor and fishing does not normally take place near dumping grounds (Bucke & Waterman, 1988). The fact that 10% or so of fish from the North Sea show overt lesions and deformities is enough to convince many people that some form of pollution is to blame, especially as fish from other, less polluted, areas do not show such high incidence rates.

The application of organic materials to both estuarine and coastal ecosystems in modest quantities has a fertilising effect, increasing invertebrate productivity and leading to an increase in their predators, including both fish and birds. However, this increase in productivity may be accompanied by a reduction in the species diversity of invertebrate communities. This has happened on the sea bed of the northern Adriatic (Vidaković, 1983) and in the mudflats of the Scheldt Estuary (van Impe, 1985), both areas that receive organic material in the form of sewage. In the Scheldt, during the same 30-year period that an increase in invertebrate productivity has been recorded, there have been increases in the numbers of wading birds feeding there, though the situation is complicated by the loss of so many mudflats elsewhere in the delta.

At high levels of organic enrichment, the water column may become sufficiently deoxygenated that fish numbers decline. This has the effect, in intertidal areas, of increasing the food supply available to waders since they often consume the same types of prey as the fish. During the last 20 years, there has been a decrease in the number of shorebirds wintering in the Clyde Estuary. It has been suggested that improved water quality has allowed fish to recolonise the estuary and so compete for food with the waders. Thus Flounders, and fish predators such as Cormorants, have increased during this period, but Dunlins, Redshanks and Lapwings have declined (Furness *et al.*, 1986). At high levels of organic enrichment, a further range of problems may develop. In estuaries, dense slimy green mats of algae such as *Enteromorpha*, may develop on the surface of intertidal sediments, especially during the summer months. Since the underlying sediment often becomes deoxygenated, invertebrates die and there is a consequent reduction in feeding opportunities for birds (Perkins & Abbott, 1972). Such nutrient enrichment may occur in the direct vicinity of sewage outfalls. Many of our rivers carry large quantities of dissolved phosphates and nitrates (derived from sewage treatment plant effluents and run-off from agricultural land) and so estuaries which look superficially unpolluted may also be affected by similar problems. The Ythan, in Scotland, and Langstone Harbour have both developed algal mats (these mats cover up to 30% of the intertidal area in the latter area) which are blamed for the reduction in numbers of some birds (Tubbs & Tubbs, 1983).

In coastal waters, especially where these are enclosed and relatively shallow, high nutrient inputs can result in blooms of algae, such as *Cladophora*, which in due course deplete oxygen levels in the water column and result in the deaths of fish and benthic invertebrates. This has happened in several parts of the Baltic and the Kattegat between Denmark and Sweden (Rosenberg, 1985). The Baltic as a whole receives almost a million tonnes of nitrogen per year, mostly from the rivers flowing into it, whilst the North Sea receives about a million and a half tonnes. In Tolo Harbour, Hong Kong, a doubling of organic input in the form of animal and human excrement has resulted in blooms of the minute toxic marine dinoflagellate *Gonyaulax polygramma* and there have consequently been large fish kills (Wu, 1988). In 1989, a ban on the human consumption of shellfish had to be introduced (Wu, 1989).

Toxic 'red tides' (so-called because the dinoflagellates are so abundant they colour the sea slightly red) occur in Britain from time to time, especially along the north-east coast. *Gonyaulax tamarensis* is usually the organism involved. In this case, there is no firm link with organic pollution but it is suspicious that no cases of paralytic shellfish poisoning (caused by eating shellfish that have themselves eaten the dinoflagellate) were recorded in humans prior to 1968. Blooms of toxic dinoflagellates now occur every year, but only occasionally at levels that pose a threat to humans, fish and birds. The years 1968 and 1975 were particularly bad, and considerable numbers of seabirds, notably Shags, were killed (Fig. 87). A total of 490 bird corpses were found along the coast of Northumberland southwards to Sunderland in 1975. The majority of these were Shags and Herring Gulls, but there were also some Cormorants, Fulmars and Eiders. Whilst some birds became poisoned by eating shellfish, others, including the Shags, presumably ate contaminated fish. Prolonged easterly winds occurred during the summers of both 1968 and 1975, and it is possible that these helped to concentrate *Gonyaulax* populations along the coast (Armstrong *et al.*, 1978). *Chrysochromulina polylepis* blooms along the west coast of Sweden have likewise been responsible for the deaths of many Eiders consuming contaminated mussels.

A relatively recent source of organic pollution of inshore waters is provided by marine fish farms. The main problem derives from the food that is fed to the fish but which they fail to eat. This, together with the high concentration of fish faeces, leads to heavy organic enrichment in the vicinity of the fish cages. In lochs and bays with relatively weak tidal currents this may enrich the sediments beneath the fish cages to such an extent that they become anoxic and outgassing of $H_2S$ takes place. The

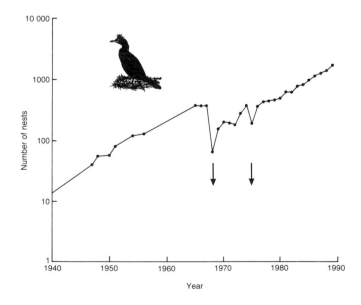

**Fig. 87.** Size of the breeding colony of Shags on the Farne Islands. The arrows indicate those years (1968 and 1975) when red tides caused reductions in numbers through the deaths of adults. Based on information in Armstrong *et al.* (1978) and P. Hawkey (personal communication).

gas is extremely toxic to fish and may be the cause of increased gill damage and mortality observed in some farms. As a result of this 'souring' of the area, the cages have to be relocated in order to achieve better fish condition and growth rates (Lumb, 1989). Given the rapid expansion in salmon farming which is currently occurring in inshore waters, the possibility that it may encourage local phytoplankton blooms also needs to be considered (Frid & Mercer, 1989) and appropriate independent monitoring, funded by the industry, should take place as a condition of granting licences. No such monitoring is taking place in Britain at the moment. One way to improve the situation would be to develop more efficient ways of feeding the fish, so as to reduce the amount of waste (Gowen & Bradbury, 1987).

Although organic pollution of the sea can sometimes be beneficial to birds, the balance between advantage and disadvantage is very delicate and the situation is one which we are doing little to monitor or control. Recent agreements to regulate pollution in the North Sea may in due course provide the means by which the large scale problems of eutrophication, especially algal blooms, may be controlled. Given the range of types of pollution entering coastal waters and their different sources, such agreements will need to cover a wide range of discharges if they are to be successful, including those into rivers as well as directly into the sea itself.

**Exploitation of fisheries**

Man's fishing activities at sea make available a good deal of food which birds would otherwise be unable to exploit. This food falls into three main categories. Firstly, there is the waste resulting from the processing at sea of fish and other marine animals. All of this material, including guts and some blubber, is jettisoned into the sea. The increase in this source of food was suggested by Fisher (1952) as being responsible for the increase and spread of the Fulmar in the twentieth century. The idea has been extended to account for increases in other species such as Gannets and gulls. However, the Fulmar population continues to expand at the present time, despite a more recent decrease in the amount of food waste available at sea (Monaghan & Zonfrillo, 1986).

The second type of food made available by fishing consists of whole fish and other species that do not normally come to the surface but are brought up during fishing activities. For example, during purse seining, some fish from the huge shoals of mid-water and deep-water fish, such as Norway Pout and Blue Whiting, escape as the net is closed. These may become available to birds at the surface and some may be taken from inside the netted area itself. Thirdly, non-target species and fish smaller than regulation size are usually rejected and become available to scavenging birds because they are moribund.

In the Irish Sea fishery for Norway Lobster (scampi), a total of 20 species of seabirds have been observed scavenging on the rejects. Herring Gulls are most abundant, constituting 66% of the attendant birds, with an average of about 550 individuals present at a boat when the catch is being sorted. Kittiwakes are next most abundant (26%), followed by Great Black-backed Gulls, Gannets, Fulmars and Lesser Black-backed Gulls. These species tend to scavenge in different ways, though Sprats and gadoids are the main fish taken by all of them. Fulmars are particularly attracted to the Norway Lobster livers discarded when these crustaceans are being gutted (Watson, 1981).

Lesser Black-backed Gulls breeding on Skokholm Island seem to be very dependent on human fishing activities with well over half of the diet consisting of fish such as Norway Pout. When fishing vessels in the area caught fewer fish in 1989 and 1990 there was a dramatic decrease in the amount of fish brought to the young and in their breeding success. The reasons for the poor fish catch are unknown, but an influx of warmer water into the area, bringing with it exotic species such as Trigger-fish (Henderson, 1990), may also have caused a change in the distribution of commercially important fish and a consequent reduction in the number of fishing vessels in the area. There was a marked improvement in breeding success in 1991.

Furness, Hudson & Ensor (1988) reviewed the use made by seabirds of human fishing activities around the British coast. Industrial fisheries, involving the processing of small fish into fish meal, produced little food for birds. Whitefish and Norway Lobster fisheries, on the other hand, produced a lot, because a significant proportion of the catch was of non-target or undersized fish and was rejected. While Norway Lobster boats are the main source of food in the Irish Sea, in the North Sea it is the whitefish vessels, and in this case fish guts are the most abundant food source. The authors estimated that up to 2.5 million seabirds (weighing 1000 g each) could probably be supported by the waste, with only about 25% of the fish discards and 10% of the offal remaining uneaten. This relatively high efficiency of scavenging is aided by the British practice of not discarding the heads with the guts, as happens in North America (the heads make the offal sink faster). Thus, for the approximately 3 million potentially scavenging seabirds around the British coast, fishing waste is clearly of major importance.

Fulmars tend to physically dominate other species and take most of the fish waste in northern Britain, whilst Herring Gulls are commoner in the south. Great and Lesser Black-backed Gulls, Kittiwakes, Gannets and Great Skuas are the next most abundant scavenging species. It has been suggested that the superior competitive ability of Fulmars, together with their increasing numbers, is beginning to exert an adverse influence on the numbers of less dominant scavengers, such as Herring and Lesser Black-backed Gulls (Hudson & Furness, 1989). Immature gulls are also less efficient than adults at handling fish of the same size (Hudson, 1989).

Even the shellfish industry, exploiting shallow and intertidal waters, can make extra food available to seabirds. For example, cockle gatherers and bait diggers in the Burry Inlet attract Common Gulls, which search spoil heaps for food as well as the wheel ruts of the horse-drawn cockle carts where crushed cockles can be found. Atlantic Brent Geese associate with clam fishermen in the USA where they take *Zostera* brought to the surface of the water during clam harvesting (Kirby & Obrecht, 1980). Eiders may also consume cultured mussels at shellfish farms and thus protective netting or some sacrificial mussels may need to be provided to protect the main stock.

The influence of human fishing activities is not entirely beneficial to seabirds, however. Methods of locating and catching fish have become more sophisticated during the last half-century or so, but the difficulties of agreeing and enforcing legislation to regulate catches and mesh sizes in international waters have remained largely unresolved. Thus there has been a monotonous pattern to the exploitation of commercial fish stocks, which ends in overfishing and at least a temporary disappearance of the fishery. This has happened with North Sea whitefish, Herring, and Mackerel; English Channel sole and plaice; South African Pilchards; South American anchovies and North Atlantic Cod (Furness, 1982; Bannister, 1983; Cooper, Williams & Britton, 1984; Furness & Ainley, 1984). Indeed, it is difficult to find a single example of a large fishery that has not been overexploited in this way, except where it occurs solely within the territorial waters of a single country.

The virtual disappearance of fish stocks through overexploitation has both direct and indirect effects on seabirds. Those species which are directly dependent on the stock obviously become reduced in numbers if they cannot find alternative sources of food. This is demonstrated by figures from the south and west coasts of Africa where guano yields can be used as an index of the changing numbers of seabirds (Fig. 88). The guano is produced mainly by Cape Gannets, Cape Cormorants and Jackass Penguins. In the early years of the Pilchard fishery there was no affect on the guano yield, even though the latter probably fluctuated in relation to the abundance of Pilchards. During the 1960s, the rapid increase in Pilchard catches began to reduce the guano yield. When the fishery crashed in the late 1960s, many boats went out of business and seabird populations recovered a little. Crashes have also occurred in the anchovy-dependent seabird colonies of Peru (also commercially important guano producers). Here Peruvian Cormorants, Peruvian Boobies and Brown Pelicans underwent a catastrophic population decline following overexploitation of the anchovy fishery. The failure of the fishery was triggered by an El Niño event (p. 14), but this was not the cause of the long-term decline in fish stocks. Overfishing might in some cases be of benefit to birds by making greater numbers of small, young fish available than would be the case in a balanced natural population (Perrins, personal communication).

Indirect effects can occur when the food resources that would normally have been consumed by the fish, or other target species, are eaten instead by other animals in the marine food chain. For example, the decline in whale and seal stocks in Antarctica due to hunting has lead to an increase in the productivity of the larger krill. This in turn has allowed considerable population increases to occur in the Chinstrap and Adélie Penguins that feed on the krill (Croxall *et al.*, 1984). Seal populations are now beginning to recover, and krill itself is being commercially exploited, so the situation may change once again. Seals may compete with birds for breeding sites as well as food, as is happening on Mercury Island off the Namibian coast where seals are re-occupying sites taken over by Jackass Penguins and Cape and Bank Cormorants (Crawford *et al.*, 1989).

In the early 1960s, a period of cool northerly winds in winter generated a mass of cooler and fresher water than normal off the Greenland coast. This mass was several hundred kilometres in diameter and

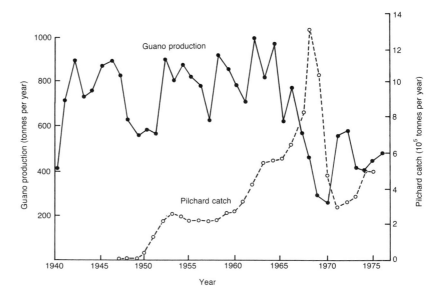

**Fig. 88.** Guano production on Bird Rock, measured by the amount harvested each year, and the size of the Pilchard catch landed at Walvis Bay, off the South African coast. Redrawn from Crawford & Shelton (1978).

400 m deep. It remained identifiable for 17 years as it drifted across the North Atlantic, passing through northern Europe as the 'great salinity anomaly' of the 1970s. Plankton productivity was reduced in the colder water and consequently recruitment to Icelandic fish stocks was poor in 1965–71 and to some North Sea fish stocks in the mid-1970s (Cushing, 1990).

Natural fluctuations of this sort are often the cause of periodic breeding failures amongst birds. Those associated with El Niño, those occurring in 1948–51 in Fig. 88, and perhaps also that which affected Puffins in west Scotland in the early 1970s are all examples of this. It has been suggested that breeding failures amongst Shetland seabirds during the 1980s were due to such a natural fluctuation (Bourne, 1989*b*), but many people believe it was the commercial exploitation of sandeels for fish meal that was responsible. This exploitation began in 1974 and reached its peak in 1982, since when a decline in stocks has occurred. The Department of Agriculture and Fisheries for Scotland pointed out that this decline occurred as a result of poor recruitment at a time when the spawning stock was large – suggesting some other cause than overfishing. Nevertheless, a close season for sandeels, lasting from July to December, was introduced with the support of local fishermen, to be followed in late 1990 by the total closure of the fishery. During the period 1981–83, it has been estimated that the catch, in combination with the numbers taken by natural predators (birds, other fish and seals), was 20% greater than the annual production of the population concerned (Furness, 1990). Such statistics make it hard to resist the conclusion that overfishing played a significant part in the decline. Similar declines in Californian sardines and Peruvian anchovies could only retrospectively be attributed to overfishing. These too are small fish with a short life span and rapid turnover, which makes population analysis difficult.

There is no doubt that the breeding failures of seabirds which occurred in Shetland were directly attributable to a decrease in the availability of sandeels. Chick production in Arctic Terns is directly related to the estimated recruitment of first-year sandeels, which form the main food for the young after they hatch. Three times as many sandeels were brought to each nest on the flourishing colony on Coquet Island off the coast of north-east England as were brought to Shetland colonies in the late 1980s (Avery & Green, 1989). 1990 was the seventh disastrous breeding season in a row for Shetland seabirds. Arctic Terns were the worst affected, with 10 000 pairs rearing less than 100 young in 1989 and only one chick known to have been reared in 1990. However, following the total ban on the sandeel fishery, 24 000 pairs raised 30 000 young in 1991. Breeding failure had previously been virtually continuous since 1984 (Monaghan & Uttley, 1989). The much smaller numbers of Common Terns that nest in the area are less dependent on sandeels, but had none the less bred poorly because of high rates of predation by the Herring Gulls, which had fewer Arctic Tern chicks to eat. Arctic Skuas, Great Skuas, Kittiwakes and Puffins have all suffered serious declines in breeding success, and Red-throated Divers, Fulmars, Black Guillemots and Razorbills have done poorly (Furness, 1989; Heubeck, 1989; Okill, 1989). All these species consume sandeels during the breeding season. In many cases, the adults laid eggs and hatched young, but these either starved to death or were eaten by predators because the parents were absent for longer than usual searching for food.

The fact that some species, such as Fulmars, seem only to have fed on sandeels relatively recently suggests that an initial increase in the size of the sandeel stocks may have been triggered by the removal of their competitors and predators, such as Herring and Mackerel (Anderson & Ursin, 1977; Furness & Todd, 1984). This may in turn have allowed some seabird populations to increase (Furness, 1982).

Changes in fish stocks other than sandeels have already been mentioned, and these too have brought about changes in seabird populations. Sprats declined in the northern part of the North Sea during the late 1970s through overfishing (Johnson, 1983), but Herrings (because they are protected) increased. These changes have been mirrored in the diet of Puffins on the Isle of May (Hislop & Harris, 1985). In the Firth of Clyde where breeding tern and Kittiwake numbers declined between 1970 and 1980, there is a significant positive correlation between the number of pairs and the number of young Herrings available to them, with signs of a recent increase in both. These surface feeding species are thus at least partly dependent on young Herrings in this area (Monaghan & Zonfrillo, 1986).

The relationship between populations of fish and those of seabirds is a two-way process and, not surprisingly, birds are often accused by fishermen of holding down fish populations, although there is little evidence to support this. In the 1920s, an increase in all five species of British breeding terns at Blakeney Point in Norfolk was blamed for the decline in flatfish in the area and there were calls for the colony to be destroyed. An investigation of the terns' diets revealed that none of them consumed any young flatfish at all (Collinge, 1925).

Attempts have been made to assess the overall impact of seabirds both on individual fisheries and throughout the whole country. Furness (1978), for example, estimated that seabirds consume 29% of the fish production within a 45 km radius of Foula in Shetland. This estimate was criticised by Bourne (1983b), who reckoned 0.25-4% was more likely because of underestimates in fish production and bird foraging range, and because birds take a lot of plankton and fish offal. Furness (1984) countered most of these points and produced an estimate of 20% for the North Sea. Both authors in this debate admit the great difficulty in making estimates of this sort, and both seem to agree that whereas human fishing activities have had a significant impact upon seabird populations, birds have had relatively little impact upon the yield of most fisheries.

Birds have also been blamed for causing declines in intertidal fisheries. In the 1970s, Oystercatchers were blamed for the decline in the cockle fishery in the Burry Inlet, which was particularly severe in 1974 and 1975. In this case, there is no doubt that the 20 000 or so wintering Oystercatchers in the area at the time did subsist largely on cockles, and so a cull was encouraged by offering a free shotgun cartridge for every Oystercatcher corpse (the law already allowed this otherwise protected species to be shot in the Burry Inlet). This resulted in some 11 000 birds being killed and the population being reduced to about 12 000. There was no resultant improvement in the cockle fishery. Today, the Oystercatcher population is about the same size as it was before the cull, yet the cockle fishery has shown a dramatic improvement. This is probably because weather, acting through spat settlement, is the major factor determining the size of the cockle population, rather than the large numbers of cockles removed each year by Oystercatchers (Franklin, 1977).

Seaweed harvesting still takes place in a few places around the British coast, either for human consumption or alginate production. Since this involves the removal of living, attached weed (unlike the collection of cast-up weed for use as a fertiliser), it presumably causes a reduction in food supply for some birds that feed on rocky shores. However, this is an effect which has never been measured and areas denuded of weed may in any case be initially colonised by invertebrates such as limpets and barnacles which birds can eat. The Outer Hebrides is one site where seaweed is harvested for alginate production. Most of the cropping takes place in the sheltered sea lochs on the east coast. If 20 cm or more of the harvested *Ascophyllum* is left attached to the rock, it grows relatively rapidly and can be cropped again after about three years (Norton & Powell, 1979).

Bait digging might once have been considered a relatively innocuous activity with relatively little effect on birds. However, its intensity in some coastal areas has become so great that it does have a significant effect on the invertebrate populations of soft sediment shores and consequently reduces the food supply of birds locally. Populations of lug-worms, for example, can be completely wiped out by using hand operated suction cylinders to extract them from the mud. A study of bait digging activities in Swansea Bay, in South Wales, showed no recovery of lugworm densities within dug areas six months after digging, even though worms were still present nearby. In the same area during the course of the year, the majority of stones on rocky parts of the shore were moved or overturned by anglers searching for moulting crabs. However, the impact of this on crab populations is difficult to assess (Cryer, Whittle & Williams, 1987). Bait digging may also liberate heavy metals and other pollutants from buried sediment layers (p. 247).

The exploitation of seabird colonies for guano, as happens at many of the huge colonies utilising the rich food resources of the Humboldt and Benguela Currents, can be done in ways that cause relatively little disturbance to the birds. Indeed it is in the obvious long-term interests of the industry to maximise the number of birds present. This has not prevented the total removal of guano down to the bedrock, and this eliminates those species which rely on burrows for nesting. Between 1880 and 1970, a 90% reduction in the population of Jackass Penguins occurred along the South African coast as a result of such guano harvesting. The construction of causeways to offshore islands, in order to facilitate guano collection, has also caused problems by allowing ready access for predators such as mongooses (Vermeer & Rankin, 1984).

### Disturbance

Many forms of human activity lead to the disturbance of coastal birds, and three main factors have intensified this during the present century. The first is the increasing size of the human population and, the second, its greater mobility because of the widespread availability of motor transport. The third is the increasing amount of leisure time available to most of us fortunate enough to belong to one of the affluent nations. This results in a greatly expanded demand for coastal leisure activities and facilities. For example, in the British Isles, these factors combined to produce a fivefold increase in the membership of the major boating and water sports organisations between 1960 and 1980 (Owen, Atkinson-Willes & Salmon, 1986).

Even seemingly innocuous activities, such as walking along the shore or bird watching, can cause disturbance. For example, walking across the firm Wadden Sea sandflats is a popular recreation in the Netherlands. Smit & Visser (1989) showed that a single person or group of people doing this causes feeding waders to vacate an area of about 0.23 km². When a group of people is noisy and active, this area increases to 1.85 km². Occasional disturbances of this sort have relatively little impact on the birds, but when they occur regularly, the costs to the birds mount up. These costs include the loss of both feeding time and the

Table 22. *Number and percentage (in brackets) of observations of disturbances of wader roosts by craft of different types in the German Waddensee. From Koepff & Dietrich (1986)*

| Type of craft | Occasions observed | Disturbances produced | Percentage disturbances per craft observation |
| --- | --- | --- | --- |
| Yacht | 34 (13) | 8 (12) | 24 |
| Motorboat | 16 (6) | 3 (5) | 19 |
| Canoe | 29 (11) | 24 (38) | 83 |
| Windsurfer | 180 (70) | 29 (45) | 16 |

energy expended in flying away and flying back again. Some areas will be deserted completely if the disturbance is frequent enough. Fortunately, since the shore is usually undisturbed at night, birds may be able to exploit the food resources of these areas then. This is certainly the case with Sanderlings on some busy beaches in Florida (Burger & Gochfield, 1991).

Roosting waders are particularly vulnerable to disturbance. Unlike wildfowl and gulls, they do not roost on the water at high tide, but instead congregate on the shore as near as possible to their preferred feeding areas (in order to minimise energy expenditure in flying back and forth). Of the various forms of disturbance at one of the major wader roosts in the Firth of Forth, the most frequent for the smaller waders was Crows and for the larger ones it was humans. Disturbance by bird watchers, fishermen and aeroplanes was more disruptive than that by Crows, causing the waders to fly around for several minutes (Furness, 1973b). As discussed already, shooting causes a considerable amount of disturbance to bird populations, including the non-target species.

The effects of water-based recreational activities on freshwater birds have been quite well documented, but less attention has been paid to coastal birds in this respect. In one comprehensive study carried out in the German part of the Wadden Sea (Koepff & Dietrich, 1986), the effects of several types of boats on roosting waders were quantified (Table 22). Canoes were the most likely form of craft to cause disturbance, because they passed closest to the shore. However, windsurfers caused the greatest total amount of disturbance because, although individually they were the least disruptive, they were present in the largest numbers at the site. There was relatively little difference in the distance at which different species took flight, although there was a lot of variability in all of the species. Most waders began to fly when approached to a distance of 400–500 m, but Oystercatchers stayed put longer – until 200 m or so. In most species, well over half the individuals had taken to the air by the time the craft had approached to 200 m, though for Oystercatchers this did not happen until 50–100 m. The average flight distances of the individual species measured were as follows: Curlew 269 m, Knot 249 m, Bar-tailed Godwit 226 m, Grey Plover 192 m, and Dunlin 122 m.

Similar values were recorded when waders were approached on foot from the land. Large flocks tended to rise earlier than small ones. Several species were more sensitive to a boat approaching directly towards them than they were to one travelling parallel to the roost. Flocks of Shelduck and Wigeon resting on the water flew away when approached to 300–500 m. Eiders and Cormorants resting on sand banks generally allowed a closer approach (100–200 m).

Breeding Common Terns at another site showed a quite different pattern of susceptibily to disturbance, allowing most vessels to approach to within 10 m of the breeding island without taking flight. The exception in this case was windsurfers, which caused most terns to depart at a distance of 40 m, with the mastheads often being attacked at a distance of 10 m. This sensitivity may be a result of the clear visibility and mobility of the human form on these craft. Occasional (illegal) egg collecting does take place at tern colonies and the strong response to humans may be a consequence of this. By contrast, the greatest threat to wintering waders and wildfowl comes from shooting, and hence wintering roosts of these species take to the air at much greater distances.

Perhaps the most telling conclusion of this German study is that the shorebird population of the area has decreased since 1975 as the amount of disturbance has increased, with the most disturbed sites being deserted first. The authors recommend a boat exclusion zone of 500 m around all shorebird winter roost sites, especially those on offshore islands which are otherwise the roosts most free from disturbance.

Many breeding colonies of birds in Britain have been lost through human disturbance, e.g. Kentish Plovers (p. 229) and Little Terns. Little Terns are particularly vulnerable because they nest in small colonies of 5–30 pairs on sand and shingle beaches. Wardening schemes can be very successful at protecting colonies and allowing them to expand. A wide range of management techniques has been developed to deal with problems such as nest flooding and predation by Foxes (Haddon & Knight, 1983; O Briain & Farrelly, 1990). Even auks nesting on cliffs can be vulnerable, losing eggs or chicks to predators such as Jackdaws and gulls when disturbance frightens the adults away. Guillemots may leave their nesting ledges so hurriedly when disturbed by boats or aircraft that their eggs roll off (Evans & Nettleship, 1985). Eider crèches in parts of Sweden disturbed by boats are much more likely to be attacked successfully by Great Black-backed Gulls. Sites with ten such boat disturbances a day had a mortality rate that was twice as high as undisturbed sites (Åhlund & Götmark, 1989). Rock climbing at breeding cliffs can be very disturbing and has had to be discouraged at vulnerable areas such as South Stack on Anglesey. Here some 2000 auks nest adjacent to a popular climbing area and a voluntary agreement with climbers to avoid certain routes proved effective in minimising disturbance (Edington & Edington, 1986). It is also interesting to note that breeding bird numbers and guano yields on Bird Island in South Africa increased significantly when visitors were banned during the breeding season (Jarvis & Cram, 1971).

## Climatic change

Of all the changes that man has brought about to habitats on earth, long-term climatic change resulting from the effects of greenhouse gases is likely to be the most far reaching for marine and coastal ecosystems. Desertification following the loss of natural vegetation and soil may devastate terrestrial ecosystems, but it only affects the sea by increasing siltation and the local supply of nutrients. A warming of the oceans, however, will not only result in shifts in the distribution of marine organisms, but may result in complex changes to the system of water circulation. Predicting what these changes will be is just as difficult as predicting what climatic changes will occur on land, and there will also be a strong interaction between the two. The El Niño effect (p. 14) demonstrates what a relatively small change in oceanic circulation can do.

A typical scenario envisages a doubling of $CO_2$ concentrations in the atmosphere by 2050, accompanied by a rise in sea level of 80 cm, a rise in mean temperature of 3 °C and a change in rainfall of ±20% (Department of the Environment, 1988). This degree of rise in sea level probably means that waders will be the group of birds most profoundly affected by climatic change (Elkins, 1990). In fact, sea levels have been rising in the south and east of Britain for at least 100 years as a result of purely natural phenomena (p. 9). In 1911, the Royal Commission on Coast Erosion reported that during the previous 30 years, 194 km$^2$ of dry land had been reclaimed from the sea, whereas only 27 km$^2$ had been lost. However, erosion had removed 283 km$^2$ of intertidal land while deposition had resulted in gains of only 93 km$^2$ (Ranwell & Boar, 1986). This net loss of intertidal habitat meant that the slope of the shore was getting steeper. Any sea level rise imposed on a shore with a concave profile will have the same effect (Fig. 89). Although reductions in steepening and increases in intertidal area are possible on convex shores, in reality sea walls are likely to be constructed in most such cases (to protect land and property) and the result will again be losses of intertidal area (Fig. 89).

In places that are already protected by sea walls, which includes most European estuaries, the result of an increase in sea level will be a reduction in the area of mud and sandflats and a decrease in the length of time for which they are exposed. This in turn is likely to affect shorebird populations. Given that the average tidal range around the British coast is about 4.1 m, and assuming that the high tide mark remains (or is constrained to remain) in the same position, a one metre rise in sea level on a shore with an even slope of 1 in 35 (a typical value for the open coast) would result in a reduction in the distance from the high tide to the low tide mark of about 35 m. This is equivalent to a loss in intertidal area of about 25%, plus a 25% reduction in exposure time for the remaining 75%. Unless new feeding areas for birds are created by allowing existing land areas to be inundated, bird populations that feed intertidally are going to decrease. Strong candidates for inundation are those parts of the East Anglian fens where poor land husbandry has allowed the wind to remove most of the fertile peat, leaving only relatively infertile clay. These parts are now also very low lying and are therefore the most expensive to continue to drain.

The costs of improving existing sea defences throughout the country have been estimated at about £5 billion, but this does not include areas that are currently protected naturally by cliffs and sand dunes, etc., and which may in future require extra defences. In estuarine areas, and

**Fig. 89.** Effects of a rise in sea level on the size of the intertidal area on different types of shore. On an evenly sloping shore (*a*), no change in area will occur, unless a new sea wall is constructed (*b*). On a concave shore (*c*) the intertidal area will decrease and on a convex one (*d*) it will increase. Derived from Smit, Lambeck & Wolff (1987).

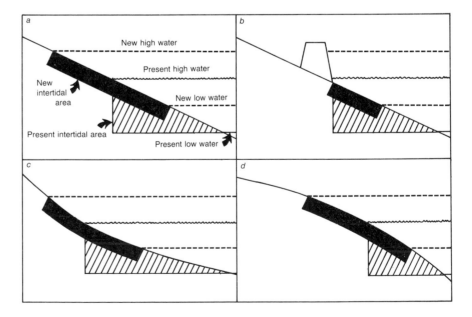

probably elsewhere, erosion will be accelerated because wave action will be concentrated for longer periods around the new high water mark. Increased average wind velocities as a result of climatic change may exacerbate this problem. As a consequence, sediments near the high tide mark will be removed and deposited in more sheltered areas further down the shore, probably resulting in further losses of intertidal area. It has been suggested, for example, that a one metre rise in sea level in Essex will reduce the area of intertidal mudflats by 30–50%, and produce an almost total loss of saltmarsh (Boorman, Goss-Custard & McGrorty, 1989).

Measures that might be undertaken to offset these changes include additional plantings of *Spartina anglica* in some sheltered areas and the use of hurdles or brushwood screens and fences to decrease water velocities and encourage siltation. However, the rate of sea level rise may be so great (greater than 10 mm per year) that sedimentation cannot keep up. Also, the sediment would have to come from somewhere and the only source in some cases is the lower shore, so the overall result would be a further steepening of the shore and a further net loss of intertidal area. Permeable or impermeable barrages across estuaries are an alternative method of protecting the land, but these do not ameliorate any of the adverse effects on shorebirds. Indeed, the impermeable barrages completely destroy intertidal habitats.

Climatic changes accompanying rising sea levels are likely to include a shrinkage in the area of arctic tundra and consequently a decline in the breeding areas of many wading birds. With a few exceptions (e.g. Fig. 90), the future therefore looks grim for them. Some species may also benefit from the creation of enlarged intertidal areas where the destruction of rainforests is leading to soil erosion. In Madagascar, for example, the Betsiboka River has a rapidly enlarging delta because of the removal of up to 25 000 tonnes of soil per square kilometre per year in its catchment (Wells, 1989).

Given the tremendous problems for humans that all these changes are bringing about, and will continue to bring about, why should we worry about the effects on birds? There are three main reasons. Firstly, it is our lack of respect for the other life-forms with which we share the planet that has led us into the present situation. Unless we replace destructive exploitation of the earth's resources by renewable exploitation, we will become victims ourselves. Secondly, bird populations have provided advanced warning to us of a number of potentially dangerous environmental contaminants, such as pesticides, PCBs and heavy metals (Furness, 1987b; Hardy, Stanley & Greig-Smith, 1987). Thirdly, birds improve the quality of life, through their direct role in recreation (birdwatching, photography, hunting) (Filion, 1987; Jacquemot & Filion,

1987), through their cultural and artistic influences (Diamond, 1987; Isack, 1987) and through their embodiment of the spirit of freedom. The latter is exemplified by the reaction of local people to the sight of large numbers of Black-headed Gulls in the Thames above London in the

**Fig. 90.** Breeding Black-winged Stilts. This species very rarely nests in Britain, but is one of the few waders that might do so in increased numbers if the climate becomes warmer. Chris Rose.

winter of 1887/88 as described, a little patronisingly, by Tristram-Valentine (cited in Hudson, 1898): 'their appearance has, from its variety, caused a corresponding excitement among Londoners, as is proved by the numbers of people that have crowded the bridges and embankments to watch their movements. To a considerable portion of these, no doubt, the marvellous flight and power of wing of the gull came as an absolute revelation'.

### The future for coastal birds

Although the future will be full of problems, there is much that we can do to conserve coastal birds. The examples of positive management given in this chapter show what can be achieved. However, there is an important step that we in Britain need to take now and that involves developing a co-ordinated policy for the use of our coast. Given that as much as 10% of the intertidal area of British estuaries is under immediate threat (Davidson, 1991), the need is urgent. What happens to the coast at present is the result of a series of *ad hoc* decisions by individuals and planning authorities, which often take no account of similar or contradictory decisions made about adjacent areas. For example, sea defences are often the responsibility of the National Rivers Authority in estuaries and local authorities along the open coast. Residents of the recently flooded North Wales town of Towyn found that responsibility for the breached sea defences rests with no less than seven separate bodies, including British Rail and the Welsh Office (Whittle, 1991).

In any list of priorities concerning what should happen to our coast, its protection from erosion may clearly need to take priority over other matters. Following this, there are a number of international treaties and conventions that are supposed to influence the way all relevant sites are developed (Table 23). Of these, the EC Wild Birds Directive incorporates a list of species under threat which includes over 50 coastal birds (see Appendix). Most of these international obligations are met through a framework of national law relating to environmental conservation (Table 24). Some sections of the coast are in the ownership of bodies whose purpose is conservation, such as the National Trust, the RSPB and local wildlife trusts. These areas at least should be secure. RSPB reserves currently include the breeding areas of more than 40% of all the Roseate Terns in the British Isles, 20–30% of the Kittiwakes, Guillemots and Razorbills and 10–20% of the Gannets, Arctic Skuas and other terns (Lloyd, 1989). This amounts in total to 35 000 breeding auks and a quarter of a million pairs of other seabirds. In winter, RSPB reserves support over 300 000 wildfowl and a similar number of waders. In the

Table 23. *International measures for conservation which can be used to restrict development in coastal areas. Based mainly on information in Grimmett & Jones (1989) and Davidson et al. (1991)*

*Biosphere Reserves*
Designated under UNESCO's Man and the Biosphere Programme, these reserves are for the conservation of natural areas and genetic diversity. Launched in 1970

*Ramsar Convention*
Full title: Convention on Wetlands of International Importance Especially as Waterfowl Habitat. Came into force in 1975. About 300 sites designated so far (18 of these, and a further 43 awaiting designation, cover parts of British estuaries). 24 European contracting parties, and many in other parts of the world. Promotes the wise use of all wetlands and the designation and protection of internationally important ones

*World Heritage Convention*
Full title: Convention concerning the Protection of the World Cultural and Natural Heritage. Came into force in 1975, but very few sites as yet specified. Identifies and protects sites of natural heritage offering outstanding universal value from any of several points of view, including conservation, science and natural beauty

*Biogenetic Reserves*
Full title: European Network of Biogenetic Reserves. Established in 1976, with well over 100 sites now included. Intended to encourage individual EC members to provide national legal protection to areas with typical, unique, rare or threatened habitats so that they could be included in the network

*EC Wild Birds Directive*
Full title: Directive on the Conservation of Wild Birds (79/409/EEC). Came into force in 1981. Requires the preservation of a sufficient diversity of habitats to ensure the maintenance of natural populations of wild birds, with special measures required for those on Annex 1 (see Appendix)

*SPAs*
Full title: Special Protection Areas. Established under the EC Wild Birds Directive as the main means of achieving the latter's objectives. 761 sites now designated as SPAs (14 covering parts of British estuaries)

*Barcelona Convention*
Full title: Convention for the Protection of the Mediterranean Sea against Pollution. Established in 1976, with a Protocol on Mediterranean Specially Protected Areas adopted in 1982. There are now 41 of the latter, which can include sites of biological, ecological, scientific, aesthetic, historical, archaeological, cultural or educational interest

*Berne Convention*
Full title: Council of Europe Convention on the Conservation of European Wildlife and Natural Habitats. Came into force in 1982. Requires the conservation of both wild species and natural habitats, including most species of birds, and their breeding, wintering and moulting areas, as well as sites used during migration

Table 23—*contd.*

*Bonn Convention*
Full title: Convention on the Conservation of Migratory Species of Wild
Animals. Came into force in 1983. 12 European signatories and others in the
rest of the world. Aims to provide strict protection for globally threatened
species and to conserve and manage those which would benefit from
international co-operation

*ESAs*
Full title: Environmentally Sensitive Areas. Established in 1985. Allows special
payments to be made to farmers in the EC who forego agricultural
intensification in these designated areas of landscape, wildlife or
archaeological importance. 18 SPAs designated so far in the UK

*Important Bird Areas*
An inventory of sites in Europe which in the opinion of ICBP and IWRB (as
set out in Grimmett & Jones, 1989) require effective legal protection under
one or more of the above conventions

latter case this amounts to some 3% of the East Atlantic Flyway population (Everett, Cadbury & Dawson, 1988).

As has already been discussed in this chapter, there is a whole range
of legitimate interest groups wishing to use the coast for other purposes
– recreation, tourism, housing, industry, etc. Bodies concerned with sea
defences are currently foremost amongst those calling for a nationally
integrated policy to resolve and co-ordinate these competing interests. A
Standing Conference on the Problems Associated with the Coastline
(SCOPAC) has been set up by local authorities responsible for the coast
between Weymouth and Worthing to try and integrate the needs of sea
defence, development and conservation. The latter is an example of a
local mechanism, but it still lacks broad nationally agreed policy
guidelines. The Countryside Commission, the Marine Conservation
Society and the RSPB have all called for a national review of coastal
policy, the latter suggesting the formulation of a National Coastal Policy
(Rothwell & Housden, 1990). A conference of planning officers and
other experts concerned with coastal matters have produced an 'Agenda
for action' which again emphasises the need for national planning
guidelines (Houston & Jones, 1990). Despite this, the Government
remains opposed at the moment to the development of a national policy
for the coast.

How would such a coastal policy work if it ever comes about? Firstly,
all areas of the coast would have to conform to certain national and
regional criteria setting down limits to the amount of various types of
development that would be permitted. Specifically, these developments
would include housing, tourism, caravan parks, recreational facilities,

Table 24. *Land ownership and management categories of conservation and amenity areas, including coastal ones, in the United Kingdom. Updated from Grimmett & Jones (1989) and Nature Conservancy Council (1989). The Countryside Commission now has responsibility only for England. The Nature Conservancy Council has been replaced in England by English Nature. Scottish Natural Heritage and the Countryside Council for Wales combine these responsibilities in Scotland and Wales respectively*

| Date of legislation | Type | Purpose | Nature and control over development | Number |
|---|---|---|---|---|
| *Conservation* | | | | |
| 1949 | NNR[a] National Nature Reserve | Conservation of flora, geological or physiographical features | Managed by the Nature Conservancy Council | 234 |
| 1949 | LNR[a] Local Nature Reserve | Nature conservation of some sort | Some owned by local authorities, others privately. Controlled by by-laws | 176 |
| 1981 | MNR[b] Marine Nature Reserve | Conservation of flora, fauna, geological or physiographical features | Managed by the Nature Conservancy Council | 2 |
| 1981 | SSSI[b] Site of Special Scientific Interest (Area of Special Scientific Interest in Northern Ireland) | As above | Largely privately owned. Damaging operations restricted on a statutary basis for 4 months, then voluntarily. If not, compulsory purchase can be used as a last resort, but this is very rare | 5184 |

| Amenity | | | | |
|---|---|---|---|---|
| 1949 | National Park[a] | Conservation of landscape and amenity in England and Wales | Very little, but Countryside Commission promotes them and may urge planning authorities to prevent certain types of development | 10 |
| 1949 | AONB[a] Area of Outstanding Natural Beauty | Conservation of landscape in England and Wales | Access orders and management agreements can be made | 48 |
| 1968 | Country Parks | To cater for more intensive recreational activities in a countryside environment, usually close to large centres of population | Controlled by by-laws | 170 |
| 1973 | Heritage Coasts | To conserve and protect areas of attractive coastline and to provide recreational and interpretive facilities | Very little | 43 |
| 1980 | National Scenic Area | Conservation of landscape and amenity in Scotland | As for AONBs, but promoted by the Countryside Commission for Scotland | 40 |

[a]Established under the 1949 National Parks and Access to the Countryside Act.
[b]Established under the 1981 Wildlife and Countryside Act.

industrial and commercial development, fisheries, mineral extraction
and waste discharges. Targets could also be set for the conservation of
the coast, free from any sort of building development. All developments
and activities deterimental to wildlife in areas of designated national or
international importance should be prevented. Some conservation areas
could, however, be zoned to permit a range of different activities in the
most suitable sections, such as rock climbing, water skiing, power boating, scuba diving and bird watching. While such restrictions may sound
draconian, they are an inevitable consequence of the increasing pressures from all sides. There are plenty of examples of highly successful
conservation policies. At one marine nature reserve near Auckland in
New Zealand, fishermen were initially opposed to its declaration, but
when catches increased dramatically outside the reserve they became
strong supporters. Exactly the same thing has happened at a Spanish
reserve on the Medas Islands off the Costa Brava (McDonald, 1990).

Development, conservation and zoning proposals need to be set out
in a structure plan, similar to the one that local authorities are obliged to
produce at present for the whole of their planning areas. The coastal
structure plan could indeed form part of the overall structure plan as it
does at present. The difference would be that it would have to conform
to certain national guidelines and would only be approved when it did
so. The activities of the National Rivers Authority, and others with legal
rights and obligations in the coastal zone, would have to be regulated in
a similar way.

What happens on dry land is reasonably amenable to planning controls because all that is required is a modification of existing procedures.
The littoral zone and the open sea are much more difficult areas, not
least because of the international rights that exist over the latter. Planning controls already exist over such activities as aggregate dredging
and fishing. Whilst a national strategy is certainly needed for the former,
the latter merely requires firm government controls on catch limits to
maintain maximum sustainable yields (based on independent scientific
advice). The latter sounds simple, yet governments have repeatedly
lacked the will to achieve any worthwhile limits until after a fishery has
collapsed. In international waters, international agreements and controls are clearly required. In this case, the difficulty is in finding and
applying effective sanctions against those countries who do not conform
to such agreements. The successes and failures of the International
Whaling Commission are well known and illustrate the problems perfectly. On the whole, the international scene is encouraging and new
developments, especially those relating to environmental pollution (e.g.
GESAMP, 1991), are acting as a spur to individual countries to improve
their national legislation on environmental conservation.

This final section has had little to say about birds despite the fact that it is supposed to be concerned with their future. This is because their future depends on decisions made by politicians about the rules and regulations which govern our lives. It is up to us to let them know what we think are the right decisions, and to take them to task when they make the wrong ones.

# APPENDIX

## Scientific names

## Birds

Asterisks indicate EC Wild Birds Directive Annex 1 listed coastal species which occur regularly in the British Isles, or which are supported in internationally significant breeding or wintering numbers. Based on Cramp & Simmons (1977, 1983), Grimmett & Jones (1989), Batten *et al.* (1990) and Lloyd, Tasker & Partridge (1991).

| Order | Family | | | Status of coastal species | | |
|---|---|---|---|---|---|---|
| | | | | EC Annex I | Non-breeding | Breeding |
| Sphenisciformes | Spheniscidae | King Penguin | *Aptenodytes patagonica* | | | |
| | | Adélie Penguin | *Pygoscelis adeliae* | | | |
| | | Chinstrap Penguin | *P. antarctica* | | | |
| | | Rockhopper Penguin | *Eudyptes chrysocome* | | | |
| | | Jackass Penguin | *Spheniscus demersus* | | | |
| Gaviiformes | Gaviidae | Red-throated Diver | *Gavia stellata* | * | * | * |
| | | Black-throated Diver | *G. arctica* | * | * | |
| | | Great Northern Diver | *G. immer* | * | * | |
| Podicipediformes | Podicipedidae | Little Grebe | *Tachybaptus ruficollis* | | | |
| | | Great Crested Grebe | *Podiceps cristatus* | | | |
| | | Red-necked Grebe | *P. grisegena* | | | |
| | | Slavonian Grebe | *P. auritus* | * | | |
| | | Black-necked Grebe | *P. nigricollis* | | | |
| Procellariiformes | Diomedeidae | Black-browed Albatross | *Diomedea melanophris* | | | |
| | | Wandering Albatross | *D. exulans* | | | |
| | | Black-footed Albatross | *D. nigripes* | | | |
| | | Laysan Albatross | *D. immutabilis* | | | |
| | | Sooty Albatross | *Phoebetria fusca* | | | |
| | Procellariidae | Fulmar | *Fulmarus glacialis* | | * | * |
| | | Snow Petrel | *Pagodroma nivea* | | | |
| | | Soft-plumaged Petrel | *Pterodroma mollis* | | | |
| | | Dark-rumped Petrel | *P. phaeopygia* | | | |
| | | Bonin Petrel | *P. hypoleuca* | | | |
| | | Broad-billed Prion | *Pachyptila vittata* | | | |
| | | White-chinned Petrel | *Procellaria aequinoctialis* | | | |
| | | Great Shearwater | *Puffinus gravis* | | | |
| | | Sooty Shearwater | *P. griseus* | | | |
| | | Manx Shearwater | *P. puffinus* | | | * |
| | | Mediterranean Shearwater | *P. yelkouan* | | | |
| | | Short-tailed Shearwater | *P. tenuirostris* | | | |
| | Oceanitidae | Wilson's Petrel | *Oceanites oceanicus* | | | |
| | | Storm Petrel | *Hydrobates pelagicus* | * | | * |
| | | Leach's Petrel | *Oceanodroma leucorrhoa* | * | | * |

280

Appendix—*cont.*

| Order | Family | | | Status of coastal species | | |
|---|---|---|---|---|---|---|
| | | | | EC Annex I | Non-breeding | Breeding |
| Procellariiformes *cont.* | | Guadelupe Storm-petrel | *O. macrodactyla* | | | |
| | Pelecanoididae | Common Diving Petrel | *Pelecanoides urinatrix* | | | |
| Pelecaniformes | Phaethontidae | Red-billed Tropicbird | *Phaethon aethereus* | | | |
| | | White-tailed Tropicbird | *P. lepturus* | | | |
| | Sulidae | Gannet | *Sula bassana* | | * | * |
| | | Cape Gannet | *S. capensis* | | | |
| | | Australian Gannet | *S. serrator* | | | |
| | | Blue-footed Booby | *S. nebouxii* | | | |
| | | Peruvian Booby | *S. variegata* | | | |
| | Phalacrocoracidae | Cormorant | *Phalacrocorax carbo* | | * | * |
| | | Shag | *P. aristotelis* | | * | * |
| | | Socotra Cormorant | *P. nigrogularis* | | | |
| | | Cape Cormorant | *P. capensis* | | | |
| | | Bank Cormorant | *P. neglectus* | | | |
| | | Peruvian Cormorant | *P. bougainvillii* | | | |
| | Pelecanidae | White Pelican | *Pelecanus onocrotalus* | * | | |
| | | Dalmatian Pelican | *P. crispus* | | | |
| | | Brown Pelican | *P. occidentalis* | | | |
| | Fregatidae | Magnificent Frigatebird | *Fregata magnificens* | | | |
| Ciconiiformes | Ardeidae | Bittern | *Botaurus stellaris* | * | | |
| | | Green Heron | *Butarides striatus* | | | |
| | | Cattle Egret | *Bubulcus ibis* | | | |
| | | Little Egret | *Egretta garzetta* | | | |
| | | Grey Heron | *Ardea cinerea* | | | |
| | | Purple Heron | *A. purpurea* | | | |
| | | Great Blue Heron | *A. herodias* | | | |
| | Ciconiidae | White Stork | *Ciconia ciconia* | | | |
| | Threskiornithidae | Spoonbill | *Platalea leucorodia* | | | |
| | | Roseate Spoonbill | *Ajaia ajaja* | | | |
| Phoenicopteriformes | Phoenicopteridae | Greater Flamingo | *Phoenicopterus ruber* | | | |
| | | Chilean Flamingo | *P. chilensis* | | | |
| | | Lesser Flamingo | *Phoeniconaias minor* | | | |
| Anseriformes | Anatidae | Mute Swan | *Cygnus olor* | | * | * |
| | | Bewick's Swan | *C. columbianus* | * | * | |
| | | Whooper Swan | *C. cygnus* | * | * | |
| | | Bean Goose | *Anser fabalis* | | * | |
| | | Pink-footed Goose | *A. brachyrhynchus* | | * | |
| | | White-fronted Goose | *A. albifrons* | * | * | |
| | | Greylag Goose | *A. anser* | | * | |
| | | Snow Goose | *A. caerulescens* | | | |
| | | Canada Goose | *Branta canadensis* | | | |
| | | Barnacle Goose | *B. leucopsis* | * | * | |
| | | Brent Goose (Brant) | *B. bernicla* | | * | |
| | | Egyptian Goose | *Alopochen aegyptiacus* | | | |
| | | Shelduck | *Tadorna tadorna* | | * | * |
| | | Wigeon | *Anas penelope* | | * | |
| | | Gadwall | *A. strepera* | | * | |
| | | Teal | *A. crecca* | | * | |
| | | Mallard | *A. platyrhynchos* | | * | |
| | | Pintail | *A. acuta* | | * | |
| | | Garganey | *A. querquedula* | | | |
| | | Cinnamon Teal | *A. cyanoptera* | | | |
| | | Shoveler | *A. clypeata* | | * | |
| | | Pochard | *Aythya ferina* | | * | |
| | | Tufted Duck | *A. fuligula* | | | |
| | | Scaup | *A. marila* | | * | |
| | | Lesser Scaup | *A. affinus* | | | |

Appendix—*cont.*

| Order | Family | | | Status of coastal species | | |
|---|---|---|---|---|---|---|
| | | | | EC Annex I | Non-breeding | Breeding |
| Anseriformes *cont.* | | Eider | *Somateria mollissima* | | * | * |
| | | Long-tailed Duck | *Clangula hyemalis* | | * | |
| | | Common Scoter | *Melanitta nigra* | | * | |
| | | Velvet Scoter | *M. fusca* | | * | |
| | | Goldeneye | *B. clangula* | | * | |
| | | Smew | *Mergus albellus* | | | |
| | | Red-breasted Merganser | *M. serrator* | | * | * |
| | | Goosander | *M. merganser* | | | |
| Falconiformes | Accipitridae | Red Kite | *Milvus milvus* | | | |
| | | White-tailed Eagle | *Haliaeetus albicilla* | * | | |
| | | Hen Harrier | *Circus cyaneus* | * | | |
| | | Goshawk | *Accipiter gentilis* | | | |
| | | Sparrowhawk | *A. nisus* | | | |
| | | Golden Eagle | *Aquila chrysaetos* | | | |
| | Pandionidae | Osprey | *Pandion haliaetus* | * | | |
| | Falconidae | Merlin | *Falco columbarius* | * | | |
| | | Gyrfalcon | *F. rusticolus* | | | |
| | | Peregrine | *F. peregrinus* | * | | * |
| Galliformes | Tetraonidae | Red (Willow) Grouse | *Lagopus lagopus* | | | |
| | Phasianidae | Grey Partridge | *Perdix perdix* | | | |
| | | Jungle Fowl | *Gallus gallus* | | | |
| | | Pheasant | *Phasianus colchicus* | | | |
| Gruiformes | Rallidae | Water Rail | *Rallus aquaticus* | | | |
| | | Corncrake | *Crex crex* | * | | |
| | | Moorhen | *Gallinula chloropus* | | | |
| | | Coot | *Fulica atra* | | | |
| | Otididae | Great Bustard | *Otis tarda* | | | |
| Charadriiformes | Rostratulidae | Painted Snipe | *Rostratula benghalensis* | | | |
| | Haematopodidae | Oystercatcher | *Haematopus ostralegus* | | * | * |
| | Recurvirostridae | Black-winged Stilt | *Himantopus himantopus* | * | | |
| | | Avocet | *Recurvirostra avosetta* | * | * | * |
| | Burhinidae | Stone Curlew | *Burhinus oedicnemus* | | | |
| | Dromadidae | Crab Plover | *Dromas ardeola* | | | |
| | Glareolidae | Egyptian Plover | *Pluvianus aegyptius* | | | |
| | Charadriidae | Little Ringed Plover | *Charadrius dubius* | | | |
| | | Ringed Plover | *C. hiaticula* | | * | * |
| | | Kittlitz's Sand Plover | *C. pecuarius* | | | |
| | | Kentish Plover | *C. alexandrinus* | | | |
| | | Dotterel | *C. morinellus* | | | |
| | | Golden Plover | *Pluvialis apricaria* | * | * | |
| | | Grey Plover | *P. squatarola* | | * | |
| | | Lapwing | *Vanellus vanellus* | | * | |
| | Scolopacidae | Knot | *Calidris canutus* | | * | |
| | | Sanderling | *C. alba* | | * | |
| | | Semipalmated Sandpiper | *C. pusilla* | | | |
| | | Little Stint | *C. minuta* | | | |
| | | Temminck's Stint | *C. temminckii* | | | |
| | | Curlew Sandpiper | *C. minutilla* | | | |
| | | Purple Sandpiper | *C. maritima* | | * | |
| | | Dunlin | *C. alpina* | | * | |
| | | Ruff | *Philomachus pugnax* | * | | |
| | | Jack Snipe | *Lymnocryptes minimus* | | | |
| | | Snipe | *Gallinago gallinago* | | | |
| | | Woodcock | *Scolopax rusticola* | | | |
| | | Short-billed Dowitcher | *Limnodromus griseus* | | | |
| | | Black-tailed Godwit | *Limosa limosa* | | * | |
| | | Bar-tailed Godwit | *L. lapponica* | | * | |

Appendix—*cont.*

| Order | Family | | | Status of coastal species | | |
|---|---|---|---|---|---|---|
| | | | | EC Annex I | Non-breeding | Breeding |
| Charadriiformes cont. | | Slender-billed Curlew | *Numenius tenuirostris* | * | | |
| | | Whimbrel | *N. phaeopus* | | * | |
| | | Bristle-thighed Curlew | *N. tahitiensis* | | | |
| | | Curlew | *N. arquata* | | * | * |
| | | Spotted Redshank | *Tringa erythropus* | | | |
| | | Redshank | *T. totanus* | | * | * |
| | | Greenshank | *T. nebularia* | | * | |
| | | Green Sandpiper | *T. ochropus* | | | |
| | | Wood Sandpiper | *T. glareola* | * | | |
| | | Common Sandpiper | *Actitis hypoleucos* | | | |
| | | Turnstone | *Arenaria interpres* | | * | |
| | | Wilson's Phalarope | *Phalaropus tricolor* | | | |
| | | Red-necked Phalarope | *P. lobatus* | * | | |
| | | Grey Phalarope | *P. fulicarius* | | | |
| | Stercorariidae | Pomarine Skua | *Stercorarius pomarinus* | | | |
| | | Arctic Skua | *S. parasiticus* | | * | |
| | | Long-tailed Skua | *S. longicaudus* | | | |
| | | Great Skua | *S. skua* | | | * |
| | Laridae | Mediterranean Gull | *Larus melanocephalus* | * | | |
| | | Laughing Gull | *L. atricilla* | | | |
| | | Little Gull | *L. minutus* | | | |
| | | Sabine's Gull | *L. sabini* | | | |
| | | Black-headed Gull | *L. ridibundus* | | * | * |
| | | Bonaparte's Gull | *L. philadelphia* | | | |
| | | Ring-billed Gull | *L. delawarensis* | | | |
| | | Common Gull | *L. canus* | | * | * |
| | | Lesser Black-backed Gull | *L. fuscus* | | | * |
| | | Herring Gull | *L. argentatus* | | * | * |
| | | Yellow-legged Herring Gull | *L. cachinnans* | | | |
| | | California Gull | *L. californicus* | | | |
| | | Western Gull | *L. occidentalis* | | | |
| | | Glaucous Gull | *L. hyperboreus* | | | |
| | | Great Black-backed Gull | *L. marinus* | | * | * |
| | | Glaucous-winged Gull | *L. glaucescens* | | | |
| | | Swallow-tailed Gull | *L. furcatus* | | | |
| | | Ross's Gull | *Rhodostethia rosea* | | | |
| | | Kittiwake | *Rissa tridactyla* | | * | * |
| | | Ivory Gull | *Pagophila eburnea* | | | |
| | Sternidae | Caspian Tern | *Sterna caspia* | | | |
| | | Sandwich Tern | *S. sandvicensis* | * | | * |
| | | Roseate Tern | *S. dougallii* | * | | * |
| | | Common Tern | *S. hirundo* | * | | * |
| | | Arctic Tern | *S. paradisaea* | * | | * |
| | | Sooty Tern | *S. fuscata* | | | |
| | | Little Tern | *S. albifrons* | * | | * |
| | | Black Tern | *Chlidonias niger* | | | |
| | | Fairy Tern | *Gygis alba* | | | |
| | Alcidae | Guillemot | *Uria aalge* | | * | * |
| | | Brünnich's Guillemot | *U. lomvia* | | | |
| | | Razorbill | *Alca torda* | | * | * |
| | | Great Auk | *Pinguinus impennis* | | | |
| | | Black Guillemot | *Cepphus grylle* | | * | * |
| | | Little Auk | *Alle alle* | | | |
| | | Cassin's Auklet | *Ptychoramphus aleuticus* | | | |
| | | Puffin | *Fratercula arctica* | | | * |
| | | Tufted Puffin | *Lunda cirrhata* | | | |

Appendix—*cont.*

| Order | Family | | | EC Annex I | Status of coastal species Non-breeding | Breeding |
|---|---|---|---|---|---|---|
| Columbiformes | Columbidae | Rock Dove | *Columbia livia* | | | |
| | | Woodpigeon | *C. palumbus* | | | |
| Cuculiformes | Cuculidae | Cuckoo | *Cuculus canorus* | | | |
| Strigiformes | Tytonidae | Barn Owl | *Tyto alba* | | | |
| | Strigidae | Eagle Owl | *Bubo bubo* | | | |
| | | Little Owl | *Athene noctua* | | | |
| | | Tawny Owl | *Strix aluco* | | | |
| | | Short-eared Owl | *Asio flammeus* | * | | |
| Caprimulgiformes | Caprimulgidae | Nightjar | *Caprimulgus europaeus* | | | |
| Apodiformes | Apodidae | Swift | *Apus apus* | | | |
| Coraciiformes | Alcedinidae | Kingfisher | *Alcedo atthis* | | | |
| Passeriformes | Alaudidae | Skylark | *Alauda arvensis* | | | |
| | | Shore Lark | *Eremophila alpestris* | | | |
| | Hirundinidae | Swallow | *Hirundo rustica* | | | |
| | | House Martin | *Delichon urbica* | | | |
| | Motacillidae | Meadow Pipit | *Anthus pratensis* | | | |
| | | Rock Pipit | *A. spinoletta* | | | |
| | | Yellow Wagtail | *Motacilla flava* | | | |
| | | Pied Wagtail | *M. alba* | | | |
| | Cinclidae | Dipper | *Cinclus cinclus* | | | |
| | Troglodytidae | Wren | *Troglodytes troglodytes* | | | |
| | Prunellidae | Dunnock | *Prunella modularis* | | | |
| | Turdidae | Robin | *Erithacus rubecula* | | | |
| | | Black Redstart | *Phoenicurus ochruros* | | | |
| | | Stonechat | *Saxicola torquata* | | | |
| | | Wheatear | *Oenanthe oenanthe* | | | |
| | | Blackbird | *Turdus merula* | | | |
| | | Song Thrush | *T. philomelos* | | | |
| | Sylviidae | Grasshopper Warbler | *Locustella naevia* | | | |
| | | Sedge Warbler | *Acrocephalus schoenobaenus* | | | |
| | | Reed Warbler | *A. scirpaceus* | | | |
| | | Whitethroat | *Sylvia communis* | | | |
| | | Chiffchaff | *Phylloscopus collybita* | | | |
| | | Willow Warbler | *P. trochilus* | | | |
| | | Firecrest | *Regulus ignicapillus* | | | |
| | Paridae | Great Tit | *Parus major* | | | |
| | Corvidae | Magpie | *Pica pica* | | | |
| | | Chough | *Pyrrhocorax pyrrhocorax* | * | | |
| | | Jackdaw | *Corvus monedula* | | | |
| | | Carrion Crow | *C. corone* | | | |
| | | Raven | *C. corax* | | | |
| | Sturnidae | Starling | *Sturnus vulgaris* | | | |
| | Passeridae | House Sparrow | *Passer domesticus* | | | |
| | Fringillidae | Chaffinch | *Fringilla coelebs* | | | |
| | | Greenfinch | *Carduelis chloris* | | | |
| | | Goldfinch | *C. carduelis* | | | |
| | | Linnet | *C. cannabina* | | | |
| | | Twite | *C. flavirostris* | | * | * |
| | | Redpoll | *C. flammea* | | | |
| | Emberizidae | Lapland Bunting | *Calcarius lapponicus* | | | |
| | | Snow Bunting | *Plectrophenax nivalis* | | | |
| | | Cirl Bunting | *Emberiza cirlus* | | | |
| | | Reed Bunting | *E. schoeniclus* | | | |
| | | Corn Bunting | *Miliaria calandra* | | | |

## Plants

| | |
|---|---|
| Annual Sea-blite | *Suaeda maritima* |
| Beaked Tasselweed | *Ruppia maritima* |
| Blackthorn | *Prunus spinosus* |
| Bladder Wrack | *Fucus vesiculosus* |
| Channel Wrack | *Pelvetia canaliculata* |
| Common Chickweed | *Stellaria media* |
| Common Nettle | *Urtica dioica* |
| Common Saltmarsh-grass | *Puccinellia maritima* |
| Common Sorrel | *Rumex acetosa* |
| Creeping Bent | *Agrostis stolonifera* |
| Creeping Willow | *Salix repens* |
| Curled Dock | *Rumex crispus* |
| Elder | *Sambucus nigra* |
| Flat Wrack | *Fucus spiralis* |
| Gorse | *Ulex europaeus* |
| Knotted Wrack | *Ascophyllum nodosum* |
| Marram | *Ammophila arenaria* |
| Perennial Ryegrass | *Lolium perenne* |
| Red Fescue | *Festuca rubra* |
| Saltmarsh Rush | *Juncus gerardii* |
| Sea Arrowgrass | *Triglochin maritima* |
| Sea Aster | *Aster tripolium* |
| Sea-buckthorn | *Hippophae rhamnoides* |
| Sea Club-rush | *Scirpus maritimus* |
| Sea Mayweed | *Tripleurospermium maritimum* |
| Sea Plantain | *Plantago maritima* |
| Sea Purslane | *Halimione portulacoides* |
| Sea Wheat Grass | *Agropyron junceum* |
| Small Nettle | *Urtica urens* |
| Spear-leaved Orache | *Atriplex prostrata* |
| Sycamore | *Acer pseudoplatanus* |
| Thongweed | *Himanthalia elongata* |
| Tree Mallow | *Lavatera arborea* |
| Thrift | *Armeria maritima* |
| White Clover | *Trifolium repens* |
| White Poplar | *Populus alba* |
| Wild Privet | *Ligustrum vulgare* |
| Yorkshire-fog | *Holcus lanatus* |

## Invertebrates

| | |
|---|---|
| Common Cockle | *Cerastoderma edule* |
| Common Mussel | *Mytilus edulis* |
| Common Sea Urchin | *Echinus esculentus* |
| Common Starfish | *Asterias rubens* |
| Dogwhelk | *Nucella lapillus* |
| Edible Crab | *Carcinus pagurus* |
| Horseshoe Crab | *Limulus polyphemus* |
| Laver Spire Shell | *Hydrobia ulvae* |
| Lugworm | *Arenicola marina* |
| Norway Lobster | *Nephrops norvegicus* |
| Opossum Shrimp | *Neomys integer* |
| Peppery Furrow Shell | *Scrobicularia plana* |
| Ragworm | *Nereis diversicolor* |
| Razor Shell | *Ensis solen* |
| Sand Gaper | *Mya arenaria* |
| Sand Mason | *Lanice conchilega* |
| Sea Slater | *Ligia oceanica* |
| Shorecrab | *Carcinus maenus* |
| Shrimp | *Crangon vulgaris* |
| Spindle Shell | *Neptunea antiqua* |
| Thick-shelled Dogwhelk | *Nassarius reticulatus* |
| Yellow Meadow Ant | *Lasius flavus* |

## Reptiles

| | |
|---|---|
| Adder | *Vipera berus* |
| Grass Snake | *Natrix natrix* |

## Fish

| | |
|---|---|
| Bass | *Dicentrarchus labrax* |
| Bleak | *Alburnus alburnus* |
| Blue Whiting | *Micromesistius poutassou* |
| Bream | *Abramis brama* |
| Butterfish | *Pholis gunnellus* |
| Cod | *Gadus morhua* |
| Eel | *Anguilla anguilla* |
| Flounder | *Platichthys flesus* |
| Herring | *Clupea harengus* |
| Mackerel | *Scomber scombrus* |
| Norway Pout | *Trisopterus esmarkii* |
| Perch | *Perca fluviatilis* |
| Pike | *Esox lucius* |
| Pilchard | *Sardinia pilchardus* |
| Pollack | *Pollachius pollachius* |
| Roach | *Rutilus rutilus* |
| Salmon | *Salmo salar* |
| Sardine (young Pilchard) | *Sardinia pilchardus* |
| Sea Scorpion | *Taurulus bubalis* |
| Sprat | *Sprattus sprattus* |
| Tench | *Tinca tinca* |
| Trigger-fish | *Balistes carolinensis* |
| Yellowfin Tuna | *Thunnus albacares* |

## Mammals

| | |
|---|---|
| Badger | *Meles meles* |
| Black Rat | *Rattus rattus* |
| Cave Lion | *Panthera leo* |
| Common Seal | *Phoca vitulina* |
| Ferret | *Mustela furo* |
| Fox | *Vulpes vulpes* |
| Grey Seal | *Halichoerus grypus* |
| Grey Squirrel | *Sciurus carolinensis* |
| Hedgehog | *Erinaceus europaeus* |
| House Mouse | *Mus domesticus* |
| Humpback Whale | *Megaptera novaeangliae* |
| Mink | *Mustela vison* |
| Norway Rat | *Rattus norvegicus* |
| Otter | *Lutra lutra* |
| Pine Marten | *Martes martes* |
| Polecat | *Mustela putorius* |
| Rabbit | *Oryctolagus cuniculus* |
| Red Deer | *Cervus elephas* |
| Red Squirrel | *Sciurus vulgaris* |
| Sea Otter | *Enhydra lutris* |

**Mammals**—*cont.*

| | | | |
|---|---|---|---|
| Stoat | *Mustela erminea* | Wildcat | *Felis silvestris* |
| Weasel | *Mustela nivalis* | Woolly Mammoth | *Mammuthus primigenius* |
| | | Woolly Rhinoceros | *Coelodonta antiquitatus* |

# REFERENCES

Abramson, M. (1979). Vigilance as a factor influencing flock formation among Curlews *Numenius arquata*. *Ibis* **121**, 213–216.

Åhlund, M. & Götmark, F. (1989). Gull predation on eider ducklings *Somateria mollissima*: effects of human disturbance. *Biological Conservation* **48**, 115–127.

Ainley, D.G. (1977). Feeding methods in seabirds: a comparison of polar and tropical nesting communities in the eastern Pacific Ocean. In *Adaptations within Antarctic ecosystems*. (ed. G.A. Llano), pp. 669–685. Washington, D.C.: Smithsonian Institution.

Alder, L.P. (1982). Feeding behaviour of Skylarks in hard winters. *British Birds* **75**, 33.

Alerstam, T. (1981). The course and timing of bird migration. In *Animal migration*. (ed. D.J. Aidley), pp. 9–54. Cambridge: Cambridge University Press.

Allaine, D. & Lebreton, J.-D. (1990). The influence of age and sex on wing-tip pattern in adult Black-headed Gulls *Larus ridibundus*. *Ibis* **132**, 560–567.

Allen, J.R.L. (1987). Late Flandrian shoreline oscillations in the Severn Estuary: The Rumney Formation at its typesite (Cardiff area). *Philosophical Transactions of the Royal Society of London* **315B**, 157–174.

Allport, G., O'Brien, M. & Cadbury, C.J. (1986). *Survey of Redshank and other breeding birds on saltmarshes in Britain 1985*. Report to the Nature Conservancy Council. Sandy: Royal Society for the Protection of Birds.

Altenburg, W. (1987). *Waterfowl in West African coastal wetlands*. Zeist: Dutch Working Group for International Wader- and Waterfowl Research, Report 15.

Altenburg, W. & van der Kamp, J. (1988). Coastal waders in Guinea. *Wader Study Group Bulletin* **54**, 33–35.

Alzieu, C. & Heral, M. (1984). Ecotoxicological effects of organotin compounds on oyster culture. In *Ecotoxicological testing for the marine environment. Vol. 1*. (ed. G. Persoone, E. Jaspers & C. Claus), pp. 187–196. Ghent: State University of Ghent and Institute for Marine Scientific Research, Bredene.

Amlaner, C.J. & McFarland, D.J. (1981). Sleep in the herring gull (*Larus argentatus*). *Animal Behaviour* **29**, 551–556.

Andersen, K.P. & Ursin, E. (1977). A multispecies extension to the Beverton and Holt theory of fishing, with accounts of phosphorus circulation and primary production. *Meddelelser fra Danmarks Fiskeri- og Havundersøgelser* **7**, 319–435.

Annett, C. & Pierotti, R. (1989). Chick hatching as a trigger for dietary switching in the Western Gull. *Colonial Waterbirds* **12**, 4–11.

Anonymous. (1989). Nets kill seabirds. *Scottish Bird News* **15**, 2.

Appelquist, H., Drabaek, I. & Asbirk, S. (1985). Variation in mercury content of guillemot feathers over 150 years. *Marine Pollution Bulletin* **16**, 244–248.

Armstrong, I.H., Coulson, J.C., Hawkey, P. & Hudson, M.J. (1978). Further mass seabird deaths from paralytic shellfish poisoning. *British Birds* **71**, 58–68.

Ashmole, N.P. (1971). Sea bird ecology and the marine environment. In *Avian biology. Vol. I.* (ed. D.S. Farner & J.R. King), pp. 223–286. London: Academic Press.

Au, D.W. & Pitman, R.L. (1988). Seabird relationships with tropical tunas and dolphins. In *Seabirds & other marine vertebrates. Competition, predation and other interactions.* (ed. J. Burger), pp. 174–212. New York: Columbia University Press.

Avery, M. & Green, R. (1989). Not enough fish in the sea. *New Scientist* **123 (22 July)**, 28–29.

Bannister, R.C.A. (1983). Too much fishing on North Sea flatfish. *Fishing Prospects* **1983**, 1–4.

Barnard, C.J. & Thompson, D.B.A. (1985). *Gulls and plovers. The ecology and behaviour of mixed-species feeding groups.* London: Croom Helm.

Barnes, R.S.K. (1989). The coastal lagoons of Britain: an overview and conservation appraisal. *Biological Conservation* **49**, 295–313.

Barnes, R.S.K. & Hughes, R.N. (1986). *An introduction to marine ecology.* Oxford: Blackwell Scientific Publications.

Barrett, J.H. & Yonge, C.M. (1958). *Collins pocket guide to the sea shore.* London: Collins.

Barrett, R.T. & Furness, R.W. (1990). The prey and diving depths of seabirds on Hornøy, North Norway after a decrease in the Barents Sea capelin stocks. *Ornis Scandinavica* **21**, 179–186.

Batten, L.A., Bibby, C.J., Clement, P., Elliott, G.D. & Porter, R.F. (eds) (1990). *Red data birds in Britain.* London: T & AD Poyser.

Batten, L.A. & Swift, J.A. (1982). British criteria for calling a ban on wildfowling in severe weather. In *Proceedings of the Second Technical Meeting on Western Palaearctic Migratory Bird Management.* (ed. D.A. Scott & M. Smart), pp. 181–189. Slimbridge: International Waterfowl Research Bureau.

Baudinette, R.R., Loveridge, J.P., Wilson, K.J., Mills, C.D. & Schmidt-Nielsen, K. (1976). Heat loss from feet of Herring Gulls at rest and during flight. *American Journal of Physiology* **230**, 920–924.

Bazely, D.R., Ewins, P.J. & McCleery, R.H. (1991). Possible effects of local enrichment by gulls on feeding-site selection by wintering Barnacle Geese *Branta leucopsis*. *Ibis* **133**, 111–114.

Bergman, G. (1982). Why are the wings of *Larus f. fuscus* so dark? *Ornis Fennica* **59**, 77–83.

Berry, R.J. (1985). *The natural history of Orkney.* London: Collins.

Bertram, B.C.R. (1978). Living in groups: predators and prey. In *Behavioural ecology: an evolutionary approach.* (ed. J.R. Krebs & N.B. Davies), pp. 64–96. Oxford: Blackwell Scientific Publications.

Bickmore, D.P. & Shaw, M.A. (1963). *The atlas of Britain and northern Ireland.* Oxford: Clarendon Press.

Bird, E.C.F. (1984) (3rd edn). *Coasts.* Cambridge, Massachusetts: Massachusetts Institute of Technology Press.

Birkhead, T.R. (1977). The effect of habitat and density on breeding success in the Common Guillemot *Uria aalge*. *Journal of Animal Ecology* **46**, 751–764.

Birkhead, T.R. & Furness, R.W. (1985). Regulation of seabird numbers. In *Behavioural ecology. Ecological consequences of adaptive behaviour.* (ed. R.M. Sibly & R.H. Smith), pp. 145–167. Oxford: Blackwell Scientific Publications.

Birkhead, T.R. & Harris, M.P. (1985). Ecological adaptations for breeding in the Atlantic Alcidae. In *The Atlantic Alcidae*. (ed. D.N. Nettleship & T.R. Birkhead), pp. 205–231. London: Academic Press.

Blake, B.F., Tasker, M.L., Hope Jones, P., Dixon, T.J., Mitchell, R. & Langslow, D.R. (1984). *Seabird distribution in the North Sea*. Huntingdon: Nature Conservancy Council.

Blomqvist, S. & Elander, M. (1981). Sabine's gull (*Xema sabini*), Ross's gull (*Rhodostethia rosea*) and Ivory gull (*Pagophila eburnea*) gulls in the Arctic: a review. *Arctic* **34**, 122–132.

Blomqvist, S. & Peterz, M. (1984). Cyclones and pelagic seabird movements. *Marine Ecology – Progress Series* **20**, 85–92.

Boer, G. de (1988). History of the Humber coastline. In *A dynamic estuary: man, nature and the Humber*. (ed. N.V. Jones), pp. 16–30. Hull: Hull University Press.

Boer W.F. de & Drent, R.H. (1989). A matter of eating or being eaten? The breeding performance of arctic geese and its implications for waders. *Wader Study Group Bulletin* **55**, 11–17.

Boorman, L.A., Goss-Custard, J.D. & McGrorty, S. (1989). *Climatic change, rising sea level and the British coast*. London: HMSO.

Boudewijn, T. (1984). The role of digestibility in the selection of spring feeding sites by Brent Geese. *Wildfowl* **35**, 97–105.

Bourne, W.R.P. (1983*a*). Auk wrecks and industrial fisheries. *Marine Pollution Bulletin* **14**, 159.

Bourne, W.R.P. (1983*b*). Birds, fish and offal in the North Sea. *Marine Pollution Bulletin* **14**, 294–296.

Bourne, W.R.P. (1984). Seabird wrecks. *Marine Pollution Bulletin* **15**, 165.

Bourne, W.R.P. (1989*a*). The organization of seabird research. *Marine Pollution Bulletin* **20**, 158–163.

Bourne, W.R.P. (1989*b*). Shetland's sand eels. *New Scientist* **123 (30 Sept.)**, 71–72.

Bourne, W.R.P. (1990). Population fluctuations in seabirds. *BTO News* **171**, 16–17.

Bourne, W.R.P. & Clark, G.C. (1984). The occurrence of birds and garbage at the Humboldt front off Valparaiso, Chile. *Marine Pollution Bulletin* **15**, 343–344.

Bourne, W.R.P. & Imber, M.J. (1982). Plastic pellets collected by a prion on Gough Island, central South Atlantic ocean. *Marine Pollution Bulletin* **13**, 20–21.

Bourne, W.R.P., Mackrill, E.J., Paterson, A.M. & Yésou, P. (1988). The Yelkouan Shearwater *Puffinus (puffinus?) yelkouan*. *British Birds* **81**, 306–319.

Bowes, A., Lack, P.C. & Fletcher, M.R. (1984). Wintering gulls in Britain, January 1983. *Bird Study* **31**, 161–170.

Boyce, D.A. (1985). Merlins and the behavior of wintering shorebirds. *Raptor Research* **19**, 94–96.

Boyd, H. (1964). Wildfowl and other water-birds found dead in England and Wales in January-March 1963. *Wildfowl Trust Annual Report* **15**, 20–22.

Boyd, H. (1987). Do June temperatures affect the breeding success of Dark-bellied Brent Geese *Branta b. bernicla*? *Bird Study* **34**, 155–159.

Boyd, H. & Maltby, L.S. (1979). The Brant of the western Queen Elizabeth Islands, N.W.T. In *Management and biology of Pacific Flyway geese*. (ed. R.L. Jarvis & J.C. Bartonek), pp. 5–21. Corvallis: O.S.U. Book Stores.

Bradstreet, M.S.W. & Brown, R.G.B. (1985). Feeding ecology of the Atlantic Alcidae. In *The Atlantic Alcidae*. (ed. D.N. Nettleship & T.R. Birkhead), pp. 263–318. London: Academic Press.

Bretagnolle, A., Carruthers, M., Cubitt, M., Bioret, F. & Guillandre, J.-P. (1991).

Six captures of a dark-rumped, fork-tailed, storm-petrel in the north-eastern Atlantic. *Ibis* **133**, 351–356.

Britton, R.H. & Johnson, A.R. (1987). An ecological account of a Mediterranean salina: the Salin de Giraud, Camargue (S.France). *Biological Conservation* **42**, 185–230.

Brooke, M. de L. (1989). Determination of the absolute visual threshold of a nocturnal seabird, the Common Diving Petrel *Pelecanoides urinatrix*. *Ibis* **131**, 290–294.

Brooke, M. de L. (1990). *The Manx Shearwater*. London: T & AD Poyser.

Brown, R.G.B. (1980). Seabirds as marine animals. In *Behavior of marine animals. Vol. 4. Marine birds*. (ed. J. Burger, B.L. Olla & H.E. Winn), pp. 1–39. New York: Plenum Press.

Brown, R.G.B., Barker, S.P., Gaskin, D.E. & Sandeman, M.R. (1981). The foods of Great and Sooty Shearwaters *Puffinus gravis* and *P. griseus* in eastern Canadian waters. *Ibis* **123**, 19–30.

Bruijns, M.F.M. (1963). Bird mortality in the Netherlands in the spring of 1960, due to the use of pesticides in agriculture. *Bulletin of the International Council for Bird Preservation* **9**, 70–75.

Bryant, D.M. & Leng, J. (1975). Feeding distribution and behaviour of Shelduck in relation to food supply. *Wildfowl* **26**, 20–30.

Buchanan, J.B., Schick, C.T., Brennan, L.A. & Herman, S.G. (1988). Merlin predation on wintering Dunlins: hunting success and Dunlin escape tactics. *Wilson Bulletin* **100**, 108–118.

Bucke, D. & Watermann, B. (1988). Effects of pollutants on fish. In *Pollution of the North Sea: an assessment*. (ed. W. Salomons, B.L. Bayne, E.K. Duursma & V. Förstner), pp. 612–623. Berlin: Springer-Verlag.

Bull, K.R., Every, W.J., Freestone, P., Hall, J.R., Osborn, D., Cooke, A.S. & Stowe, T. (1983). Alkyl lead pollution and bird mortalities on the Mersey Estuary, UK, 1979–1981. *Environmental Pollution* **31A**, 239–259.

Bullock, I.D., Drewett, D.R. & Mickleburgh, S.P. (1983). The Chough in Britain and Ireland. *British Birds* **76**, 377–401.

Burd, F. (1989). *The saltmarsh survey of Great Britain. An inventory of British saltmarshes*. Peterborough: Nature Conservancy Council, Research and Survey in Nature Conservation No. 17.

Burger, A.E. & Simpson, M. (1986). Diving depths of Atlantic Puffins and Common Murres. *Auk* **103**, 828–830.

Burger, J. (1980). The transition to independence and postfledging parental care in seabirds. In *Behavior of marine animals. Vol.4. Marine birds*. (ed. J. Burger, B.L. Olla & H.E. Winn), pp. 367–447. New York: Plenum Press.

Burger, J. & Gochfield, M. (1991). Human activity influence and diurnal and nocturnal foraging of Sanderlings (*Calidris alba*). *Condor* **93**, 259–265.

Burton, R. (1985). *Bird behaviour*. London: Granada Publishing.

Buxton, N.E. (1982). Wintering coastal waders of Lewis and Harris. *Scottish Birds* **12**, 38–43.

Cadbury, C.J., Berry, R., Chapman, B., Collier, R.V., Parker, A., Redman, B., Wilkinson, R.B. & Wilson, P. (1975). *Breeding birds of the Wash*. Report to the Institute of Terrestrial Ecology. Sandy: Royal Society for the Protection of Birds.

Cadbury, C.J., Green, R.E. & Allport, G. (1987). Redshanks and other breeding waders of British saltmarshes. *RSPB Conservation Review* **1**, 37–40.

Cadbury, C.J., Hill, D., Partridge, J. & Sorenson, J. (1989). The history of the

Avocet population and its management in England since recolonisation. *RSPB Conservation Review* **3**, 9–13.

Cain, K. (1984). Functions of metallothionein. In *Metals in animals*. (ed. D. Osborn), pp. 55–61. Cambridge: Institute of Terrestrial Ecology.

Campbell, B. & Lack, E. (eds) (1985). *A dictionary of birds*. Calton: T & AD Poyser.

Campbell, L.H. (1978). Patterns of distribution and behaviour of flocks of seaducks wintering at Leith and Musselburgh, Scotland. *Biological Conservation* **14**, 111–124.

Campbell, L.H. (1989). The importance of breeding waders on croft and farmland in Shetland. *RSPB Conservation Review* **3**, 75–78.

Campbell, L.H., Barrett, J. & Barrett, C.F. (1986). Seaducks in the Moray Firth: a review of their current status and distribution. *Proceedings of the Royal Society of Edinburgh* **91B**, 105–112.

Campbell, L.H. & Milne, H. (1983). Moulting eiders in eastern Scotland. *Wildfowl* **34**, 105–107.

Carpenter, E.J., Anderson, S.J., Harvey, G.R., Miklas, H.P. & Peck, B.B. (1972). Polystyrene spherules in coastal waters. *Science, New York* **178**, 749–750.

Carter, R.W.G. (1989). *Coastal environments: an introduction to the physical, ecological and cultural systems of coastlines*. London: Academic Press.

Castro, G., Stoyan, N. & Myers, J.P. (1989). Assimilation efficiency in birds: a function of taxon or food type? *Comparative Biochemistry and Physiology* **92A**, 271–278.

Castro, M. (1984). Auks drown in Spanish nets. *BTO News* **132**, 1.

Chapman, A. (ed.) (1986). *RSPB reserves. Report for 1985*. Sandy: Royal Society for the Protection of Birds.

Chapman, V.J. (ed.) (1977). *Ecosystems of the world. 1. Wet coastal ecosystems*. Amsterdam: Elsevier.

Chardine, J.W. (1987). The influence of pair-status on the breeding behaviour of the Kittiwake *Rissa tridactyla* before egg-laying. *Ibis* **129**, 515–526.

Charman, K. (1979). Feeding ecology and energetics of the dark-bellied brent goose (*Branta bernicla bernicla*) in Essex and Kent. In *Ecological processes in coastal environments*. (ed. R.L. Jefferies & A.J. Davy), pp. 451–465. Oxford: Blackwell Scientific Publications.

Charman, K. & Macey, A. (1978). The winter grazing of saltmarsh vegetation by Dark-bellied Brent Geese. *Wildfowl* **29**, 153–162.

Charnov, E.L., Orians, G.H. & Hyatt, K. (1976). Ecological implications of resource depression. *American Naturalist* **110**, 247–259.

Clafton, F.R. (1976). Portland Bill. In *Bird observatories in Britain and Ireland*. (ed. R. Durman), pp. 191–205. Berkhamsted: T & AD Poyser.

Clark, N.A. (1983). The ecology of Dunlin (*Calidris alpina* L.) wintering on the Severn Estuary. Ph.D. thesis: University of Edinburgh.

Clark, N.A. & Davidson, N.C. (1986). WSG project on the effects of severe weather on waders: sixth progress report. *Wader Study Group Bulletin* **46**, 7–8.

Coles, B. & Coles, J. (1986). *Sweet Track to Glastonbury: the Somerset Levels in prehistory*. London: Thames and Hudson.

Coles, B. & Coles, J. (1989). *People of the wetlands: bogs, bodies and lake-dwellers*. London: Guild Publishing.

Coles, S.M. (1979). Benthic microalgal populations on intertidal sediments and their role as precursors to salt marsh development. In *Ecological processes in coastal environments*. (ed. R.L. Jefferies & A.J. Davy), pp. 25–42. Oxford: Blackwell Scientific Publications.

Collinge, W.E. (1925). An investigation of the food of terns at Blakeney Point, Norfolk. *Transactions of the Norfolk and Norwich Naturalists' Society* **12**, 35–53.

Colman, J. (1940). On the faunas inhabiting intertidal seaweeds. *Journal of the Marine Biological Association of the United Kingdom* **24**, 129–183.

Cooke, A.S., Bell, A.A. & Haas, M.B. (1982). *Predatory birds, pesticides and pollution*. Cambridge: Institute of Terrestrial Ecology.

Cooper, J., Williams, A.J. & Britton, P.L. (1984). Distribution, population sizes and conservation of breeding seabirds in the Afrotropical region. In *Status and conservation of the world's seabirds*. (ed. J.P. Croxall, P.G.H. Evans & R.W. Schreiber), pp. 403–419. Cambridge: International Council for Bird Preservation, Technical Publication No.2.

Cormack, D. (1984). Seabirds and oil. *Marine Pollution Bulletin* **15**, 345–347.

Cott, H.B. (1947). The edibility of birds: illustrated by five years' experiments and observations (1941–1946) on the food preferences of the hornet, cat and man; and considered with special reference to the theories of adaptive coloration. *Proceedings of the Zoological Society of London* **116**, 371–524.

Coulson, J.C., Monaghan, P., Butterfield, J.E.L., Duncan, N., Ensor, K., Sheddon, C. & Thomas, C. (1984). Scandinavian Herring Gulls wintering in Britain. *Ornis Scandinavica* **15**, 79–88.

Coulson, J.C. & Porter, J.M. (1985). Reproductive success of the Kittiwake *Rissa tridactyla*: the roles of clutch size, chick growth rates and parental quality. *Ibis* **127**, 450–464.

Coulson, J.C. & Thomas, C. (1985). Differences in the breeding performance of individual kittiwake gulls, *Rissa tridactyla* (L.). In *Behavioural ecology. Ecological consequences of adaptive behaviour*. (ed. R.M. Sibly & R.H. Smith), pp. 489–503. Oxford: Blackwell Scientific Publications.

Coulson, J.C. & Wooller, R.D. (1984). Incubation under natural conditions in the kittiwake gull, *Rissa tridactyla*. *Animal Behaviour* **32**, 1204–1215.

Craik, J.C.A. (1990). The price of mink. *Scottish Bird News* **19**, 4–5.

Craik, J.C.A. (1991). More serious than Shetland? *Seabird Group Newsletter* **60**, 2–4.

Craik, K.J.W. (1944). White plumage of sea-birds. *Nature, London* **153**, 288.

Cramp, S. (ed.) (1985). *The birds of the western Palearctic, Vol. IV*. Oxford: Oxford University Press.

Cramp, S., Bourne, W.R.P. & Saunders, D. (1974). *The seabirds of Britain and Ireland*. London: Collins.

Cramp, S. & Simmons, K.E.L. (eds.). (1977). *The birds of the western Palearctic, Vol.I*. Oxford: Oxford University Press.

Cramp, S. & Simmons, K.E.L. (eds.). (1983). *The birds of the western Palearctic, Vol.III*. Oxford: Oxford University Press.

Crawford, R.J.M., David, J.H.M., Williams, A.J. & Dyer, B.M. (1989). Competition for space: recolonising seals displace endangered, endemic seabirds off Namibia. *Biological Conservation* **48**, 59–72.

Crawford, R.J.M. & Shelton, P.A. (1978). Pelagic fish and seabird interrelationships off the coasts of South West and South Africa. *Biological Conservation* **14**, 85–901.

Croxall, J.P., Evans, P.G.H. & Schreiber, R.W. (eds.) (1984). *Status and conservation of the world's seabirds*. Cambridge: International Council for Bird Preservation, Technical Publication No. 2.

Croxall, J.P. & Prince, P.A. (1990). Recoveries of Wandering Albatrosses *Diomedia exulans* ringed at South Georgia 1958–1986. *Ringing & Migration* **11**, 43–51.

Croxall, J.P., Prince, P.A., Hunter, I., McInnes, S.J. & Copestake, P.G. (1984). The seabirds of the Antarctic Peninsula, islands of the Scotia Sea, and Antarctic Continent between 80°W and 20°W: their status and conservation. In *Status and conservation of the world's seabirds*. (ed. J.P. Croxall, P.G.H. Evans & R.W. Schreiber), pp. 637–666. Cambridge: International Council for Bird Preservation, Technical Publication No. 2.

Cryer, M., Whittle, N. & Williams, R. (1987). The impact of bait collection by anglers on marine intertidal invertebrates. *Biological Conservation* **42**, 83–93.

Cudworth, J. (1976). Spurn. In *Bird observatories in Britain and Ireland*. (ed. R. Durman), pp. 232–250. Berkhamsted: T & AD Poyser.

Cullen, E. (1957). Adaptations in the Kittiwake to cliff-nesting. *Ibis* **99**, 275–302.

Curtis, D.J., Galbraith, C.G., Smyth, J.C. & Thompson, D.B.A. (1985). Seasonal variations in prey selection by estuarine black-headed gulls (*Larus ridibundus*). *Estuarine, Coastal and Shelf Science* **21**, 75–89.

Cushing, D.H. (1990). Plankton production and year-class strength in fish populations: an update of the match/mismatch hypothesis. *Advances in Marine Biology* **26**, 249–293.

Cutts, L.L. (1985). Defense as a factor in bill-tucking. *Florida Field Naturalist* **13**, 58–63.

Daniels, D., Heath, J. & Rawson, W. (1984). A declaration of intent in the kittiwake gull (*Rissa tridactyla*). *Animal Behaviour* **32**, 1151–1156.

Davidson, N.C. (1981). Survival of shorebirds (Charadrii) during severe weather: the role of nutritional reserves. In *Feeding and survival strategies of estuarine organisms*. (ed. N.V. Jones & W.J. Wolff), pp. 231–249. New York: Plenum Press.

Davidson, N.C. (1982). Changes in the body-condition of Redshanks during mild winters: an inability to regulate reserves? *Ringing & Migration* **4**, 51–62.

Davidson, N.C. (1984). How valid are flight range estimates for waders? *Ringing & Migration* **5**, 49–64.

Davidson, N.C. (1991). Breeding waders on British wet grasslands. *Wader Study Group Bulletin* **61 Supplement**, 36–41.

Davidson, N.C. & Clark, N.A. (1985). The effects of severe weather in January and February 1985 on waders in Britain. *Wader Study Group Bulletin* **44**, 10–16.

Davidson, N.C. & Evans, P.R. (1986). The role and potential of man-made and man-modified wetlands in the enhancement of the survival of overwintering shorebirds. *Colonial Waterbirds* **9**, 176–188.

Davidson, N.C., Laffoley, D.d'A., Doody, J.P., Way, L.S., Gordon, J., Key, R., Drake, C.M., Pienkowski, M.W., Mitchell, R. & Duff, K.L. (1991). *Nature conservation and estuaries in Great Britain*. Peterborough: Nature Conservancy Council.

Davidson, N.C. & Pienkowski, M.W. (eds.) (1987). The conservation of international flyway populations of waders. *Wader Study Group Bulletin* **49 Supplement**, 1–151.

Davies, M. (1988). The importance of Britain's twites. *RSPB Conservation Review* **2**, 91–94.

Davis, P. & Moss, D. (1984). *Spartina* and waders – the Dyfi estuary. In Spartina anglica *in Great Britain*. (ed. P. Doody), pp. 37–40. Huntingdon: Nature Conservancy Council, Research and Survey in Nature Conservation No. 5.

Dean, T. (1991). *The natural history of Walney Island*. Burnley: Faust Publications.

DeGange, A.R. & Day, R.H. (1991). Mortality of seabirds in the Japanese land-based gillnet fishery for salmon. *Condor* **93**, 251–258.

Dennis, R.N. (1925). Notes on Sussex ornithology. In *Extracts from the diaries (1845–1869) of Robert Nathaniel Dennis*. (ed. W.H. Mullens & N.F. Ticehurst). London: H.F. & G. Witherby.

Dennis, R. (1991). Dark skies and darker water. *BTO News* **174**, 14–15.

Department of the Environment. (1981). *Report of the sub-committee on the disposal of sewage sludge to land*. London: National Water Council, Standing Technical Committee Reports No. 20.

Department of the Environment. (1988). *Possible impact of climate change on the natural environment in the United Kingdom*. London: Department of the Environment.

Department of Energy. (1989). *Taking power from water*. Harwell: Department of Energy.

Dethlefsen, V. (1984). Diseases in North Sea fishes. *Helgolander Meeresunters* **37**, 353–374.

Dhondt, A.A. (1987). Cycles of lemmings and Brent Geese *Branta b. bernicla*: a comment on the hypothesis of Roselaar and Summers. *Bird Study* **34**, 151–154.

Diamond, A.W. (1987). A global view of cultural and economic uses of birds. In *The value of birds*. (ed. A.W. Diamond & F.L. Filion), pp. 99–109. Cambridge: International Council for Bird Preservation, Technical Publication No. 6.

Dick, W.J.A. & Pienkowski, M.W. (1979). Autumn and early winter weights of waders in north-west Africa. *Ornis Scandinavica* **10**, 117–123.

Dick, W.J.A., Piersma, T. & Prokosch, P. (1987). Spring migration of the Siberian Knots *Calidris canutus canutus*: results of a co-operative Wader Study Group project. *Ornis Scandinavica* **18**, 5–16.

Dony, J.G., Jury, S.L. & Perring, F.H. (1986) (2nd edn). *English names of wild flowers*. Peterborough: Botanical Society of the British Isles.

Doody, J.P. (1985). The conservation of sand dunes in Great Britain – a review. In *Sand dunes and their management*. (ed. J.P. Doody), pp. 43–50. Peterborough: Nature Conservancy Council, Research and Survey in Nature Conservation No. 13.

Doody, J.P. & Barnett, B. (1987). Introduction. In *The Wash and its environment*. 1–10. Peterborough: Nature Conservancy Council, Research and Survey in Nature Conservation, No. 7.

Drent, R.H. (1970). Functional aspects of incubation in the Herring Gull. *Behaviour*, **Supplement 17**, 1–132.

Drent, R.H., Ebbinge, B. & Weijand, B. (1979). Balancing the energy budgets of arctic-breeding geese throughout the annual cycle: a progress report. *Verhandlungen der Ornithologischen Gesellschaft in Bayern* **23**, 239–264.

Drijver, C.A. (1983). Coastal protection and land reclamation in the Wadden Sea area. In *Nature conservation, nature management and physical planning in the Wadden Sea area*, Report 11 of the Wadden Sea Working Group. (ed. M.F. Mörzer Bruyns & W.J. Wolff), pp. 130–144. Rotterdam: A.A. Balkema.

Druehl, L. (1973). Marine transplantation. *Science, New York* **179**, 12.

Drury, W.H. (1965). Results of a study of Herring Gull populations and movements in southeastern New England. In *Le Problème des Oiseaux sur les Aérodromes*. (ed. R.G. Busnel & J. Giban), pp. 207–219. Paris: Institut National de la Recherche Agronomique.

Dugan, P.J. (1981). The importance of nocturnal foraging in shorebirds: a consequence of increased invertebrate prey activity. In *Feeding and survival strategies of estuarine organisms*. (ed. N.V. Jones & W.J. Wolff), pp. 251–260. New York: Plenum Press.

Dugan, P.J., Evans, P.R., Goodyer, L.R. & Davidson, N.C. (1981). Winter fat reserves in shorebirds: disturbance of regulated levels by severe weather conditions. *Ibis* **123**, 359–363.

Duinker, J.C., Schultz, D.E. & Petrick, G. (1988). Selection of chlorinated biphenyl congeners for analysis in environmental samples. *Marine Pollution Bulletin* **19**, 19–25.

Dunn, E.K. (1973). Robbing behaviour of Roseate Terns. *Auk* **90**, 641–651.

Dunnet, G.M. & Ollason, J.C. (1982). The feeding dispersal of Fulmars *Fulmarus glacialis* in the breeding season. *Ibis* **124**, 359–361.

Durell, S.E.A. le V. dit & Goss-Custard, J.D. (1984). Prey selection within a size-class of mussels, *Mytilus edulis*, by oystercatchers, *Haematopus ostralegus*. *Animal Behaviour* **32**, 1197–1203.

Ebbinge, B.S. (1989). A multifactorial explanation for variation in breeding performance of Brent Geese *Branta bernicla*. *Ibis* **131**, 196–204.

Ebbinge, B.S., Canters, K. & Drent, R. (1975). Foraging routines and estimated daily food intake in Barnacle Geese wintering in the northern Netherlands. *Wildfowl* **26**, 5–19.

Edington, J.M. & Edington, M.A. (1986). *Ecology, recreation and tourism*. Cambridge: Cambridge University Press.

Elkins, N. (1983). *Weather and bird behaviour*. Calton: T & AD Poyser.

Elkins, N. (1990). What does global warming mean for our birds? *Scottish Bird News* **18**, 4–5.

Ens, B.J. & Goss-Custard, J.D. (1984). Interference among oystercatchers, *Haematopus ostralegus*, feeding on mussels, *Mytilus edulis*, on the Exe Estuary. *Journal of Animal Ecology* **53**, 217–231.

Erichsen, J.T. (1985). Vision. In *A dictionary of birds*. (ed. B. Campbell & E. Lack), pp. 623–629. Calton: T & AD Poyser.

Erwin, R.M. (1980). Censusing waterbird colonies: some sampling experiments. *Transactions of the Linnaean Society of New York* **9**, 77–86.

Erwin, R.M. (1981). Censusing wading bird colonies: an update on the 'flight-line' count method. *Colonial Waterbirds* **4**, 91–95.

Erwin, R.M. (1982). Observer variability in estimating numbers: an experiment. *Journal of Field Ornithology* **53**, 159–167.

Erwin, R.M. & Custer, T.W. (1982). Estimating reproductive success in colonial waterbirds: an evaluation. *Colonial Waterbirds* **5**, 49–56.

Etheridge, B. (1982). Distribution of Dunlin *Calidris alpina* nests on an area of South Uist machair. *Bird Study* **29**, 239–243.

Evans, A. (1987). Relative availability of the prey of wading birds by day and by night. *Marine Ecology – Progress Series* **37**, 103–107.

Evans, C.D.R. (1980). *Report on the geological and geophysical investigations conducted in the Severn Estuary during 1979*. Leeds: Institute of Geological Sciences, Report No. STP 23.

Evans, G.E. & Collins, M. (1987). Sediment supply and deposition in the Wash. In *The Wash and its environment*. (ed. P. Doody & B. Barnett), pp. 48–63. Peterborough: Nature Conservancy Council, Research and Survey in Nature Conservation No. 7.

Evans, M.E., Wood, N.A. & Kear, J. (1973). Lead shot in Bewick's Swans. *Wildfowl* **24**, 56–60.

Evans, P.G.H. (1981). Ecology and behaviour of the Little Auk *Alle alle* in west Greenland. *Ibis* **123**, 1–18.

Evans, P.G.H. (1984). The seabirds of Greenland: their status and conservation.

In *Status and conservation of the world's seabirds*. (ed. J.P. Croxall, P.G.H. Evans & R.W. Schreiber), pp. 49–84. Cambridge: International Council for Bird Preservation, Technical Publication No. 2.

Evans, P.G.H. & Nettleship, D.N. (1985). Conservation of the Atlantic Alcidae. In *The Atlantic Alcidae*. (ed. D.N. Nettleship & T.R. Birkhead), pp. 427–488. London: Academic Press.

Evans, P.R. (1979). Reclamation of intertidal land: some effects on Shelduck and wader populations in the Tees estuary. *Verhandlungen der Ornithologischen Gesellschaft in Bayern* **23**, 147–168.

Evans, P.R. (1986). Use of the herbicide 'Dalapon' for control of *Spartina* encroaching on intertidal mudflats: beneficial effects on shorebirds. *Colonial Waterbirds* **9**, 171–175.

Evans, P.R. & Pienkowski, M.W. (1982). Behaviour of shelducks *Tadorna tadorna* in a winter flock: does regulation occur? *Journal of Animal Ecology* **44**, 241–262.

Evans, P.R., Uttley, J.D., Davidson, N.C. & Ward, P. (1987). Shorebirds (S.Os Charadrii and Scolopaci) as agents of transfer of heavy metals within and between estuarine ecosystems. In *Pollutant transport and fate in ecosystems*. (ed. P.J. Coughtrey, M.H. Martin & M.H. Unsworth), pp. 337–352. Oxford: Blackwell Scientific Publications.

Evans, S. (1986). The mainland coast. In *The nature of West Wales*. (ed. D. Saunders), pp. 89–104. Buckingham: Barracuda Books.

Everett, M.J. (1987). The Elmley experiment. *RSPB Conservation Review* **1**, 31–33.

Everett, M.J., Cadbury, C.J. & Dawson, L. (1988). The importance of RSPB reserves for wintering and migrant wildfowl and waders. *RSPB Conservation Review* **2**, 57–63.

Everett, M.J., Hepburn, I.R., Ntiamoa-Baidu, Y. & Thomas, G.J. (1987). Roseate Terns in Britain and West Africa. *RSPB Conservation Review* **1**, 56–58.

Ewins, P.J. (1985). Growth, diet and mortality of Arctic Tern *Sterna paradisaea* chicks in Shetland. *Seabird* **8**, 59–68.

Ewins, P.J. (1989). The breeding biology of Black Guillemots *Cepphus grylle* in Shetland. *Ibis* **131**, 507–520.

Ewins, P.J., Birkhead, T.R. & Partridge, K.E. (1988). Auks die in Cornish fishing nets. *BTO News* **156**, 1.

Falandysz, J., Jakuczun, B. & Mizera, T. (1988). Metals and organochlorines in four female White-tailed eagles. *Marine Pollution Bulletin* **19**, 521–526.

Fauchald, K. & Jumars, P.A. (1979). The diet of worms: a study of polychaete feeding guilds. *Oceanography and Marine Biology, an Annual Review* **17**, 193–284.

Feare, C.J. (1971). Predation of limpets and dogwhelks by Oystercatchers. *Bird Study* **18**, 121–129.

Feare, C.J. (1974). The utilization of mangroves by Seychelles birds. *Ibis* **116**, 543–545.

Feare, C.J. (1976). The breeding of the Sooty tern *Sterna fuscata* in the Seychelles and the effects of experimental removal of its eggs. *Journal of Zoology, London* **179**, 317–360.

Feare, C.J. (1984). Human exploitation. In *Status and conservation of the world's seabirds*. (ed. J.P. Croxall, P.G. H. Evans & R.W. Schreiber), pp. 691–699. Cambridge: International Council for Bird Preservation, Technical Publication No. 2.

Feare, C.J. & Summers, R.W. (1985). Birds as predators on rocky shores. In *The ecology of rocky coasts*. (ed. P.G. Moore & R. Seed), pp. 249–264. London: Hodder and Stoughton.

Feduccia, A. (1980). *The age of birds.* Cambridge, Massachusetts: Harvard University Press.

Ferns, P.N. (1987). The Taff Estuary and its birds. *Transactions of the Cardiff Naturalists' Society* **100**, 13–25.

Ferns, P.N. (1989). Current prospects for the Severn Barrage. *Biologist* **36**, 201–206.

Ferns, P.N. & Mudge, G.P. (1981). Accuracy of nest counts at a mixed colony of Herring and Lesser Black-backed Gulls. *Bird Study* **28**, 244–246.

Filion, F.L. (1987). Birds as a socio-economic resource: a strategic concept in promoting conservation. In *The value of birds.* (ed. A.W. Diamond & F.L. Filion), pp. 7–14. Cambridge: International Council for Bird Preservation, Technical Publication No. 6.

Fisher, J. (1952). *The Fulmar.* London: Collins.

Fisher, J. & Lockley, R.M. (1954). *Sea-birds.* London: Collins.

Fisher, J. & Vevers, H.G. (1951). The present population of the the North Atlantic Gannet (*Sula bassana*). *Proceedings of the International Ornithological Congress* **10**, 463–467.

Fogg, G.E. (1985). Biological activities at a front in the western Irish Sea. In *Proceedings of the 19th European Marine Biology Symposium.* (ed. P.E. Gibbs), pp. 87–95. Cambridge: Cambridge University Press.

Franklin, A. (1977). The Burry Inlet cockle fishery. In *Problems of a small estuary.* (ed. A. Nelson-Smith & E.M. Bridges), pp. 3:1/1–3:1/5. Swansea: Institute of Marine Studies, University College of Swansea.

Frazer, D. (1983). *Reptiles and amphibians in Britain.* London: Collins.

Frid, C.L.J. & Mercer, T.S. (1989). Environmental monitoring of caged fish farming in macrotidal environments. *Marine Pollution Bulletin* **20**, 379–383.

Fuchs, E. (1977). Kleptoparasitism of Sandwich Terns *Sterna sandvicensis* by Black-headed Gulls *Larus ridibundus*. *Ibis* **119**, 183–190.

Fuller, R.J. (1982). *Bird habitats in Britain.* Calton: T & AD Poyser.

Fuller, R.J., Reed, T.M., Buxton, N.E., Webb, A., Williams, T.D. & Pienkowski, M.W. (1986). Populations of breeding waders Charadrii and their habitats on the crofting lands of the Outer Hebrides, Scotland. *Biological Conservation* **37**, 333–361.

Furness, B.L. (1983). Plastic particles in three procellariiform seabirds from the Benguela Current, South Africa. *Marine Pollution Bulletin* **14**, 307–308.

Furness, R.W. (1973a). Wader populations at Musselburgh. *Scottish Birds* **7**, 275–281.

Furness, R.W. (1973b). Roost selection by waders. *Scottish Birds* **7**, 281–287.

Furness, R.W. (1978). Energy requirements of seabird communities: a bioenergetics model. *Journal of Animal Ecology* **47**, 39–53.

Furness, R.W. (1982). Competition between fisheries and seabird communities. *Advances in Marine Biology* **20**, 225–307.

Furness, R.W. (1984). Seabird biomass and food consumption in the North Sea. *Marine Pollution Bulletin* **15**, 244–248.

Furness, R.W. (1985). Plastic particle pollution: accumulation by procellariiform seabirds at Scottish colonies. *Marine Pollution Bulletin* **16**, 103–106.

Furness, R.W. (1987a). *The skuas.* Calton: T & AD Poyser.

Furness, R.W. (1987b). Seabirds as monitors of the marine environment. In *The value of birds.* (ed. A.W. Diamond & F.L. Filion), pp. 217–230. Cambridge: International Council for Bird Preservation, Technical Publication No. 6.

Furness, R.W. (1988). Predation on ground-nesting seabirds by island populations of red deer *Cervus elephas* and sheep *Ovis*. *Journal of Zoology, London* **216**, 565–573.

Furness, R.W. (1989). Changes in diet and breeding ecology of seabirds on Foula, 1971–88. In *Seabirds and sandeels*. (ed. M. Heubeck), pp. 22–26. Lerwick: Shetland Bird Club.

Furness, R.W. (1990). A preliminary assessment of the quantities of Shetland sandeels taken by seabirds, seals, predatory fish and the industrial fishery in 1981–83. *Ibis* **132**, 205–217.

Furness, R.W. & Ainley, D.G. (1984). Threats to seabird populations presented by commercial fisheries. In *Status and conservation of the world's seabirds*. (ed. J.P. Croxall, P.G.H. Evans & R.W. Schreiber), pp. 701–708. Cambridge: International Council for Bird Preservation, Technical Publication No.2.

Furness, R.W., Galbraith, H., Gibson, I.P. & Metcalfe, N.B. (1986). Recent changes in numbers of waders on the Clyde Estuary, and their significance for conservation. *Proceedings of the Royal Society of Edinburgh* **90B**, 171–184.

Furness, R.W., Hudson, A.V. & Ensor, K. (1988). Interactions between scavenging seabirds and commercial fisheries around the British Isles. In *Seabirds & other marine vertebrates. Competition, predation and other interactions*. (ed. J. Burger), pp. 240–268. New York: Columbia University Press.

Furness, R.W. & Monaghan, P. (1987). *Seabird ecology*. Glasgow: Blackie.

Furness, R.W. & Todd, C.M. (1984). Diets and feeding of Fulmars *Fulmarus glacialis* during the breeding season: a comparison between St Kilda and Shetland colonies. *Ibis* **126**, 379–387.

GESAMP (1991). *The state of the marine environment*. Oxford: Blackwell Scientific Publications.

Gardiner, J. (1974). The chemistry of cadmium in natural water – II. The adsorption of cadmium on river muds and naturally occurring solids. *Water Research* **8**, 157–164.

Gardner, L.L. (1925). The adaptive modifications and the taxonomic value of the tongue in birds. *Proceedings of the United States National Museum* **67**, 1–49.

Gaston, A.J. (1985). Development of the young in the Atlantic Alcidae. In *The Atlantic Alcidae*. (ed. D.N. Nettleship & T.R. Birkhead), pp. 319–354. London: Academic Press.

Gaston, A.J. & Nettleship, D.N. (1981). *The Thick-billed Murres of Prince Leopold Island*. Ottawa: Canadian Wildlife Service, Monograph Series, No.6.

Gerritsen, A.F.C. & Meiboom, A. (1986). The role of touch in prey density estimation by *Calidris alba*. *Netherlands Journal of Zoology* **36**, 530–561.

Gerritsen, A.F.C. & Sevenster, J.G. (1985). Foraging behaviour and bill anatomy in sandpipers. *Fortschrifte der Zoologie* **30**, 237–239.

Gibb, J. (1956). Food, feeding habits and territory of the Rock Pipit *Anthus spinoletta*. *Ibis* **98**, 506–530.

Goethe, F. (1981). Common gull (*Larus canus* L.). In *Birds of the Wadden Sea*, Report 6 of the Wadden Sea Working Group. (ed. C.J. Smit & W.J. Wolff), pp. 229–238. Rotterdam: A.A.Balkema.

Goodall, A. (1988). Life in the Humber: (D) Birds. In *A dynamic estuary: man, nature and the Humber*. (ed. N.V. Jones), pp. 83–97. Hull: Hull University Press.

Gordon, N. (1976). The Isle of May. In *Bird observatories in Britain and Ireland*. (ed. R. Durman), pp. 176–190. Berkhamsted: T & AD Poyser.

Gorman, M.L. (1974). The significance of habitat selection during nesting of the Eider *Somateria mollissima mollissima*. *Ibis* **116**, 152–154.

Goss-Custard, J.D. (1970). Feeding dispersion in some overwintering wading birds. In *Social behaviour in birds and mammals*. (ed. J.H. Crook), pp. 3–35. London: Academic Press.

Goss-Custard, J.D. & Durell, S.E.A. le V. dit (1987). Age-related effects in oystercatchers, *Haematopus ostralegus*, feeding on mussels, *Mytilus edulis*. II. Aggression. *Journal of Animal Ecology* **56**, 537–548.

Goss-Custard, J.D., Jenyon, R.A., Jones, R.E., Newberry, P.E. & Williams, R. le B. (1977). The ecology of the Wash. II. Seasonal variation in the feeding conditions of wading birds (Charadrii). *Journal of Applied Ecology* **14**, 701–719.

Goss-Custard, J.D., McGrorty, S., Reading, C.J. & Durell, S.E. A. le V. dit (1980). Oystercatchers and mussels on the Exe Estuary. *Devon Association Special Volume* **2**, 161–185.

Goss-Custard, J.D. & Moser, M.E. (1988). Rates of change in the numbers of dunlin, *Calidris alpina*, wintering in British estuaries in relation to the spread of *Spartina anglica*. *Journal of Applied Ecology* **25**, 95–109.

Goss-Custard, J.D., Yates, M.G., McGrorty, S., Moy, I., Durell, S.E.A. le V. dit, Lakhani, K., Rispin, E. & Yates, T. (1987). The wading birds (Charadrii) of the Wash, and the possible effect of further land-claim on their numbers. In *The Wash and its environment*. (ed. P. Doody & B. Barnett), pp. 173–185. Peterborough: Nature Conservancy Council, Research and Survey in Nature Conservation No. 7.

Götmark, F. (1987). White underparts in gulls function as hunting camouflage. *Animal Behaviour* **35**, 1786–1792.

Götmark, F. (1989). Costs and benefits to Eiders nesting in gull colonies: a field experiment. *Ornis Scandinavica* **20**, 283–288.

Gowen, R.J. & Bradbury, N.B. (1987). The ecological impact of salmonid farming in coastal waters: a review. *Oceanography and Marine Biology. An Annual Review* **2**, 563–575.

Gray, P.W.G. & Jones, E.B.G. (1977). The attempted clearance of *Sargassum muticum* from Britain. *Environmental Conservation* **4**, 303–308.

Green, G. & Cade, M. (1989). *Where to watch birds in Dorset, Hampshire & the Isle of Wight*. London: Christopher Helm.

Green, G.H., Greenwood, J.J.D. & Lloyd, C.S. (1977). The influence of snow conditions on the date of breeding of wading birds in north-east Greenland. *Journal of Zoology, London* **183**, 311–328.

Green, R. (1987). Breeding birds of the Wash saltmarshes. In *The Wash and its environment*. (ed. P. Doody & B. Barnett), pp. 133–137. Peterborough: Nature Conservancy Council, Research and Survey in Nature Conservation No. 7.

Green, R.E. (1988). Effects of environmental factors on the timing and success of breeding of common snipe *Gallinago gallinago* (Aves: Scolopacidae). *Journal of Applied Ecology* **25**, 79–93.

Green, R.E. & Cadbury, C.J. (1987). Breeding waders of lowland wet grasslands. *RSPB Conservation Review* **1**, 10–13.

Green, R.E., Cadbury, C.J. & Williams, G. (1987). Floods threaten Black-tailed Godwits breeding at the Ouse Washes. *RSPB Conservation Review* **1**, 14–16.

Greenhalgh, M.E. (1975). *Wildfowl of the Ribble Estuary*. Chester: Wildfowlers' Association of Great Britain and Ireland.

Greenwood, P.J. (1980). Mating systems, philopatry and dispersal in birds and mammals. *Animal Behaviour* **28**, 1140–1162.

Gribble, F.C. (1962). Census of Black-headed Gull colonies in England and Wales, 1958. *Bird Study* **9**, 56–71.

Grieve, S. (1885). *The great auk, or garefowl* (Alca impennis, Linn.) *its history, archaeology and remains*. London: T.C. Jack.

Grimmett, R.F.A. & Jones, T.A. (1989). *Important bird areas in Europe*. Cambridge: International Council for Bird Preservation, Technical Publication No. 9.

Grubb, T.C. (1974). Olfactory navigation to the nesting burrow in Leach's Petrel (*Oceanodroma leucorrhoa*). *Animal Behaviour* **22**, 192–202.

Gudmundsson, G.A., Lindström, Å. & Alerstam, T. (1991). Optimal fat loads and long-distance flights by migrating Knots *Calidris canutus*, Sanderlings *C. alba* and Turnstones *Arenaria interpres*. *Ibis* **133**, 140–152.

Gurney, J.H. (1913). *The Gannet, a bird with a history*. London: Witherby.

Gwinner, E. (1975). Circadian and circannual rhythms in birds. In *Avian biology. Vol. V.* (ed. D.S. Farner & J.R. King), pp. 221–285. London: Academic Press.

Haddon, P.C. & Knight, R.C. (1983). *A guide to Little Tern conservation*. Sandy: Royal Society for the Protection of Birds.

Hainsworth, F.R. (1988). Induced drag savings from ground effect and formation flight in brown pelicans. *Journal of experimental Biology* **135**, 431–444.

Hale, W.G. (1980). *Waders*. London: Collins.

Haney, J.C. (1986*a*). Seabird affinities for Gulf Stream frontal eddies: responses of mobile marine consumers to episodic upwelling. *Journal of Marine Research* **44**, 361–384.

Haney, J.C. (1986*b*). Offshore feeding by gulls (*Larus*) at ocean fronts in the northeast Gulf of Mexico. *Northeast Gulf Science* **8**, 167–172.

Haney, J.C. & Stone, A.E. (1988). Seabird foraging tactics and water clarity: are plunge divers really in the clear? *Marine Ecology – Progress Series* **49**, 1–9.

Hanson, H.C. (1962). The dynamics of condition factors in Canada Geese and their relation to seasonal factors. *Arctic Institute of North America, Technical Paper* **12**, 1–68.

Hardy, A.R., Stanley, P.I. & Greig-Smith, P.W. (1987). Birds as indicators of the intensity of use of agricultural pesticides in the UK. In *The value of birds*. (ed. A.W. Diamond & F.L. Filion), pp. 119–132. Cambridge: International Council for Bird Preservation, Technical Publication No. 6.

Harper, P.C. (1987). Feeding behaviour and other notes on 20 species of Procellariiformes at sea. *Notornis* **34**, 169–192.

Harradine, J. (1985). Duck shooting in the United Kingdom. *Wildfowl* **36**, 81–94.

Harris, M.P. (1966). Breeding biology of the Manx Shearwater *Puffinus puffinus*. *Ibis* **108**, 17–33.

Harris, M.P. (1970*a*). Rates and causes of increases of some British gull populations. *Bird Study* **17**, 325–335.

Harris, M.P. (1970*b*). Abnormal migration and hybridization of *Larus argentatus* and *L. fuscus* after interspecies cross-fostering experiments. *Ibis* **112**, 488–498.

Harris, M.P. (1981). Age determination and first breeding of British Puffins. *British Birds* **74**, 246–256.

Harris, M.P. (1984). *The Puffin*. Calton: T & AD Poyser.

Harris, M.P. (1989). Recent breeding success of seabirds around the British Isles. In *Seabirds and sandeels*. (ed. M. Heubeck), pp. 19–21. Lerwick: Shetland Bird Club.

Harris, M.P. & Birkhead, T.R. (1985). Breeding ecology of the Atlantic Alcidae. In *The Atlantic Alcidae*. (ed. D.N. Nettleship & T.R. Birkhead), pp. 155–204. London: Academic Press.

Harris, M.P. & Lloyd, C.S. (1977). Variation in counts of seabirds from photographs. *British Birds* **70**, 200–205.

Harris, M.P., Morley, C. & Green, G.H. (1978). Hybridization of Herring and Lesser Black-backed Gulls in Britain. *Bird Study* **25**, 161–166.

Harris, M.P. & Osborn, D. (1981). Effect of a polychlorinated biphenyl on the survival and breeding of puffins. *Journal of Applied Ecology* **18**, 471–479.

Harris, M.P., Towll, H., Russell, A.F. & Wanless, S. (1990). Maximum dive depths attained by auks feeding young on the Isle of May. *Scottish Birds* **16**, 25–28.

Harris, M.P. & Wanless, S. (1988). The breeding biology of Guillemots *Uria aalge* on the Isle of May over a six year period. *Ibis* **130**, 172–192.

Harris, M.P., Webb, A. & Tasker, M.L. (1991). Growth of young Guillemots *Uria aalge* after leaving the colony. *Seabird* **13**, 40–44.

Harrison, P. (1983). *Seabirds: an identification guide*. Beckenham, Kent: Croom Helm.

Haynes, F.N. (1984). *Spartina* in Langstone Harbour, Hampshire. In Spartina anglica *in Great Britain*. (ed. P. Doody), pp. 5–10. Huntingdon: Nature Conservancy Council, Research and Survey in Nature Conservation No. 5.

Hays, H. & Cormons, G. (1974). Plastic particles found in tern pellets, on coastal beaches and at factory sites. *Marine Pollution Bulletin* **5**, 44–46.

Hatchwell, B.J. (1991). An experimental study of the effects of timing of breeding on the reproductive success of common guillemots (*Uria aalge*). *Journal of Animal Ecology* **60**, 721–736.

Henderson, P.A. (1990). Wildlife reports. Marine and freshwater life. *British Wildlife* **2**, 120–121.

Hendry, G. (1988). Where does all the green go? *New Scientist* **120 (5 Nov.),** 38–42.

Hensler, G.L. & Nichols, J.D. (1981). The Mayfield method of estimating nest success: a model, estimators and simulation results. *Wilson Bulletin* **93**, 42–53.

Heubeck, M. (1989). Breeding success of Shetland's seabirds: Arctic Skua, Kittiwake, Guillemot, Razorbill and Puffin. In *Seabirds and sandeels*. (ed. M. Heubeck), pp. 11–18. Lerwick: Shetland Bird Club.

Heubeck, M., Harvey, P.V. & Okill, J.D. (1991). Changes in the Shetland Guillemot *Uria aalge* population and the pattern of recoveries of ringed birds, 1959–1990. *Seabird* **13**, 3–21.

Hill, D.A. (1989). Manipulating water habitats to optimise wader and wildfowl populations. In *Biological habitat reconstruction*. (ed. G.P. Buckley), pp. 328–343. London: Belhaven Press.

Hill, D.I. (1988). *Saltmarsh vegetation of the Wash. An assessment of change from 1971 to 1985*. Peterborough: Nature Conservancy Council, Research and Survey in Nature Conservation No. 13.

Hill, K.J. (1985). Extrarenal excretion. In *A dictionary of birds*. (ed. B. Campbell & E. Lack), pp. 196. Calton: T & AD Poyser.

Hislop, J.R.G. & Harris, M.P. (1985). Recent changes in the food of young Puffins *Fratercula arctica* on the Isle of May in relation to fish stocks. *Ibis* **127**, 234–239.

Hislop, J.R.G. & MacDonald, W.S. (1989). Damage to fish by seabirds in the Moray Firth. *Scottish Birds* **15**, 151–155.

Hodges, A.F. (1974). The orientation of adult Kittiwakes *Rissa tridactyla* at the nest site in Northumberland. *Ibis* **117**, 235–240.

Hogg, R.A., White, V.J. & Smith, G.R. (1990). Suspected botulism in cattle associated with poultry litter. *Veterinary Record* **126**, 476–479.

Holden, A.V. (1981). Organochlorines – an overview. *Marine Pollution Bulletin* **12**, 110–115.

Holdgate, M.W.(ed.). (1971). *The sea bird wreck in the Irish Sea autumn 1969.* London: Natural Environment Research Council.

Holley, A.J.F. (1982). Post-fledging interactions on the territory between parents and young Herring Gulls *Larus argentatus. Ibis* **124**, 198–203.

Holligan, P.M., Aarup, T. & Groom, S.B. (1989). The North Sea satellite colour atlas. *Continental Shelf Research* **9**, 665- 765.

Hope Jones, P. & Kinnear, P.K. (1979). Moulting Eiders in Orkney and Shetland. *Wildfowl* **30**, 109–113.

Hope Jones, P., Monnat, J.-Y., Cadbury, C.J. & Stowe, T.J. (1978). Birds oiled during the *Amoco Cadiz* incident – an interim report. *Marine Pollution Bulletin* **9**, 307–310.

Houston, J. & Jones, C. (eds) (1990). *Planning and management of the coastal heritage.* Southport: Sefton Metropolitan Borough Council.

Howell, R. (1985). The effect of bait-digging on the bioavailability of heavy metals from surficial intertidal marine sediments. *Marine Pollution Bulletin* **16**, 292–295.

Hudson, A.V. (1989). Interspecific and age-related differences in the handling time of discarded fish by scavenging seabirds. *Seabird* **12**, 40–44.

Hudson, A.V. & Furness, R.W. (1989). The behaviour of seabirds foraging at fishing boats around Shetland. *Ibis* **131**, 225–237.

Hudson, P.J. (1982). Nest site characteristics and breeding success in the Razorbill *Alca torda. Ibis* **124**, 355–359.

Hudson, W.H. (1898). *Birds in London.* London: Longmans, Green and Co.

Hughes, R.G. (1988). Dispersal by benthic invertebrates: the *in situ* swimming behaviour of the amphipod *Corophium volutator. Journal of the Biological Association of the United Kingdom* **68**, 565–579.

Hulscher, J.B. (1976). Localisation of cockles (*Cardium edule* L.) by the Oystercatcher (*Haematopus ostralegus* L.) in darkness and daylight. *Ardea* **64**, 292–310.

Hulscher, J.B. (1982). The Oystercatcher *Haematopus ostralegus* as a predator of the bivalve *Macoma balthica* in the Dutch Wadden Sea. *Ardea* **70**, 89–152.

Hulscher, J.B. (1985). Growth and abrasion of the Oystercatcher bill in relation to dietary switches. *Netherlands Journal of Zoology* **35**, 124–154.

Hutchinson, L.V., Wenzel, B.M., Stager, K.E. & Tedford, B.L. (1984). Further evidence for olfactory foraging by Sooty Shearwaters and Northern Fulmars. In *Marine birds: their feeding ecology and commercial fisheries relationships.* (ed. D.N. Nettleship, G.A. Sanger & P.F. Springer), pp. 72–77. Ottawa: Canadian Wildlife Service.

Hutton, M. & Symon, C. (1987). Sources of cadmium discharge to the UK environment. In *Pollutant transport and fate in ecosystems.* (ed. P.J. Coughtrey, M.H. Martin & M.H. Unsworth), pp. 223–237. Oxford: Blackwell Scientific Publications.

Imber, M.J. (1973). The food of Grey-faced Petrels *Pterodroma macroptera gouldi* with special reference to diurnal vertical migration of their prey. *Journal of Animal Ecology* **42**, 645–662.

Ingold, P. (1980). Anpassungen der Eier und des Brutverhaltens von Trottellummen (*Uria aalge aalge* Pont.) an das Brüten auf Felssimsen. *Zeitschrift für Tierpsychologie* **53**, 341–388.

Irving, L. (1966). Adaptations to cold. *Scientific American* **214 (Jan.)**, 94–101.

Isack, H.A. (1987). The cultural and economic importance of birds among the Boran people of northern Kenya. In *The value of birds*. (ed. A.W. Diamond & F.L. Filion), pp. 89–98. Cambridge: International Council for Bird Preservation, Technical Publication No. 6.

Jackson, D.B. (1988). Habitat selection and breeding ecology of three species of waders in the Western Isles of Scotland. Ph.D. thesis: University of Durham.

Jacob, J. (1982). Stomach oils. In *Avian biology. Vol. VI*. (ed. D.S. Farner, J.R. King & K.C. Parkes), pp. 325–340. London: Academic Press.

Jacob, J. & Ziswiler, V. (1982). The uropygial gland. In *Avian biology. Vol. VI*. (ed. D.S. Farner, J.R. King & K.C. Parkes), pp. 199–324. London: Academic Press.

Jacquemot, A. & Filion, F.L. (1987). The economic significance of birds in Canada. In *The value of birds*. (ed. A.W. Diamond & F.L. Filion), pp. 15–21. Cambridge: International Council for Bird Preservation, Technical Publication No. 6.

James, I.D. (1981). Fronts and shelf-circulation models. *Philosophical Transactions of the Royal Society of London* **A 302,** 597–604.

Jarvis, M.J.F. & Cram, D.L. (1971). Bird Island, Lamberts Bay, South Africa: an attempt at conservation. *Biological Conservation* **3**, 269–272.

Jefferies, D.J. & Parslow, J.L.F. (1976). The genetics of bridling in guillemots from a study of hand-reared birds. *Journal of Zoology, London* **179**, 411–420.

Jehl, J.R. & Murray, B.G. (1986). The evolution of normal and reverse sexual size dimorphism in shorebirds and other birds. In *Current ornithology. Vol. 3*. (ed. R.F. Johnston), pp. 1–86. New York: Plenum Press.

Jenkins, D., Murray, M.G. & Hall, P. (1975). Structure and regulation of a shelduck (*Tadorna tadorna* (L.)) population. *Journal of Animal Ecology* **44**, 201–231.

Johnson, A.R. (1975). Camargue flamingos. In *Flamingos*. (ed. J. Kear & N. Duplaix-Hall), pp. 17–25. Berkhamsted: T & AD Poyser.

Johnson, O.W. & Morton, M.L. (1976). Fat content and flight range in shorebirds summering on Enewetak Atoll. *Condor* **78**, 144–145.

Johnson, P.O. (1983). The English North Sea sprat fisheries. *Fishing Prospects* **1983,** 19–24.

Johnston, D.W. & McFarlane, R.W. (1967). Migration and bioenergetics of flight in the Pacific Golden Plover. *Condor* **69**, 156–168.

Jones, L.L. & DeGange, A.R. (1988). Interactions between seabirds and fisheries in the north Pacific Ocean. In *Seabirds & other marine vertebrates. Competition, predation and other interactions*. (ed. J. Burger), pp. 269–291. New York: Columbia University Press.

Jönsson, P.E. (1987). Sexual size dimorphism and disassortative mating in the Dunlin *Calidris alpina schinzii* in southern Sweden. *Ornis Scandinavica* **18**, 257–264.

Jury, J.A. (1986). Razorbill swimming at depth of 140 m. *British Birds* **79**, 339.

Kadlec, J.A. & Drury, W.H. (1968). Aerial estimation of the size of gull breeding colonies. *Journal of Wildlife Management* **32**, 287–293.

Kalejta, B. (1984). The winter bird populations on the Taff/Ely saltmarsh. B.Sc. thesis: University of Wales.

Kear, J. (1964). Colour preference in young Anatidae. *Ibis* **106**, 361–369.

Kear, J. (1990). *Man and wildfowl*. London: T & AD Poyser.

Kenward, R.E. (1978). Hawks and doves: factors affecting success and selection in goshawk attacks on woodpigeons. *Journal of Animal Ecology* **47**, 449–460.

Kilpi, M. (1989). The effect of varying pair numbers on reproduction and use of space in a small Herring Gull *Larus argentatus* colony. *Ornis Scandinavica* **20**, 204–210.

Kirby, J.S., Waters, R.J. & Prys-Jones, R.P. (1991). *Wildfowl and wader counts 1989–90*. Slimbridge: Wildfowl & Wetlands Trust.

Kirby, R.E. & Obrecht, H.H. (1980). Atlantic Brant – human commensalism on eelgrass beds in New Jersey. *Wildfowl* **31**, 158–160.

Kirkham, I.R. & Nisbet, I.C.T. (1987). Feeding techniques and field identification of Arctic, Common and Roseate Terns. *British Birds* **80**, 41–47.

Klaassen, M. (1990). Short note on the possible occurrence of heat stress in roosting waders on the Banc d'Arguin, Mauritania. *Ardea* **78**, 63–65.

Klaassen, M. & Ens, B.J. (1990). Is salt stress a problem for waders wintering on the Banc d'Arguin, Mauritania? *Ardea* **78**, 67–74.

Koepff, C. & Dietrich, K. (1986). Störungen von Küstenvögeln durch Wasserfahrzeuge. *Die Vogelwarte* **33**, 232–248.

Krebs, J.R., MacRoberts, M.H. & Cullen, J.M. (1972). Flocking and feeding in the Great Tit *Parus major* – an experimental study. *Ibis* **114**, 507–530.

Kruuk, H., Nolet, B. & French, D. (1988). Fluctuations in numbers and activity of inshore demersal fishes in Shetland. *Journal of the Marine Biological Association of the United Kingdom* **68**, 601–617.

Lack, P.C. (ed.) (1986). *The atlas of wintering birds in Britain and Ireland*. Calton: T & AD Poyser.

Lack, D. & Venables, L.S.V. (1937). The heathland birds of South Haven Peninsula, Studland Heath, Dorset. *Journal of Animal Ecology* **6**, 62–72.

Lambeck, R.H.D, Sandee, A.J.J. & Wolff, de L. (1989). Long-term patterns in the wader usage of an intertidal flat in the Oosterschelde (SW Netherlands) and the impact of the closure of an adjacent estuary. *Journal of Applied Ecology* **26**, 419–431.

Lange, G. (1968). Über Nahrung, Nahrungsaufnahme und Verdauungstrakt mitteleuropäischer Limikolen. *Beiträge Vogelkunde* **13**, 225–334.

Laursen, K., Gram, I. & Frikke, J. (1984). Traekkende vandfugle ved det fremskudte dige ved Højer, 1982. *Danske Vildtundersøgelser* **37**, 1–36.

Leewis, R.J., Baptist, H.J.M. & Meininger, P.L. (1984). The Dutch Delta Area. In *Coastal waders and wildfowl in winter*. (ed. P.R. Evans, J.D. Goss-Custard & W.G. Hale), pp. 253–260. Cambridge: Cambridge University Press.

Legg, R. (1978). The arrival of Muntjac. In *Steep Holm – a case history in the study of evolution*. (ed. Kenneth Allsop Trust & J. Fowles), pp. 133–142. Sherborne: Dorset Publishing Company.

Lendrem, D.W. (1983). Sleeping and vigilance in birds. I. Field observations of the mallard (*Anas platyrhynchos*). *Animal Behaviour* **31**, 532–538.

Levy, E.M. (1980). Oil pollution and seabirds: Atlantic Canada 1976–77 and some implications for northern environments. *Marine Pollution Bulletin* **11**, 51–56.

Lewis, J. (1986). Origins and influences. In *The nature of West Wales*. (ed. D. Saunders), pp. 13–20. Buckingham: Barracuda Books.

Lewis, W.V. (1932). The formation of Dungeness foreland. *Geographical Journal* **80**, 309–324.

Lifjeld, J.T. (1984). Prey selection in relation to body size and bill length of five species of waders feeding in the same habitat. *Ornis Scandinavica* **15**, 217–226.

Lima, S.L. (1986). Predation risk and unpredictable feeding conditions: determinants of body mass in birds. *Ecology* **67**, 377–385.

Lipinski, M.R. & Jackson, S. (1989). Surface-feeding on cephalopods by procellariiform seabirds in the southern Benguela region, South Africa. *Journal of Zoology, London* **218**, 549–563.

Little, D.I., Howells, S.E., Abbiss, T.P. & Rostron, D. (1987). Some factors affecting the fate of estuarine sediment hydrocarbons and trace metals in Milford Haven 1978–82. In *Pollutant transport and fate in ecosystems*. (ed. P.J. Coughtrey, M.H. Martin & M.H. Unsworth), pp. 55–87. Oxford: Blackwell Scientific Publications.

Lloyd, C.S. (1989). The importance of RSPB reserves for breeding seabirds. *RSPB Conservation Review* **3**, 35–40.

Lloyd, C.S., Tasker, M.L. & Partridge, K.E. (1991). *The status of seabirds in Britain and Ireland*. London: T & AD Poyser.

Lockley, R.M. (1983). *Flight of the storm petrel*. Newton Abbot: David & Charles.

Long, S.P. & Mason, C.F. (1983). *Saltmarsh ecology*. Glasgow: Blackie.

Loonen, M. & van Eerden, M. (1989). The significance of water as a key trigger enabling Wigeon *Anas penelope* to feed on grass leaves of *Agrostis stolonifera* in a wet meadow system in Lake Lauwersmeer, The Netherlands. *Wader Study Group Bulletin* **57**, 16.

Lotem, A., Schechtman, E. & Katzir, G. (1991). Capture of submerged prey by little egrets, *Egretta garzetta garzetta*: strike depth, strike angle and the problem of light refraction. *Animal Behaviour* **42**, 341–346.

Love, J.A. (1983). *The return of the Sea Eagle*. Cambridge: Cambridge University Press.

Lucas, F.A. (1890). The expedition to Funk Island, with observations upon the history and anatomy of the Great Auk. *Report of the Smithsonian Institution* **1887-'88,** 493–529.

Lumb, C.M. (1989). Self-pollution by Scottish salmon farms? *Marine Pollution Bulletin* **20**, 375–379.

MacDonald, J.D. (1960). Secondary external nares of the gannet. *Proceedings of the Zoological Society of London* **135**, 357–363.

McDonald, K. (1990). In fish heaven. *Diver* **(July, 1990),** 60.

McFarland, D. (1989). *Problems of animal behaviour*. Harlow: Longman Scientific & Technical.

McGrath, S.P. (1987). Long-term studies of metal transfers following application of sewage sludge. In *Pollutant transport and fate in ecosystems*. (ed. P.J. Coughtrey, M.H. Martin & M.H. Unsworth), pp. 301–317. Oxford: Blackwell Scientific Publications.

McLannahan, H.M.C. (1973). Some aspects of the ontogeny of cliff nesting behaviour in the Kittiwake (*Rissa tridactyla*) and the Herring Gull (*Larus argentatus*). *Behaviour* **44**, 36–88.

MacLennan, A.S. (1986). Oil pollution in the Cromarty Firth and inshore Moray Firth. *Proceedings of the Royal Society of Edinburgh* **91B**, 275–282.

McLusky, D.S. (1981). *The estuarine ecosystem*. Glasgow: Blackie.

McVicar, A.H., Bruno, D.W. & Fraser, C.O. (1988). Fish diseases in the North Sea in relation to sewage sludge dumping. *Marine Pollution Bulletin* **19**, 169–173.

Madsen, J. (1988). Autumn feeding ecology of herbivorous wildfowl in the Danish Wadden Sea, and impact of food supplies and shooting on movement. *Danish Review of Game Biology* **13 (4),** 1–32.

Madsen, J. (1989). Spring feeding ecology of Brent Geese *Branta bernicla*: annual variation in salt marsh food supplies and effects of grazing on growth of vegetation. *Danish Review of Game Biology* **13 (7),** 1–16.

Makepeace, M. & Patterson, I.J. (1980). Duckling mortality in the Shelduck, in relation to density, aggressive interaction and weather. *Wildfowl* **31**, 57–72.

Mallory, E.P. & Schneider, D.C. (1979). Agonistic behavior in Short-billed Dowitchers feeding on a patchy resource. *Wilson Bulletin* **91**, 271–278.

Marchant, J.H., Hudson, R., Carter, S.P. & Whittington, P. (1990). *Population trends in British breeding birds*. Tring: British Trust for Ornithology.

Markus, M.B. (1985). Number of feathers. In *A dictionary of birds*. (ed. B. Campbell & E. Lack), pp. 209–210. Calton: T & AD Poyser.

Martin, A.P. & Baird, D. (1988). Lemming cycles – which Palearctic migrants are affected? *Bird Study* **35**, 143–145.

Martin, G. (1990). Designer eyes for seabirds of the night. *New Scientist* **128 (3 Nov.),** 46–48.

Mason, A.Z., Jenkins, K.D. & Sullivan, P.A. (1988). Mechanisms of trace metal accumulation in the polychaete *Neanthes arenaceodentata*. *Journal of the Marine Biological Association of the United Kingdom* **68**, 61–80.

Mason, C.F. (1986). Invertebrate populations and biomass over four years in a coastal, saline lagoon. *Hydrobiologia* **133**, 21–29.

Mason, C.F. (1988). Parallel changes in numbers of waders and geese. *Bird Study* **36**, 80–82.

Mattocks, J.G. (1971). Goose feeding and cellulose digestion. *Wildfowl* **22**, 107–113.

May, V.J. (1985). Geomorphological aspects of Dungeness. In *Dungeness. Ecology and conservation*. (ed. B. Ferry & S. Waters), pp. 2–12. Peterborough: Nature Conservancy Council, Research and Survey in Nature Conservation No. 12.

Mayfield, H.F. (1975). Suggestions for calculating nest success. *Wilson Bulletin* **87**, 456–466.

Mead, C. (1974). The results of ringing auks in Britain and Ireland. *Bird Study* **21**, 45–86.

Mead, C. (1983). *Bird migration*. Feltham, Middlesex: Country Life Books.

Mead, C. (1989). Auk netfax. *BTO News* **163**, 8.

Mead, C. (1991). Wildlife reports. Birds. *British Wildlife* **2**, 239–241.

Mearns, R. (1982). Winter occupation of breeding territories and winter diet of Peregrines in South Scotland. *Ornis Scandinavica* **13**, 79–83.

Meire, P.M., Seys, J., Ysebaert, T., Meininger, P.L. & Baptist, H.J.M. (1989). A changing Delta: effects of large coastal engineering works on feeding ecological relationships as illustrated by waterbirds. In *Hydro-ecological relations in the Delta Waters of the South-West Netherlands*. (ed. J.C. Hooghart & C.W.S. Posthumus), pp. 109–145. Rotterdam: TNO Committee on Hydrological research.

Meltofte, H. (1978). A breeding association betweeen Eiders and tethered huskies in North-east Greenland. *Wildfowl* **29**, 45–54.

Mendenhall, V.M. & Milne, H. (1985). Factors affecting duckling survival of Eiders *Somateria mollissima* in northeast Scotland. *Ibis* **127**, 148–158.

Metcalfe, N.B. (1984). The effects of mixed-species flocking on the vigilance of shorebirds: who do they trust? *Animal Behaviour* **32**, 986–993.

Metcalfe, N.B. (1985). Prey detection by intertidally feeding lapwing. *Zeitschrift für Tierpsychologie* **67**, 45–57.

Metcalfe, N.B. (1986). Variation in winter flocking associations and dispersion patterns in the turnstone *Arenaria interpres*. *Journal of Zoology, London* **209**, 385–403.

Metcalfe, N.B. (1989). Flocking preferences in relation to vigilance benefits and aggression costs in mixed-species shorebird flocks. *Oikos* **56**, 91–98.

Midtgård, U. (1980). Heat loss from the feet of Mallards *Anus platyrhynchos* and artero-venous heat exchange in the *rete tibiotarsale*. *Ibis* **122**, 354–359.

Millard, A.V. & Evans, P.R. (1984). Colonisation of mudflats by *Spartina anglica* ; some effects on invertebrate and shorebird populations at Lindisfarne. In *Spartina anglica in Great Britain*. (ed. P. Doody), pp. 41–48. Huntingdon: Nature Conservancy Council, Research and Survey in Nature Conservation No. 5.

Milne, H. (1974). Breeding numbers and reproductive rate of Eiders at the Sands of Forvie National Nature Reserve, Scotland. *Ibis* **116**, 135–154.

Milsom, T.P. (1984). Diurnal behaviour of Lapwings in relation to moon phase during winter. *Bird Study* **31**, 117–120.

Monaghan, P. & Coulson, J.C. (1977). Status of large gulls nesting on buildings. *Bird Study* **24**, 89–104.

Monaghan, P. & Uttley, J. (1989). Breeding success of Shetland's seabirds: Arctic Tern and Common Tern. In *Seabirds and sandeels*. (ed. M. Heubeck), pp. 3–5. Lerwick: Shetland Bird Club.

Monaghan, P. & Zonfrillo, B. (1986). Population dynamics of seabirds in the Firth of Clyde. *Proceedings of the Royal Society of Edinburgh* **90B**, 363–375.

Moors, P.J. & Atkinson, I.A.E. (1984). Predation on seabirds by introduced animals, and factors affecting its severity. In *Status and conservation of the world's seabirds*. (ed. J.P. Croxall, P.G.H. Evans & R.W. Schreiber), pp. 667–690. Cambridge: International Council for Bird Preservation, Technical Publication No.2.

Morgan, R.A. (1978). Changes in the breeding bird community at Gibralter Point, Lincolnshire, between 1965 and 1974. *Bird Study* **25**, 51–58.

Mork, M. (1981). Circulation phenomena and frontal dynamics of the Norwegian coastal current. *Philosophical Transactions of the Royal Society of London* **A 302,** 635–647.

Moser, M.E., Broad, R.A., Dennis, R.H. & Madders, M. (1986). The distribution and abundance of some coastal birds on the west and north-west coasts of Scotland in winter. *Scottish Birds* **14**, 61–67.

Moser, M.E. & Carrier, M. (1983). Patterns of population turnover in Ringed Plovers and Turnstones during their spring passage through the Solway Firth in 1983. *Wader Study Group Bulletin* **39**, 37–41.

Moser, M.E. & Summers, R.W. (1987). Wader populations on the non-estuarine coasts of Britain and Northern Ireland: results of the 1984–85 Winter Shorebird Count. *Bird Study* **34**, 71–81.

Moss, R., Watson, A. & Parr, R. (1975). Maternal nutrition and breeding success in red grouse (*Lagopus lagopus scoticus*). *Journal of Animal Ecology* **44**, 233–244.

Mudge, G.P. (1978). Ecological studies of Herring Gulls (*Larus argentatus* Pont.) and other Larini, in an urban environment. Ph.D. thesis: University of Wales.

Mudge, G.P. (1986). Trends of population change at colonies of cliff-nesting seabirds in the Moray Firth. *Proceedings of the Royal Society of Edinburgh* **91B**, 73–80.

Mudge, G.P. & Ferns, P.N. (1982*a*). Breeding populations of gulls in the inner Bristol Channel, 1980. *Seabird Report* **6**, 48–49.

Mudge, G.P. & Ferns, P.N. (1982*b*). The feeding ecology of five species of gulls (Aves: Larini) in the inner Bristol Channel. *Journal of Zoology, London* **197**, 497–510.

Muirhead, S.J. & Furness, R.W. (1988). Heavy metal concentrations in the tissues of seabirds from Gough Island, South Atlantic Ocean. *Marine Pollution Bulletin* **19**, 278–283.

Munro, C.A. (1984). Roof nesting Oystercatchers. *Bird Study* **31**, 148.

Myers, J.P. (1984). Spacing behavior of nonbreeding shorebirds. In *Behavior of marine animals. Vol. 6. Shorebirds. Migration and foraging behavior.* (ed. J. Burger & B.L. Olla), pp. 271–321. New York: Plenum Press.

Myers, J.P., Connors, P.G. & Pitelka, F.A. (1979). Territory size in wintering Sanderlings: the effects of prey abundance and intruder density. *Auk* **96**, 551–561.

Nairn, R.G.W. & Sheppard, J.R. (1985). Breeding waders of sand dune machair in north-west Ireland. *Irish Birds* **3**, 53–70.

Nairn, R.G.W. & Whatmough, J.A. (1978). Breeding bird communities of a sand dune system in north-east Ireland. *Irish Birds* **1**, 160–170.

Natural Environment Research Council. (1983). *Contaminants in marine top predators*. Swindon: Natural Environment Research Council.

Nature Conservancy Council. (1989). *Fifteenth Report covering the period 1 April 1988–31 March 1989*. Peterborough: Nature Conservancy Council.

Nelson, J.B. (1964). Factors influencing clutch-size and chick growth in the North Atlantic Gannet *Sula bassana*. *Ibis* **106**, 63–77.

Nelson, J.B. (1975). The breeding biology of frigatebirds – a comprehensive review. *Living Bird* **14**, 113–156.

Nelson, J.B. (1978*a*). *The gannet*. Berkhamsted: T & AD Poyser.

Nelson, J.B. (1978*b*). *The Sulidae: gannets and boobies*. Oxford: Oxford University Press.

Nelson, J.B. (1980). *Seabirds: their biology and ecology*. London: Hamlyn.

Newton, I., Bogan, J.A. & Haas, M.B. (1989). Organochlorines and mercury in the eggs of British Peregrines *Falco peregrinus*. *Ibis* **131**, 355–376.

Newton, I. & Haas, M.B. (1984). The return of the Sparrowhawk. *British Birds* **77**, 47–70.

Nicoll, M., Summers, R.W., Underhill, L.G., Brockie, K. & Rae, R. (1988). Regional, seasonal and annual variations in the structure of Purple Sandpiper *Calidris maritima* populations in Britain. *Ibis* **130**, 221–233.

Nilsson, L. (1989). Feeding habits and exploitation of feeding areas by resting and wintering geese in southernmost Sweden. *Wader Study Group Bulletin* **57**, 18.

Noer, H. (1979). Speeds of migrating waders *Charadriidae*. *Dansk ornithologisk Forenings Tidsskrift* **73**, 215–224.

Norman, R.K. & Saunders, D.R. (1969). Status of Little Terns in Great Britain and Ireland in 1967. *British Birds* **62**, 4–13.

North, F.J. (1955). *The evolution of the Bristol Channel with special reference to the coast of South Wales*. Cardiff: National Museum of Wales.

North, J.J. (1987). The economic importance of reclamation. In *The Wash and its environment*. (ed. P. Doody & B. Barnett), pp. 163–164. Peterborough: Nature Conservancy Council, Research and Survey in Nature Conservation No. 7.

Norton, T.A. & Powell, H.T. (1979). Seaweeds and rocky shores of the Outer Hebrides. *Proceedings of the Royal Society of Edinburgh* **77B**, 141–153.

O Briain, M. & Farrelly, P. (1990). Breeding biology of Little Terns at Newcastle, Co. Wicklow and the impact of conservation action, 1985–1990. *Irish Birds* **4**, 149–168.

O'Brien, M. (1990). Breeding Waders of Wet Meadows survey – 1989. *BTO News* **168**, 6.

O'Connor, R.J. (1984). *The growth and development of birds*. Chichester: John Wiley & Sons.

O'Donald, P. (1983). *The Arctic Skua*. Cambridge: Cambridge University Press.

Ohlendorf, H.M., Klaas, E.E. & Kaiser, T.E. (1978). Organochlorine residues and eggshell thinning in Anhingas and waders. *Proceedings of the 1977 Conference of the Colonial Waterbirds Group*, 185–195.

Okill, D. (1989). Breeding success of Shetland's seabirds: Red-throated Diver, Fulmar, Gannet, Cormorant and Shag. In *Seabirds and sandeels*. (ed. M. Heubeck), pp. 5–10. Lerwick: Shetland Bird Club.

Olney, P.J.S. (1965). The autumn and winter feeding biology of certain sympatric ducks. *Transactions of the International Union of Game Biologists* **6**, 309–320.

Olson, S.L. (1977). Additional notes on subfossil bird remains from Ascension Island. *Ibis* **119**, 37–43.

Olsthoorn, J.C.M. & Nelson, J.B. (1990). The availability of breeding sites for some British seabirds. *Bird Study* **37**, 145–164.

Opp, M.R. & Ball, N.J. (1985). Bioenergetic consequences of sleep based on time budgets of free ranging glaucous-winged gulls. *Sleep Research* **14**, 21.

Osborn, D. & Harris, M.P. (1984). Organ weights and body composition in three seabird species. *Ornis Scandinavica* **15**, 95–97.

Owen, M. (1972). Movements and feeding ecology of white-fronted geese at the New Grounds, Slimbridge. *Journal of Applied Ecology* **9**, 385–398.

Owen, M. (1973). The winter feeding ecology of Wigeon at Bridgwater Bay, Somerset. *Ibis* **115**, 227–243.

Owen, M. (1976). The selection of winter food by whitefronted geese. *Journal of Applied Ecology* **13**, 715–729.

Owen, M. (1979). Food selection in geese. *Verhandlungen der Ornithologischen Gesellschaft in Bayern* **23**, 169–176.

Owen, M. (1980a). *Wild geese of the world*. London: B.T. Batsford.

Owen, M. (1980b). Abdominal profile - a condition index for wild geese in the field. *Journal of Wildlife Management* **45**, 227- 230.

Owen, M. (1987). Brent Goose *Branta b. bernicla* breeding and lemmings – a re-examination. *Bird Study* **34**, 147–149.

Owen, M. (1990). The damage-conservation interface illustrated by geese. *Ibis* **132**, 238–252.

Owen, M., Atkinson-Willes, G.L. & Salmon, D.G. (1986). *Wildfowl in Great Britain*. Cambridge: Cambridge University Press.

Owen, M. & Black, J.M. (1991). Geese and their future fortune. *Ibis* **133 Supplement,** 28–35.

Owen, M. & Kerbes, R.H. (1971). On the autumn food of Barnacle Geese at Caerlaverock National Nature Reserve. *Wildfowl* **22**, 114–119.

Owen, M., Nugent, M. & Davies, N. (1977). Discrimination between grass species and nitrogen-fertilized vegetation by young Barnacle Geese. *Wildfowl* **28**, 21–26.

Owen, M. & Williams, G. (1976). Winter distribution and habitat requirements of Wigeon in Britain. *Wildfowl* **27**, 83–90.

Owen, R.W. (1981). Fronts and eddies in the sea: mechanisms, interactions and biological effects. In *Analysis of marine ecosystems*. (ed. A.R. Longhurst), pp. 197–233. London: Academic Press.

Pain, S. (1989). Alaska has its fill of oil. *New Scientist* **123 (12 Aug.),** 34–40.

Palmer, R.S. (ed.) (1976a). *Handbook of North American birds. Vol. 1. Loons through flamingos.* New Haven: Yale University Press.

Palmer, R.S. (ed.) (1976b). *Handbook of North American birds. Vol. 2. Waterfowl (Part 1).* New Haven: Yale University Press.

Parkin, D.T., Ewing, A.W. & Ford, H.A. (1970). Group diving in the Blue-footed Booby *Sula nebouxii. Ibis* **112**, 111–112.

Parslow, J.L.F. & Jefferies, D.J. (1972). Elastic thread pollution of puffins. *Marine Pollution Bulletin* **3**, 43–45.

Patterson, A. (1989). Second MEDMARAVIS Symposium. *Seabird Group Newsletter* **54**, 7–9.

Patterson, I.J. & Makepeace, M. (1979). Mutual interference during nest-prospecting in the shelduck, *Tadorna tadorna. Animal Behaviour* **27**, 522–535.

Peakall, D.B., Hallett, D., Miller, D.S., Butler, R.G. & Kinter, W.B. (1980). Effects of ingested crude oil on black guillemots: a combined field and laboratory study. *Ambio* **9**, 28–30.

Pearson, T.H. (1968). The feeding biology of sea-bird species breeding on the Farne Islands, Northumberland. *Journal of Animal Ecology* **37**, 521–552.

Pedersen, M.B. (1988). Anti-predator responses by Jack Snipe to human interference. *Wader Study Group Bulletin* **54**, 28.

Pennycuick, C.J. (1975). Mechanics of flight. In *Avian biology. Vol. V.* (ed. D.S. Farner & J.R. King), pp. 1–75. London: Academic Press.

Pennycuick, C.J. (1982). The flight of petrels and albatrosses (Procellariiformes), observed in South Georgia and its vicinity. *Philosophical Transactions of the Royal Society of London* **B 300,** 75–106.

Pennycuick, C.J. (1987). Flight of seabirds. In *Seabirds: feeding ecology and role in marine ecosystems.* (ed. J.P. Croxall), pp. 43–62. Cambridge: Cambridge University Press.

Perkins, E.J. & Abbott, O.J. (1972). Nutrient enrichment and sand flat fauna. *Marine Pollution Bulletin* **3**, 70–72.

Perrins, C.M. (1967). The number of Manx Shearwaters on Skokholm. *Skokholm Bird Observatory Report* **1967,** 23–29.

Perrins, C.M. (1970). The timing of birds' breeding seasons. *Ibis* **112**, 242–255.

Perrins, C.M., Harris, M.P. & Britton, C.K. (1973). Survival of Manx Shearwaters *Puffinus puffinus. Ibis* **115**, 535–548.

Perrins, C.M. & Ogilvie, M.A. (1981). A study of the Abbotsbury Mute Swans. *Wildfowl* **32**, 35–47.

Pettitt, C. (1975). A review of the predators of *Littorina*, especially those of *L. saxatilis* (Olivi) (Gastropoda: Prosobranchia). *Journal of Conchology, London* **28**, 343–357.

Phillips, J.G., Butler, P.J. & Sharp, P.J. (1985). *Physiological strategies in avian biology.* Glasgow: Blackie.

Phillips, N.J. (1987). The breeding biology of White-tailed Tropicbirds *Phaethon lepturus* at Cousin Island, Seychelles. *Ibis* **129**, 10–24.

Piatt, J.F. & Nettleship, D.N. (1987). Incidental catch of marine birds and mammals in fishing nets off Newfoundland, Canada. *Marine Pollution Bulletin* **18**, 344–349.

Pienkowski, M.W. (1979). Differences in habitat requirements and distribution patterns of plovers and sandpipers as investigated by studies of feeding behaviour. *Verhandlungen der Ornithologischen Gesellschaft in Bayern* **23**, 105–124.

Pienkowski, M.W. (1981). How foraging plovers cope with environmental effects on invertebrate behaviour and availability. In *Feeding and survival strate-*

*gies of estuarine organisms.* (ed. N.V. Jones & W.J. Wolff), pp. 179–192. New York: Plenum Press.

Pienkowski, M.W. (1982). Diet and energy intake of Grey and Ringed plovers, *Pluvialis squatarola* and *Charadrius hiaticula*, in the non-breeding season. *Journal of Zoology, London* **197**, 511–549.

Pienkowski, M.W. (1983*a*). Surface activity of some intertidal invertebrates in relation to temperature and the foraging behaviour of their shorebird predators. *Marine Ecology – Progress Series* **11**, 141–150.

Pienkowski, M.W. (1983*b*). Changes in the foraging pattern of plovers in relation to environmental factors. *Animal Behaviour* **31**, 244–264.

Pienkowski, M.W. (1991). Using long-term ornithological studies in setting targets for conservation in Britain. *Ibis* **133 Supplement,** 62–75.

Pienkowski, M.W. & Evans, P.R. (1982*a*). Clutch parasitism and nesting interference between Shelducks at Aberlady Bay. *Wildfowl* **33**, 159–163.

Pienkowski, M.W. & Evans, P.R. (1982*b*). Breeding behaviour, productivity and survival of colonial and non-colonial Shelducks *Tadorna tadorna*. *Ornis Scandinavica* **13**, 101–116.

Pienkowski, M.W., Ferns, P.N., Davidson, N.C. & Worrall, D.H. (1984). Balancing the budget: measuring the energy intake and requirements of shorebirds in the field. In *Coastal waders and wildfowl in winter*. (ed. P.R. Evans, J.D. Goss-Custard & W.G. Hale), pp. 29–56. Cambridge: Cambridge University Press.

Pienkowski, M.W. & Pienkowski, A.E. (1989). Limitation by nesting habitat of the density of breeding Ringed Plovers *Charadrius hiaticula*: a natural experiment. *Wildfowl* **40**, 115–126.

Pierotti, R. (1987). Isolating mechanisms in seabirds. *Evolution* **41**, 559–570.

Pierotti, R. (1988). Associations between marine birds and mammals in the northwest Atlantic Ocean. In *Seabirds & other marine vertebrates. Competition, predation and other interactions.* (ed. J. Burger), pp. 31–58. New York: Columbia University Press.

Pierotti, R. & Annett, C. (1987). Reproductive consequences of dietary specialization and switching in an ecological generalist. In *Foraging behavior*. (ed. A.C. Kamil, J.R. Krebs & H.R. Pulliam), pp. 417–442. New York: Plenum Press.

Pierotti, R. & Annett, C. (1991). Diet choice in the Herring Gull: constraints imposed by reproductive and ecological factors. *Ecology* **72**, 319–328.

Piersma, T. (1986). Breeding waders in Europe. A review of population size estimates and a bibliography of information sources. *Wader Study Group Bulletin* **48 Supplement**, 1–116.

Piersma, T. & Engelmoer, M. (1982). Waders and their food resources; general discussion. In *Wintering waders on the Banc d'Arguin*. (ed. W. Altenburg, M. Engelmoer, R. Mes & T. Piersma), pp. 162–164. Leiden: Stichting Veth tot steun aan Waddenonderzoek.

Piersma, T. & Jukema, J. (1990). Budgeting the flight of a long-distance migrant: changes in nutrient reserve levels of Bar-tailed Godwits at successive spring staging sites. *Ardea* **78**, 315–337.

Piersma, T. & van Eerden, M.R. (1989). Feather eating in Great Crested Grebes *Podiceps cristatus*: a unique solution to the problems of debris and gastric parasites in fish-eating birds. *Ibis* **131**, 477–486.

Pilcher, R.E.M. (1964). Effects of the cold winter of 1962–63 on birds of the north coast of the Wash. *Wildfowl Trust Annual Report* **15**, 23–26.

Pingree, R.D. & Griffiths, D.K. (1978). Tidal fronts on the shelf seas around the British Isles. *Journal of Geophysical Research* **83**, 4615–4622.

Pingree, R.D. & Mardell, G.T. (1981). Slope turbulence, internal waves and phytoplankton growth at the Celtic Sea shelf-break. *Philosophical Transactions of the Royal Society of London* **A 302,** 663–682.

Place, A.R. & Jackson, S. (1990). Chitin digestion in seabirds. *Acta Congressus Internationalis Ornithologici* **20**, 289.

Place, A.R., Sievert, P. & Butler, R.G. (1991). The volume of stomach oils increases during prefledging weight loss in Leach's Storm-Petrel (*Oceanodroma leucorrhoa*) chicks. *Auk* **108**, 709–711.

Player, P.V. (1971). Food and feeding habits of the Common Eider at Seafield, Edinburgh, in winter. *Wildfowl* **22**, 100–106.

Poole, A.F. (1989). *Ospreys: a natural and unnatural history.* Cambridge: Cambridge University Press.

Potts, G.R., Coulson, J.C. & Deans, I.R. (1980). Population dynamics and breeding success of the shag, *Phalacrocorax aristotelis*, on the Farne Islands, Northumberland. *Journal of Animal Ecology* **49**, 465–484.

Poulin, J.M. (1968). Croissance du jeune fou de bassan (*Sula bassana*) pendant sa période pré-envol. *Naturaliste Canadien, Quebec* **95**, 1131–1143.

Pounder, B. (1974). Wildfowl and pollution in the Tay estuary. *Marine Pollution Bulletin* **5**, 35–38.

Powers, K.D. (1982). A comparison of two methods of counting birds at sea. *Journal of Field Ornithology* **53**, 209–222.

Prater, A.J. (1972). The ecology of Morecambe Bay. III. The food and feeding habits of knot (*Calidris canutus* L.) in Morecambe Bay. *Journal of Applied Ecology* **9**, 179–194.

Prater, A.J. (1975). The wintering population of the Black-tailed Godwit. *Bird Study* **22**, 169–176.

Prater, A.J. (1976). Breeding population of the Ringed Plover in Britain. *Bird Study* **23**, 155–161.

Prater, A.J. (1979). Trends in accuracy of counting birds. *Bird Study* **26**, 198–200.

Prater, A.J. (1981). *Estuary birds of Britain and Ireland.* Calton: T & A D Poyser.

Prater, A.J. (1989). Ringed Plover *Charadrius hiaticula* breeding population of the United Kingdom in 1984. *Bird Study* **36**, 154–159.

Prendergast, E.D.V. & Boys, J.V. (1983). *The birds of Dorset.* Newton Abbot: David & Charles.

Prestrud, P., Black, J.M. & Owen, M. (1989). The relationship between an increasing Barnacle Goose *Branta leucopsis* population and the number and size of colonies in Svalbard. *Wildfowl* **40**, 32–38.

Prop, J., van Eerden, M.R. & Drent, R.H. (1984). Reproductive success of the Barnacle Goose *Branta leucopsis* in relation to food exploitation on the breeding grounds, western Spitsbergen. *Norsk Polarinstitutt Skrifter* **181**, 87–117.

Purseglove, J. (1989). *Taming the flood.* Oxford: Oxford University Press.

Puttick, G.M. (1980). Energy budgets of curlew sandpipers at Langebaan Lagoon, South Africa. *Estuarine and Coastal Marine Science* **11**, 207–215.

Quinn, J.S. & Morris, R.D. (1986). Intraclutch egg-weight apportionment and chick survival in Caspian terns. *Canadian Journal of Zoology* **64**, 2116–2122.

Rahn, H., Greene, D.G., Tøien, Ø., Krog, J. & Mehlum, F. (1984). Estimated laying dates and eggshell conductance of the Fulmar and Brünnich's Murre in Spitsbergen. *Ornis Scandinavica* **15**, 110–114.

Ranwell, D.S. & Boar, R. (1986). *Coast dune management guide*. Huntingdon: Natural Environment Research Council.

Ranwell, D.S. & Downing, B.M. (1959). Brent goose (*Branta bernicla* (L.)) winter feeding pattern and *Zostera* resources at Scolt Head Island, Norfolk. *Animal Behaviour* **7**, 42–56.

Ratcliffe, D.A. (1970). Changes attributable to pesticides in egg breakage frequency and eggshell thickness in some British birds. *Journal of Applied Ecology* **7**, 67–115.

Reader's Digest Association. (1965). *Complete atlas of the British Isles*. London: The Reader's Digestion Association.

Reading, C.J. & McGrorty, S. (1978). Seasonal variations in the burying depth of *Macoma balthica* (L.) and its accessibility to wading birds. *Estuarine and Coastal Marine Science* **6**, 135–144.

Reay, P. (1988). *The Tamar avocets*. Plymouth: Caradon Field and Natural History Club & Plymouth Polytechnic.

Reay, P. (1989). *The Tamar avocets: Supplement 1 (1988/89)*. Plymouth: Caradon Field and Natural History Club & Plymouth Polytechnic.

Recher, H.F. & Recher, J.A. (1969). Some aspects of the ecology of migrant shorebirds. II. Aggression. *Wilson Bulletin* **81**, 140–154.

Reynolds, C.M. (1979). The heronries census: 1972–1977 population changes and a review. *Bird Study* **26**, 7–12.

Ricklefs, R.E., Place, A.R. & Anderson, D.J. (1987). An experimental investigation of the influence of diet quality on growth in Leach's storm-petrel. *American Naturalist* **130**, 300–305.

Ridley, M.W. & Percy, R. (1958). *The exploitation of sea birds in Seychelles*. London: HMSO, Colonial Research Studies No.25.

Rijke, A.M., Jesser, W.A. & Mahoney, S.A. (1989). Plumage wettability of the African Darter *Anhinga melanogaster* compared with the Double-crested Cormorant *Phalacrocorax auritus*. *Ostrich* **60**, 128–132.

Ritchie, W. (1979). Machair development and chronology in the Uists and adjacent islands. *Proceedings of the Royal Society of Edinburgh* **77B**, 107–122.

Robert, M. & McNeil, R. (1989). Comparative day and night feeding strategies of shorebird species in a tropical environment. *Ibis* **131**, 69–79.

Roberts, P. (1983). Feeding habitats of the Chough on Bardsey Island (Gwynedd). *Bird Study* **30**, 67–72.

Robertson, H.G. (1981). Annual summer and winter fluctuations of Palaearctic and resident waders (Charadrii) at Langebaan Lagoon, South Africa, 1975–1979. In *Proceedings of the Symposium on Birds of the Sea and Shore*. (ed. J. Cooper), pp. 335–345. Cape Town: African Seabird Group.

Robertson, J.G. (1982). *Machair under threat*. Portree, Isle of Skye: Habitat Scotland.

Rogers, C.M. (1987). Predation risk and fasting capacity: do wintering birds maintain optimal body mass? *Ecology* **68**, 1051–1061.

Roselaar, C.S. (1979). Fluctuaties in aantallen krombekstrandlopers *Calidris ferruginea*. *Watervogels* **4**, 202–210.

Roselaar, C.S. (1983). Subspecies recognition in Knot *Calidris canutus* and occurrence of races in Western Europe. *Beaufortia* **33**, 97–109.

Rosenberg, R. (1985). Eutrophication – the future marine coastal nuisance? *Marine Pollution Bulletin* **16**, 227–231.

Ross, A. (1989). Nuvan use in salmon farming. The antithesis of the precautionary principle. *Marine Pollution Bulletin* **20**, 372–374.

Rothwell, P. & Housden, S. (1990). *Turning the tide. A future for estuaries*. Sandy: Royal Society for the Protection of Birds.

Round, P. (1982). Inland feeding by brent geese *Branta bernicla* in Sussex, England. *Biological Conservation* **23**, 15–32.

Rufino, R., Araujo, A., Pina, J.P. & Miranda, P.S. (1984). The use of salinas by waders in the Algarve, south Portugal. *Wader Study Group Bulletin* **42**, 41–42.

Rüppell, G. (1977). *Bird flight*. New York: Van Nostrand Reinhold Company.

Ryan, P.G. (1988). Effects of ingested plastic on seabird feeding: evidence from chickens. *Marine Pollution Bulletin* **19**, 125–128.

Ryan, P.G. & Avery, G. (1987). Wreck of juvenile Blackbrowed Albatrosses on the west coast of South Africa during storm weather. *Ostrich* **58**, 139–140.

Ryan, P.G., Connell, A.D. & Gardner, B.D. (1988). Plastic ingestion and PCBs in seabirds: is there a relationship? *Marine Pollution Bulletin* **19**, 174–176.

Ryan, P.G. & Cooper, J. (1991). Rockhopper penguins and other marine life threatened by driftnet fisheries at Tristan da Cunha. *Oryx* **25**, 76–79.

Safina, C. & Burger, J. (1988). Ecological dynamics among prey fish, bluefish, and foraging common terns in an Atlantic coastal system. In *Seabirds & other marine vertebrates. Competition, predation and other interactions*. (ed. J. Burger), pp. 95–173. New York: Columbia University Press.

Sagar, P.M. & Sagar, J.L. (1989). The effects of wind and sea on the feeding of Antarctic Terns at the Snares Islands, New Zealand. *Notornis* **36**, 171–182.

Salmon, D.G. (1988). The numbers and distribution of scaup *Aythya marila* in Britain and Ireland. *Biological Conservation* **43**, 267–278.

Salmon, D.G., Moser, M.E. & Kirby, J.S. (1987). *Wildfowl and wader counts 1985–86*. Slimbridge: Wildfowl Trust.

Salmon, D.G., Prys-Jones, R.P. & Kirby, J.S. (1988). *Wildfowl and wader counts 1987–88*. Slimbridge: Wildfowl Trust.

Salmon, D.G., Prys-Jones, R.P. & Kirby, J.S. (1989). *Wildfowl and wader counts 1988–89*. Slimbridge: Wildfowl & Wetlands Trust.

Salmon, H.M. & Lockley, R.M. (1933). The Grassholm gannets – a survey and a census. *British Birds* **27**, 142–152.

Samstag, T. (1988). *For love of birds*. Sandy: Royal Society for the Protection of Birds.

Saunders, D. (1986). Islands. In *The nature of West Wales*. (ed. D. Saunders), pp. 105–124. Buckingham: Barracuda Books.

Schmidt-Neilsen, K. (1990) (4th edn). *Animal physiology: adaptation and environment*. Cambridge: Cambridge University Press.

Schneider, D.C. (1985). Predation on the urchin *Echinometra lucunter* (Linnaeus) by migratory shorebirds on a tropical reef flat. *Journal of Experimental Marine Biology and Ecology* **92**, 19–27.

Schreiber, E.A. & Schreiber, R.W. (1989). Insights into seabird ecology from a global 'natural experiment'. *National Geographic Research* **5**, 64–81.

Schrey, E. & Vauk, G.J.M. (1987). Records of entangled Gannets (*Sula bassana*) at Helgoland, German Bight. *Marine Pollution Bulletin* **18**, 350–352.

Scott, R. (1985). Dungeness: bird populations past and present. In *Dungeness. Ecology and conservation*. (ed. B. Ferry & S. Waters), pp. 64–93. Peterborough: Nature Conservancy Council, Research and Survey in Nature Conservation No. 12.

Selman, J. & Goss-Custard, J.D. (1988). Interference between foraging Redshank *Tringa totanus*. *Animal Behaviour* **36**, 1542–1544.

Sharrock, J.T.R. (1973). Sea-watching. In *The natural history of Cape Clear Island*. (ed. J.T.R. Sharrock), pp. 141–146. Berkhamsted: T & AD Poyser.

Sharrock, J.T.R. (1976). *The atlas of breeding birds in Britain and Ireland*. Tring: British Trust for Ornithology.

Shennan, I. (1987). Holocene sea-level changes in the North Sea Region. In *Sea-level changes*. (ed. M.J. Tooley & I. Shennan), pp. 109–151. Oxford: Basil Blackwell.

Shuntov, V.P. (1974). *Sea birds and the biological structure of the ocean*. Washington, D.C.: National Technical Information Service, United States Department of Commerce.

Sibly, R.M. & McCleery, R.H. (1983). The distribution between feeding sites of herring gulls breeding at Walney Island, U.K. *Journal of Animal Ecology* **52**, 51–68.

Siman, H.Y. (1989). Feeding ecology, movements and biogeographical origins of Curlew *Numenius arquata* (L.) wintering on the Severn Estuary. Ph.D. thesis: University of Wales.

Simms, E. (1979). *The public life of the street pigeon*. London: Hutchinson.

Simpson, J.H. (1981). The shelf-sea fronts: implications of their existence and behaviour. *Philosophical Transactions of the Royal Society of London* **A 302**, 531–546.

Simpson, J.H., Allen, C.M. & Morris, N.C.G. (1978). Fronts on the continental shelf. *Journal of Geophysical Research* **83**, 4607–4614.

Sitters, H.P. (1985). Cirl Buntings in Britain in 1982. *Bird Study* **32**, 1–10.

Sivak, J.G., Bobier, W.R. & Levy, B. (1978). The refractive significance of the nictitating membrane of the bird eye. *Journal of Comparative Physiology* **A**, 125, 335–339.

Smit, C.J., Lambeck, R.H.D. & Wolff, W.J. (1987). Threats to coastal wintering and staging areas of waders. *Wader Study Group Bulletin* **49 Supplement,** 105–113.

Smit, C.J. & Piersma, T. (1989). Numbers, midwinter distribution, and migration of wader populations using the East Atlantic flyway. In *Flyways and reserve networks for water birds*. (ed. H. Boyd & J.-Y. Pirot), pp. 24–63. Slimbridge: International Waterfowl and Wetlands Research Bureau, Special Publication No. 9.

Smit, C.J. & Visser, G.H. (1989). Studies on the effects of disturbance in the Dutch Wadden Sea and Delta. *Wader Study Group Bulletin* **57**, 20–21.

Smith, B. (1989). Webs of death. *BTO News* **163**, 9.

Smith, C. (1869). *The birds of Somersetshire*. London: John van Voorst.

Smith, J.E. (ed.) (1970). *'Torrey Canyon' pollution and marine life*. Cambridge: Cambridge University Press.

Smith, K.W. (1983). The status and distribution of waders breeding on wet lowland grasslands in England and Wales. *Bird Study* **30**, 177–192.

Smith, P.C. (1975). A study of the winter feeding ecology and behaviour of the Bar-tailed Godwit (*Limosa lapponica*). Ph.D. thesis: University of Durham.

Smith, P.C. & Evans, P.R. (1973). Studies of shorebirds at Lindisfarne, Northumberland. 1. Feeding ecology and behaviour of the Bar-tailed Godwit. *Wildfowl* **24**, 135–139.

Smyth, C. & Jennings, S. (1988). Coastline changes and land management in East Sussex, southern England. *Ocean & Shoreline Management* **11**, 375–394.

Snell, R.R. (1985). Underwater flight of Long-tailed Duck (Oldsquaw) *Clangula hyemalis*. *Ibis* **127**, 267.

Sobey, D.G. & Kenworthy, J.B. (1979). The relationship between herring gulls and the vegetation of their breeding colonies. *Journal of Ecology* **67**, 469–496.

Sournia, A., Brylinski, J.-M., Dallot, S., Le Corre, P., Leveau, M., Prieur, L. & Froget, C. (1990). Fronts hydrologiques au large des côtes françaises: les sites-ateliers du programme frontal. *Oceanologica Acta* **13**, 413–438.

Spaans, A.L. (1971). On the feeding ecology of the Herring Gull *Larus argentatus* Pont. in the northern part of the Netherlands. *Ardea* **59**, 73–188.

Speakman, J.R. (1987). Apparent absorption efficiencies for redshank (*Tringa totanus* L.) and oystercatchers (*Huematopus ostralegus* L.): implications for the predictions of optimal foraging models. *American Naturalist* **130**, 677–691.

St. Joseph, A.K.M. (1979). The development of inland feeding by *Branta bernicla bernicla* in southeastern England. In *First Technical Meeting on Western Palearctic Migratory Bird Management*. (ed. M. Smart), pp. 132–145. Slimbridge: International Waterfowl Research Bureau.

Staaland, H. (1968). Excretion of salt in waders, Charadrii, after acute salt loads. *Nytt magasin for zoologi* **16**, 25–28.

Stanley, P.I. & Bunyan, P.J. (1979). Hazards to wintering geese and other wildlife from the use of dieldrin, chlorfenvinphos and carbophenothion as wheat treatments. *Proceedings of the Royal Society of London* **B 205**, 31–45.

Stanley, P.I. & Minton, C.D.T. (1972). The unprecedented westward migration of Curlew Sandpipers in autumn 1969. *British Birds* **65**, 365–380.

Stawarczyk, T. (1984). Aggression and its suppression in mixed-species wader flocks. *Ornis Scandinavica* **15**, 23–37.

Stebbing, A.R.D. (1985). Organotins and water quality – some lessons to be learned. *Marine Pollution Bulletin* **16**, 383–390.

Steers, J.A. (1954). *The sea coast*. London: Collins.

Stowe, T.J. (1982*a*). Recent population trends in cliff-breeding seabirds in Britain & Ireland. *Ibis* **124**, 502–510.

Stowe, T.J. (1982*b*). *Beached bird surveys and surveillance of cliff-breeding seabirds*. Report to the Nature Conservancy Council. Sandy: Royal Society for the Protection of Birds.

Stowe, T.J. & Hudson, A.V. (1988). Corncrake studies in the Western Isles. *RSPB Conservation Review* **2**, 38–42.

Stowe, T.J. & Underwood, L.A. (1984). Oil spillages affecting seabirds in the United Kingdom, 1966–1983. *Marine Pollution Bulletin* **15**, 147–152.

Strann, K.-B. & Summers, R.W. (1990). Diet and diurnal feeding activity of Purple Sandpipers *Calidris maritima* wintering in Northern Norway. *Fauna norvegica, Series C, Cinclus* **13**, 75–78.

Strann, K.-B., Vader, W. & Barrett, R. (1991). Auk mortality in fishing nets in north Norway. *Seabird* **13**, 22–29.

Stuart, A.J. (1982). *Pleistocene vertebrates in the British Isles*. London: Longman.

Summers, R.W. (1986). Breeding production of Dark-bellied Brent Geese *Branta bernicla bernicla* in relation to lemming cycles. *Bird Study* **33**, 105–108.

Summers, R.W. (1990). Wintering north of the Arctic Circle. *BTO News* **166**, 11–12.

Summers, R.W., Ellis, P.M. & Johnstone, J.P. (1988). Waders on the coast of Shetland in winter: numbers and habitat preferences. *Scottish Birds* **15**, 71–79.

Summers, R.W. & Smith, S.M. (1983). The diet of the Knot (*Calidris canutus*) on rocky shores of eastern Scotland in winter. *Ardea* **71**, 151–153.

Summers, R.W., Smith, S., Nicoll, M. & Atkinson, N.K. (1990). Tidal and sexual differences in the diet of Purple Sandpipers *Calidris maritima* in Scotland. *Bird Study* **37**, 187–194.

Summers, R.W. & Underhill, L.G. (1987). Factors related to breeding production of Brent Geese *Branta b. bernicla* and waders (Charadrii) on the Taimyr Peninsula. *Bird Study* **34**, 161–171.

Summers, R.W. & Underhill, L.G. (1991). The relationship between body size and time of breeding in Icelandic Redshanks *Tringa t. robusta. Ibis* **133**, 134–139.

Swennen, C. (1984). Differences in quality of roosting flocks of Oystercatchers. In *Coastal waders and wildfowl in winter.* (ed. P.R. Evans, J.D. Goss-Custard & W.G. Hale), pp. 177–189. Cambridge: Cambridge University Press.

Swennen, C., Bruijn, L.L.M. de, Duiven, P., Leopold, M.F. & Marteijn, E.C.L. (1983). Differences in bill form of the oystercatcher *Haematopus ostralegus*; a dynamic adaptation to specific foraging techniques. *Netherlands Journal of Sea Research* **17**, 57–83.

Swennen, C. & Duiven, P. (1983). Characteristics of Oystercatchers killed by cold-stress in the Dutch Wadden Sea area. *Ardea* **71**, 155–159.

Swennen, C., Leopold, M.F. & Bruijn, L.L.M. de. (1989). Time-stressed oystercatchers, *Haematopus ostralegus*, can increase their intake rate. *Animal Behaviour* **38**, 8–22.

Tamisier, A. (1985). Hunting as a key environmental parameter for the Western Palearctic duck populations. *Wildfowl* **36**, 95–103.

Tasker, M.L., Webb, A., Hall, A.J., Pienkowski, M.W. & Langslow, D.R. (1987). *Seabirds in the North Sea.* Peterborough: Nature Conservancy Council.

Tay & Orkney Ringing Groups. (1984). *The shore-birds of the Orkney Islands.* Perth: Tay Ringing Group.

Teixeira, A.M. (1985). More auk deaths in Iberian nets. *BTO News* **138**, 1.

Temple, S.A. (1987). Do predators always capture substandard individuals disproportionately from prey populations? *Ecology* **68**, 669–674.

Thomas, G.J., Allen, D.A. & Grose, M.P.B. (1981). The demography and flora of the Ouse Washes, England. *Biological Conservation* **21**, 197–229.

Thompson, P.S. & Hale, W.G. (1989). Breeding site fidelity and natal philopatry in the Redshank *Tringa totanus. Ibis* **131**, 214–224.

Tinbergen, N. (1953). *The Herring Gull's world.* London: Collins.

Tinbergen, N. (1974). *Curious naturalists.* Harmondsworth: Penguin Education.

Tooley, M.J. (1978). *Sea-level changes: North-west England during the Flandrian Stage.* Oxford: Clarendon Press.

Townshend, D.J. (1982). The Lazarus syndrome in Grey Plovers. *Wader Study Group Bulletin* **34**, 11–12.

Townshend, D.J., Dugan, P.J. & Pienkowski, M.W. (1984). The unsociable plover – use of intertidal areas by Grey Plovers. In *Coastal waders and wildfowl in winter.* (ed. P.R. Evans, J.D. Goss-Custard & W.G. Hale), pp. 140–159. Cambridge: Cambridge University Press.

Tubbs, C.R. (1977). Wildfowl and waders in Langstone Harbour. *British Birds* **70**, 177–199.

Tubbs, C.R. (1984). *Spartina* on the south coast: an introduction. In *Spartina anglica in Great Britain.* (ed. P. Doody), pp. 3–4. Huntingdon: Nature Conservancy Council, Research and Survey in Nature Conservation No. 5.

Tubbs, C.R. & Tubbs, J.M. (1983). Macroalgal mats in Langstone Harbour, Hampshire, England. *Marine Pollution Bulletin* **14**, 148–149.

Tye, A. & Tye, H. (1987). The importance of Sierra Leone for wintering waders. *Wader Study Group Bulletin* **49 Supplement,** 71–75.

Underhill, L.G. & Summers, R.W. (1990). Multivariate analyses of breeding performance in Dark-bellied Brent Geese *Branta b. bernicla. Ibis* **132**, 477–482.

Uthe, J.F., Chou, C.L., Loring, D.H., Rantala, R.T.T., Bewers, J.M., Dalziel, J.,

Yeats, P.A. & Charron, R.L. (1986). Effect of waste treatment at a lead smelter on cadmium levels in American lobster (*Homarus americanus*), sediments and seawater in the adjacent coastal zone. *Marine Pollution Bulletin* **17**, 118–123.

Uttley, J.D., Monaghan, P. & Blackwood, J. (1989). Hedgehog *Erinaceus europaeus* predation on Arctic Tern *Sterna paradisaea* eggs: the impact on breeding success. *Seabird* **12**, 3–6.

Uttley, J.D., Monaghan, P. & White, S. (1989). Differential effects of reduced sandeel availability on two sympatrically breeding species of tern. *Ornis Scandinavica* **20**, 273–277.

Uttley, J.D., Thomas, C.J., Davidson, N.C., Strann, K.-B. & Evans, P.R. (1987). The spring migration system of Nearctic Knots *Calidris canutus islandica*: a reappraisal. *Wader Study Group Bulletin* **49 Supplement,** 80–84.

van den Berg, A.B. (1988). *Moroccan Slender-billed Curlew survey, winter 1987–88.* Zeist: International Council for Bird Preservation & Dutch Working Group for International Wader- and Waterfowl Research.

van Eerden, M.R. (1984). Waterfowl movements in relation to food stocks. In *Coastal waders and wildfowl in winter.* (ed. P.R. Evans, J.D. Goss-Custard & W.G. Hale), pp. 84–100. Cambridge: Cambridge University Press.

van Franeker, J.A. (1985). Plastic ingestion in the North Atlantic fulmar. *Marine Pollution Bulletin* **16**, 367–369.

van Franeker, J.A. & Bell, P.J. (1988). Plastic ingestion by petrels breeding in Antarctica. *Marine Pollution Bulletin* **19**, 672–674.

van Halewyn, R. & Norton, R.L. (1984). The status and conservation of seabirds in the Caribbean. In *Status and conservation of the world's seabirds.* (ed. J.P. Croxall, P.G.H. Evans & R.W. Schreiber), pp. 169–222. Cambridge: International Council for Bird Preservation, Technical Publication No. 2.

van Heezik, Y.M., Gerritsen, A.F.C. & Swennen, C. (1983). The influence of chemoreception on the foraging of two species of sandpiper, *Calidris alba* and *Calidris alpina*. *Netherlands Journal of Sea Research* **17**, 47–56.

van Impe, J. (1985). Estuarine pollution as a probable cause of increase of estuarine birds. *Marine Pollution Bulletin* **16**, 271–276.

van Rhijn, J. (1985). Black-headed Gull or black-headed girl? On the advantage of concealing sex by gulls and other colonial birds. *Netherlands Journal of Zoology* **35**, 87–102.

Verbeek, N.A.M. (1988). Differential predation of eggs in clutches of Glaucous-winged Gulls *Larus glaucescens*. *Ibis* **130**, 512–518.

Vermeer, K. (1963). The breeding ecology of the Glaucous-winged gull (*Larus glaucescens*) on Mandarte Island, B.C. *Occasional Papers of the British Colombia Provincial Museum* **13**, 1–104.

Vermeer, K. (1968). Ecological aspects of ducks nesting in high densities among larids. *Wilson Bulletin* **80**, 78–83.

Vermeer, K. & Rankin, L. (1984). Influence of habitat destruction and disturbance on nesting seabirds. In *Status and conservation of the world's seabirds.* (ed. J.P. Croxall, P.G.H. Evans & R.W. Schreiber), pp. 723–736. Cambridge: International Council for Bird Preservation, Technical Publication No. 2.

Vidaković, J. (1983). The influence of raw domestic sewage on density and distribution of meiofauna. *Marine Pollution Bulletin* **14**, 84–88.

Vinogradov, M.E. (1981). Ecosystems of equatorial upwellings. In *Analysis of marine ecosystems.* (ed. A.R. Longhurst), pp. 69–93. London: Academic Press.

Viscor, G. & Fuster, J.F. (1987). Relationships between morphological parameters in birds with different flying habits. *Comparative Biochemistry and Physiology* **87A**, 231–249.

Waldichuk, M. (1990). Dioxin pollution near pulpmills. *Marine Pollution Bulletin* **21**, 365–366.

Wanless, S. (1987). *A survey of the numbers and breeding distribution of the North Atlantic gannet* Sula bassana *and an assessment of the changes which have occurred since Operation Seafarer 1969/70.* Peterborough: Nature Conservancy Council, Research and Survey in Nature Conservation No.4.

Wanless, S., Burger, A.E. & Harris, M.P. (1991). Diving depths of Shags *Phalacrocorax aristotelis* breeding on the Isle of May. *Ibis* **133**, 37–42.

Wanless, S. & Harris, M.P. (1984). Effect of date on counts of Herring and Lesser Black backed Gulls. *Ornis Scandinavica* **15**, 89–94.

Wanless, S. & Harris, M.P. (1989). Kittiwake attendance patterns during chick rearing on the Isle of May. *Scottish Birds* **15**, 156–161.

Wanless, S., Harris, M.P. & Morris, J.A. (1991). Foraging range and feeding locations of Shags *Phalacrocorax aristotelis* during chick rearing. *Ibis* **133**, 30–36.

Wanless, S. & Langslow, D.R. (1983). The effects of culling on the Abbeystead and Mallowdale gullery. *Bird Study* **30**, 17–23.

Ward, P. & Zahavi, A. (1973). The importance of certain assemblages of birds as 'information-centres' for food-finding. *Ibis* **115**, 517–534.

Watson, P.S. (1981). Seabird observations from commercial trawlers in the Irish Sea. *British Birds* **74**, 82–90.

Webb, A., Harrison, N.M., Leaper, G.M., Steele, R.D., Tasker, M.L. & Pienkowski, M.W. (1990). *Seabird distribution west of Britain.* Peterborough: Nature Conservancy Council.

Wells, G. (1989). Observing earth's environment from space. In *The fragile environment.* (ed. L. Friday & R. Laskey), pp. 148–192. Cambridge: Cambridge University Press.

Wells, M.J. (1991). Aspects of the year-round ecology and behaviour of Ringed Plovers *Charadrius hiaticula* L. on South Uist, Outer Hebrides. Ph.D. thesis: University of Wales.

Welty, J.C. (1962). *The life of birds.* Philadelphia: W.B. Saunders.

White, D.H., Mitchell, C.A. & Kaiser, T.E. (1983). Temporal accumulation of organochlorine pesticides in shorebirds wintering on the south Texas coast, 1979–80. *Archives of Environmental Contamination and Toxicology* **12**, 241–245.

Whitehead, P.E. (1989). Toxic chemicals in Great Blue Heron (*Ardea herodias*) eggs in the Strait of Georgia. In *The ecology and status of marine and shoreline birds in the Strait of Georgia, British Columbia.* (ed. K. Vermeer & R.W. Butler), pp. 177–183. Ottawa: Canadian Wildlife Service.

Whiteley, J.D., Pritchard, J.S. & Slater, P.J.B. (1990). Strategies of mussel dropping by Carrion Crows *Corvus c. corone. Bird Study* **37**, 12–17.

White-Robinson, R. (1982). Inland and saltmarsh feeding of wintering Brent Geese in Essex. *Wildfowl* **33**, 113–118.

Whitfield, D.P. (1988). Sparrowhawks *Accipiter nisus* affect the spacing behaviour of wintering Turnstone *Arenaria interpres* and Redshank *Tringa totanus. Ibis* **130**, 284–287.

Whittaker, R.H. (1975) (2nd edn). *Communities and ecosystems.* London: Macmillan.

Whittle, L. (1991). Sea defences in deep water. *New Scientist* **129 (2 March)**, 58–59.

Williams, G. & Bowers, J.K. (1987). Land drainage and birds in England and Wales. *RSPB Conservation Review* **1**, 25–30.

Williams, G. & Forbes, J.E. (1980). The habitat and dietary preferences of Dark-bellied Brent Geese and Wigeon in relation to agricultural management. *Wildfowl* **31**, 151–157.

Wilson, J.A. (1975). Sweeping flight and soaring by albatrosses. *Nature, London* **257**, 307–308.

Wilson, W.H. (1988). Methods of measuring wader abundance. *Wader Study Group Bulletin* **54**, 49–50.

Withers, P.C. (1979). Aerodynamics and hydrodynamics of the 'hovering' flight of Wilson's storm petrel. *Journal of experimental Biology* **80**, 83–91.

Wolff, W.J. & Smit, C.J. (1990). The Banc d'Arguin, Mauritania, as an environment for coastal birds. *Ardea* **78**, 17–38.

Wood, G.L. (1976) (2nd edn). *The Guiness book of animal facts and feats.* Enfield: Guiness Superlatives.

Wooller, R.D. (1979). Seasonal, diurnal and area differences in calling activity within a colony of Kittiwakes *Rissa tridactyla* (L.). *Zeitschrift für Tierpsychologie* **51**, 329–336.

Wormell, P. (1976). The Manx Shearwaters of Rhum. *Scottish Birds* **9**, 103–118.

Worrall, D.H. (1981). The feeding behaviour of the Dunlin *Calidris alpina* (L.). Ph.D. thesis: University of Wales.

Wu, R.S.S. (1988). Red tide hits Hong Kong. *Marine Pollution Bulletin* **19**, 305.

Wu, R.S.S. (1989). Toxic red tides and shellfish in Hong Kong. *Marine Pollution Bulletin* **20**, 363–364.

Wynne-Edwards, V.C. (1953). Leach's petrels stranded in Scotland in October-November, 1952. *Scottish Naturalist* **64**, 167–189.

Wynne-Edwards, V.C., Lockley, R.M. & Salmon, H.M. (1936). The distribution and numbers of breeding Gannets (*Sula bassana* L.). *British Birds* **29**, 262–276.

Yalden, D.W. (1991). History of the fauna. In *The handbook of British mammals.* (3rd edn) (ed. G.B. Corbet & S. Harris), pp. 7–18. Oxford: Blackwell Scientific Publications.

Yarrell, W. (1885). *A history of British birds.* London: Howard Saunders.

Yésou, P. (1991). The sympatric breeding of *Larus fuscus, L. cachinnans* and *L. argentatus* in western France. *Ibis* **133**, 256–263.

Ydenberg, R.C. & Prins, H.H.Th. (1981). Spring grazing and the manipulation of food quality by barnacle geese. *Journal of Applied Ecology* **18**, 443–453.

Ydenberg, R.C., Prins, H.H.Th. & van Dijk, J. (1984). A lunar rhythm in the nocturnal foraging activities of wintering Barnacle Geese. *Wildfowl* **35**, 93–96.

Zink, R.M. & Eldridge, J.L. (1980). Why does Wilson's Petrel have yellow on the webs of its feet? *British Birds* **73**, 385–387.

Zwarts, L. (1981). Habitat selection and competition in wading birds. In *Birds of the Wadden Sea*, Report 6 of the Wadden Sea Working Group. (ed. C.J. Smit & W.J. Wolff), pp. 271–279. Rotterdam: A.A. Balkema.

Zwarts, L. (1984). Wading birds in Guinea-Bissau, winter 1982/83. *Wader Study Group Bulletin* **40**, 36.

# INDEX

Principal references are printed in **bold** type.